# POSTHARVEST TECHNOLOGY OF HORTICULTURAL CROPS

**Technical Editor: Adel A. Kader**

**Publication 3311**

**UNIVERSITY OF CALIFORNIA**
**Division of Agriculture and Natural Resources**

1992

For information about ordering this publication, write to:

**Publications**
**Division of Agriculture and Natural Resources**
**University of California**
**6701 San Pablo Avenue**
**Oakland, California 94608–1239**

or telephone (510) 642–2431

Publication 3311

ISBN 0-931876-99-0

Production: Heidi Seney, Senior Editor
Jim Coats, Senior Editor
Franz Baumhackl, Senior Artist

Cover color photography: Donald C. Edwards

---

WARNING ON THE USE OF CHEMICALS

Pesticides are poisonous. Always read and carefully follow all precautions and safety recommendations given on the container label. Store all chemicals in their original labeled containers in a locked cabinet or shed, away from foods or feeds, and out of the reach of children, unauthorized persons, pets, and livestock.

Recommendations are based on the best information currently available, and treatments based on them should not leave residues exceeding the tolerance established for any particular chemical. Confine chemicals to the area being treated. THE GROWER IS LEGALLY RESPONSIBLE for residues on the grower's crops as well as for problems caused by drift from the grower's property to other properties or crops.

Consult your county agricultural commissioner for correct methods of disposing of leftover spray materials and empty containers. Never burn pesticides containers.

PHYTOTOXICITY: Certain chemicals may cause plant injury if used at the wrong stage of plant development or when temperatures are too high. Injury may also result from excessive amounts or the wrong formulation or from mixing incompatible materials. Inert ingredients, such as wetters, spreaders, emulsifiers, diluents, and solvents can cause plant injury. Since formulations are often changed by manufacturers, it is possible that plant injury may occur, even though no injury was noted in previous seasons.

---

5m–pr–11/91–HS/WJC/FB

# Authors

| Author | Mailing address | Telephone |
|---|---|---|
| M. JOSEPH AHRENS<br>Postharvest Horticulturist, CE* | Department of Vegetable Crops<br>University of California<br>Davis, CA 95616-8631 | (916) 752-1412 |
| MARY LU ARPAIA<br>Subtropical Horticulturist, CE | Department of Botany and Plant Science<br>University of California<br>Riverside, CA 92521 | (714) 787-3335 |
| MARITA CANTWELL<br>Vegetable Specialist, CE | Department of Vegetable Crops<br>University of California<br>Davis, CA 95616-8631 | (916) 752-7305 |
| PICTIAW CHEN<br>Professor, AES† | Department of Agricultural Engineering<br>University of California<br>Davis, CA 95616-5294 | (916) 752-1890 |
| ROBERTA L. COOK<br>Economist, CE | Department of Agricultural Economics<br>University of California<br>Davis, CA 95616-8512 | (916) 752-1531 |
| DONALD C. EDWARDS<br>Photographer | Department of Pomology<br>University of California<br>Davis, CA 95616-8683 | (916) 752-0932 |
| ROBERT J. FORTLAGE<br>Staff Research Associate | Department of Pomology<br>University of California<br>Davis, CA 95616-8683 | (916) 752-0908 |
| ADEL A. KADER<br>Professor and Pomologist,<br>AES and CE | Department of Pomology<br>University of California<br>Davis, CA 95616-8683 | (916) 752-0909 |
| ROBERT F. KASMIRE<br>Emeritus Vegetable Marketing<br>Specialist, CE | Department of Vegetable Crops<br>University of California<br>Davis, CA 95616-8631 | (916) 752-1410 |
| F. GORDON MITCHELL<br>Pomologist, CE and AES | Department of Pomology<br>University of California<br>Davis, CA 95616-8683 | (916) 752-0508 |
| MICHAEL S. REID<br>Professor and Postharvest<br>Physiologist, AES and CE | Department of Environmental Horticulture<br>University of California<br>Davis, CA 95616-8587 | (916) 752-7060 |
| NOEL F. SOMMER<br>Emeritus Lecturer and<br>Postharvest Pathologist, AES | Department of Pomology<br>University of California<br>Davis, CA 95616-8683 | (916) 752-0908 |
| JAMES F. THOMPSON<br>Agricultural Engineer, CE | Department of Agricultural Engineering<br>University of California<br>Davis, CA 95616-5294 | (916) 752-6167 |

*CE = Cooperative Extension    †AES = Agricultural Experiment Station

# Contents

# Preface

Postharvest Technology of Horticultural Crops is the outcome of a syllabus that was developed for a short course initiated in 1979 and offered annually since then in two modes: (1) as a regular University of California, Davis course (Plant Science 196) for advanced undergraduate and graduate students interested in postharvest biology and technology of horticultural crops, and (2) as a short course organized through University Extension for participants who are not current UCD students. The latter group usually includes research and extension workers, quality control personnel, and other persons concerned with postharvest handling of fresh horticultural crops.

The first edition (published in 1985) was well received and its distribution, of nearly 5,000 copies worldwide, exceeded our expectations. In this second edition, all chapters have been updated, many were expanded, and three new chapters added.

Emphasis is on current postharvest technology procedures for fresh fruits, vegetables, and ornamentals in California. However, all the principles discussed are applicable to postharvest handling of fresh horticultural crops worldwide.

Thirty-three chapters are included, of which 19 present various aspects of postharvest technology of horticultural commodities and eleven briefly cover postharvest handling systems for certain commodities or commodity groups. It was not possible to include every horticultural crop in these 11 chapters and keep the book to a reasonable length. The remaining 3 chapters deal with sources of information, an overview of the fresh produce industry, and extension efforts. We are continuously working on improving all aspects of this book and so we welcome comments and suggestions for incorporation into future editions.

On behalf of the authors, I wish to thank all those individuals who assisted us. I especially want to acknowledge the tireless efforts of Pamela Moyer (Department of Pomology, UC Davis) in compiling the subject index, proofreading, and organizing production of this book. Thanks are also due Marilyn Mott, Ida Fierro, Susan Place, Gloria Robles, Sharon Sloan, and Cathey Wolpert for their excellent job in typing the manuscript. Thanks are also due Don Edwards (Department of Pomology, UC Davis) and the staff of Cooperative Extension's Visual Media for their help with illustrations. Furthermore, we greatly appreciate the assistance and cooperation of Franz Baumhackl, senior artist, Jim Coats and Heidi Seney, senior editors, and other University of California Agriculture and Natural Resources Publications staff members who participated in the production of this book.

ADEL A. KADER
TECHNICAL EDITOR

# 1

# Sources of Information Related to Postharvest Biology and Technology

ADEL A. KADER

Numerous sources of information related to post-harvest biology and technology of horticultural crops can be used to supplement and expand the information in this book. The purpose of this chapter is to point out the library resources available.

## A Starting Point

Textbooks and general references related to post-harvest biology and technology of horticultural crops are listed at the end of this chapter. Furthermore, each subsequent chapter includes a list of references for further reading. The following reference provides a good starting point for developing background information.

Kader, A. A., L. L. Morris, and M. Cantwell. 1991. *Postharvest handling and physiology of horticultural crops—a list of selected references*. 8th ed., Postharvest Horticulture Series 2, Department of Pomology, University of California, Davis.

## Additional library resources

To review the published literature about a specific topic or a given commodity, use a computerized information service, if one is available, or consult one or more of the following abstracting journals:

*Biological Abstracts*
*Chemical Abstracts*
*Food Science Abstracts*
*Horticultural Abstracts*
*Postharvest News and Information* (initiated in 1990)
*Review of Plant Pathology*

Review journals, including the following, periodically contain chapters on postharvest biology and technology:

*Advances in Food Research*
*Annual Review of Phytopathology*
*Annual Review of Plant Physiology*
*Critical Reviews in Food Science and Nutrition*
*Horticultural Reviews*

To follow current publications, read the weekly publication *Current Contents: Agriculture, Biology & Environmental Sciences*, published by Institute for Scientific Information, 3501 Market St., Philadelphia, PA 19104, which includes tables of contents from many periodicals, including most of those listed below. Another current awareness publication (monthly) is *Current Advances in Plant Science*, published by Pergamon Press, Fairview Park, Elmsford, NY 10523.

Until 1990 no one scientific journal specialized in postharvest biology and technology of horticultural crops. Research reports are published in a wide range of scientific journals, including the following:

*Agricultural Engineering*
*American Potato Journal*
*Economic Botany*
*Food Technology*
*Fruits d'Outre Mer*
*HortScience*
*International Journal of Food Science and Technology*
*International Journal of Refrigeration*
*Journal of Agricultural and Food Chemistry*
*Journal of Food Biochemistry*
*Journal of Food Quality*
*Journal of Food Science*
*Journal of Horticultural Science*
*Journal of Textural Studies*
*Journal of the American Society for Horticultural Science*
*Journal of the Science of Food and Agriculture*
*Phytochemistry*
*Phytopathology*
*Plant Disease*
*Plant Physiology*
*Postharvest Biology and Technology* (starting 1991)
*Proceedings of the Florida State Society for Horticultural Science*
*Proceedings of the Tropical Region of the American Society for Horticultural Science (Interamerican Society for Tropical Horticulture)*
*Scientia Horticulturae*
*Transactions of the American Society of Agricultural Engineers*
*Tropical Agriculture*
*Tropical Science*

Semitechnical and popular periodicals include:

*American Fruit Grower*
*American Vegetable Grower*
*California Agriculture*
*Citrus and Vegetables Magazine*
*Florida Grower and Rancher*
*Florists Review*
*The Goodfruit Grower*
*The Packer* (a weekly newspaper)
*Produce Business*
*Produce Marketing Association Newsletter*
*United Fresh Fruit & Vegetable Association Newsletter*
*Western Grower and Shipper*

Newsletters published periodically by Cooperative Extension specialists in postharvest biology and technology of horticultural crops at various locations include:

*Packinghouse Newsletter* (available from W. Wardowski, University of Florida, AREC, 700 Experiment Station Rd., Lake Alfred, FL 33850)

*Perishables Handling* (available from Joe Ahrens, Mann Laboratory, University of California, Davis, CA 95616)

*Tree Fruit Postharvest Journal* (available from E. Kupferman, Tree Fruit Research and Extension Center, 1100 N. Western Ave., Wenatchee, WA 98801)

Other sources include publications of the U.S. Department of Agriculture; Agricultural Experiment Stations and Cooperative Extensions in California, Florida, New York, Michigan, Oregon, Washington, and other states; International Institute of Refrigeration (Paris, France); Natural Resources Institute (Chatham, Kent, England); Postharvest Institute for Perishables (University of Idaho, Moscow, ID 83843); and other organizations.

Industry organizations that offer publications include United Fresh Fruit and Vegetable Association (UFFVA) (722 North Washington, Alexandria, VA 22314) and Produce Marketing Association (PMA) (P.O. Box 6036, Newark, DE 19714-6036).

## Visual aids

To find out about audiovisual programs (slide sets, film strips, 16-mm movies, videotapes, and so on) that deal with various aspects of postharvest technology of horticultural crops, consult the directory published by UFFVA, the listing of slide programs available from the American Society for Horticultural Science, and the publications catalog of the Division of Agriculture and Natural Resources, University of California.

## REFERENCES

1. ASHRAE. 1990. *ASHRAE handbook and product directory, applications volume.* Atlanta, GA: Am. Soc. Heating, Refrig. Air Condit. Engineers.

2. Burton, W. G. 1982. *Postharvest physiology of food crops.* London and New York: Longman. 339 pp.

3. Debney, H. G., K. J. Blacker, B. J. Redding, and J. B. Watkins. 1980. *Handling and storage practices for fresh fruits and vegetables—product manual.* Brisbane: Austral. United Fresh Fruit & Veg. Assn.

4. Dennis, C. 1983. *Postharvest pathology of fruits and vegetables.* London: Academic Press. 264 pp.

5. Eskin, N. A. M., ed. 1989. *Quality and preservation of vegetables.* Boca Raton, FL: CRC Press. 313 pp.

6. Eskin, N. A. M., ed. 1990. *Quality and preservation of fruits.* Boca Raton, FL: CRC Press. 176 pp.

7. Friend, J., and M. J. C. Rhodes, eds. 1981. *Recent advances in the biochemistry of fruit and vegetables.* New York: Academic Press. 278 pp.

8. Hardenburg, R. E., A. E. Watada, and C. Y. Wang. 1986. *The commercial storage of fruits, vegetables, and florist and nursery stocks.* U.S. Dept. Agric. Handb. 66. 130 pp.

9. Hulme, A. C., ed. 1970. *The biochemistry of fruits and their products.* Vol. 1. New York: Academic Press. 620 pp.

10. ———. 1971. *The biochemistry of fruits and their products.* Vol. 2. New York: Academic Press. 788 pp.

11. Hultin, H. O., and M. Milner, eds. 1978. *Postharvest biology and biotechnology.* Westport, CT: Food and Nutrition Press. 460 pp.

12. Lieberman, M., ed. 1983. *Postharvest physiology and crop preservation.* New York: Plenum. 575 pp.

13. Nagy, S., and P. E. Shaw, eds. 1980. *Tropical and subtropical fruits: composition, properties, and uses.* Westport, CT: AVI Publ. Co. 570 pp.

14. O'Brien, M., B. F. Cargill, and R. B. Fridley. 1983. *Principles and practices for harvesting and handling of fruits and nuts.* Westport, CT: AVI Publ. Co. 636 pp.

15. Pantastico, E. B., ed. 1975. *Postharvest physiology, handling and utilization of tropical and subtropical fruits and vegetables.* Westport, CT: AVI Publ. Co. 560 pp.

16. Peleg, K. 1985. *Produce handling, packaging and distribution.* Westport, CT: AVI Publ. Co. 625 pp.

17. Ryall, A. L., and W. J. Lipton. 1979. *Handling, transportation and storage of fruits and vegetables.* Vol. 1, *Vegetables and melons.* 2nd ed. Westport, CT: AVI Publ. Co. 588 pp.

18. Ryall, A. L., and W. T. Pentzer. 1982. *Handling, transportation and storage of fruits and vegetables.* Vol. 2, *Fruits and tree nuts.* Westport, CT: AVI Publ. Co. 610 pp.

19. Salunkhe, D. K., N. R. Bhat, and B. B. Desai. 1990. *Postharvest biotechnology of flowers and ornamental plants.* New York: Springer-Verlag. 192 pp.

20. Salunkhe, D. K., H. R. Bolin, and N. R. Reddy. 1990a. *Storage, processing, and nutritional quality of fruits and vegetables,* 2nd edition. Volume I. Boca Raton, FL: CRC Press. 312 pp.

21. ———. 1990b. *Storage, processing, and nutritional quality of fruits and vegetables,* 2nd edition. Volume II. Boca Raton, FL: CRC Press. 208 pp.

22. Salunkhe, D. K., and B. B. Desai. 1984a. *Postharvest biotechnology of fruits.* Vol. I. Boca Raton, FL: CRC Press. 184 pp.

23. ———. 1984b. *Postharvest biotechnology of fruits.* Vol. II. Boca Raton, FL: CRC Press. 168 pp.

24. ———. 1984c. *Postharvest biotechnology of vegetables.* Vol. I. Boca Raton, FL: CRC Press. 232 pp.

25. ———. 1984d. *Postharvest biotechnology of vegetables.* Vol. II. Boca Raton, FL: CRC Press. 288 pp.

26. Snowden, A. L. 1990. *A color atlas of postharvest diseases and disorders of fruits and vegetables.* Vol. 1. *General introduction and fruits.* Boca Raton, FL: CRC Press. 302 pp.

27. Weichman, J., ed. 1987. *Postharvest physiology of vegetables.* New York: Marcel Dekker. 616 pp.

28. Wills, R. H. H., W. B. McGlasson, D. Graham, T. H. Lee, and E. G. Hall. 1989. *Postharvest: An introduction to the physiology and handling of fruit and vegetables.* New York: Van Nostrand Reinhold. 174 pp.

# 2

# The Dynamic U.S. Fresh Produce Industry: An Overview

ROBERTA L. COOK

The U.S. system for marketing fresh fruits and vegetables is complex, fragmented, and dynamic. Demand for high-quality produce continues to increase, and product form and packaging are changing as more firms introduce such value-added products as lightly processed produce. The system is evolving toward more direct sales from shippers to final buyers. Fresh produce distribution through food service channels has expanded dramatically in the 1980s. Fresh produce has become a critical element in the competitive strategy of retailers, making its year-round availability a necessity. The challenge to supply seasonal, perishable products year-round has favored imports and increased integration among shippers, nationally and internationally.

For their part, other countries are developing their horticultural industries as a means to diversify exports. Improved transportation services, along with improved temperature management and modified atmosphere technology, should facilitate expanded world trade in fruits and vegetables in the 1990s. Leading the trend toward greater international integration are the multinational food processors, which have broadened their product lines to include fresh produce, increasingly marketed on a branded basis, and requiring international sourcing to achieve a year-round market presence. This chapter provides a snapshot of the evolving fresh produce industry as it enters the 1990s.

## Fruit and Vegetable Demand

### General trends

The post-World War II era was characterized by accelerating population growth, rising affluence, and a relatively homogeneous population. Under these conditions, mass marketing strategies for food became the norm, emphasizing products that could be marketed nationwide and in large volumes. Much less variety was available than today in terms of number, form, and quality of food products.

Demographic and lifestyle trends in the 1970s and 1980s segmented the American market, causing a marked increase in the diversity of consumers and consequently the products they demand. Targeted marketing replaced mass marketing in the 1980s, and even more finely tuned segmentation strategies can be expected in the future. Development of new products occurred at a record rate in the 1980s,

reaching 12,055 new products in 1989 [Food Institute Report (FIR) Jan. 27, 1990]. Fruits and vegetables were part of this trend: 214 new products were introduced in this category in 1989, compared with 146 in 1984. Even greater diversity in both consumers and products is likely in the 1990s.

## Consumption

Declining household size and the aging of the U.S. population have favorably affected fruit and vegetable consumption and can be expected to continue. Survey data show that per-capita expenditures on fresh vegetables are 87 percent greater for one-person households than for households with more than six people (McCracken 1990). Similarly, people in the age group 45 to 64 consume 39 percent more fresh fruit and 34 percent more fresh vegetables than the national average (The Food Institute 1989). The 45 to 64 age group will be the single largest segment of the population (23 percent) by the year 2000, or 61.4 million people (The Food Institute 1989).

Two key lifestyle trends affecting food consumption are (1) the entrance of women into the work force in record numbers, increasing the demand for foods of high and predictable quality that offer convenience and variety, and (2) the growth in public knowledge about how diet and health are linked and the importance of maintaining physical fitness throughout life. These trends have influenced product mix and product form of foods consumed in the United States.

Fruit and vegetable consumption has risen as a result of greater health awareness, and fresh products are gaining relative to processed products. As shown in table 2.1, per-capita fresh vegetable consumption in 1988 was 96.4 pounds, up 26 percent from 1978. Canned vegetable consumption declined over the decade from 87.1 pounds in 1978 to 83.0 pounds in 1988, and frozen vegetable consumption grew to 18.0 pounds per capita in 1988. As shown in table 2.2, per-capita fresh fruit consumption in 1988 was 96.9 pounds, compared to 84.1 pounds in 1978, an increase of 16 percent. Total fruit consumption was 211.3 pounds in 1988 compared to 191.5 pounds in 1978.

There has been a general shift in product form toward the freshlike and "natural." Many marketers have incorporated "lite" or "natural" on their labels,

Table 2.1. Commercially produced vegetables, per-capita utilization (pounds; 1 lb = 0.4536 Kg), United States, 1978–88

| Year | Grand Total | Vegetables Total | Fresh* | Total | Canning† | Freezing‡ | Potatoes Total | Fresh | Freezing | Chips§ | Other‖ |
|------|-------------|------------------|--------|-------|----------|-----------|----------------|-------|----------|--------|--------|
| 1978 | 297.3 | 177.7 | 76.4 | 101.3 | 87.1 | 14.2 | 120.1 | 46.1 | 43.4 | 16.8 | 13.8 |
| 1979 | 303.0 | 185.4 | 79.1 | 106.3 | 91.3 | 15.0 | 117.7 | 49.6 | 38.3 | 16.9 | 12.9 |
| 1980 | 299.9 | 185.6 | 80.5 | 105.1 | 90.7 | 14.4 | 114.3 | 51.1 | 35.2 | 16.7 | 11.3 |
| 1981 | 295.4 | 179.4 | 79.2 | 100.2 | 85.5 | 14.7 | 116.0 | 45.7 | 41.2 | 16.8 | 12.3 |
| 1982 | 290.0 | 175.6 | 83.2 | 92.4 | 78.9 | 13.5 | 114.3 | 46.8 | 38.4 | 17.2 | 11.9 |
| 1983 | 292.5 | 174.4 | 80.3 | 94.1 | 79.6 | 14.5 | 118.1 | 49.7 | 38.9 | 17.9 | 11.6 |
| 1984 | 317.6 | 195.5 | 87.7 | 107.8 | 90.4 | 17.4 | 122.0 | 48.8 | 43.4 | 18.1 | 11.7 |
| 1985 | 315.2 | 192.8 | 88.1 | 104.7 | 87.6 | 17.1 | 122.3 | 46.7 | 45.1 | 17.7 | 12.8 |
| 1986 | 318.0 | 192.3 | 88.8 | 103.5 | 87.7 | 15.8 | 125.7 | 49.4 | 45.8 | 18.2 | 12.3 |
| 1987 | 322.4 | 197.0 | 93.2 | 103.8 | 87.1 | 16.7 | 125.4 | 48.2 | 47.3 | 17.7 | 12.2 |
| 1988 | 321.0 | 197.4 | 96.4 | 101.0 | 83.0 | 18.0 | 123.6 | 52.4 | 42.5 | 17.0 | 11.7 |

*Includes asparagus, broccoli, carrots, cauliflower, celery, sweet corn, lettuce, onions, tomatoes, and honeydews.
†Includes asparagus, snap beans, carrots, sweet corn, green peas, pickles, and tomatoes.
‡Includes asparagus, snap beans, broccoli, carrots, cauliflower, sweet corn, and green peas.
§Includes shoestrings.
‖Includes canning and dehydrating.
Source: *Vegetables and Specialties Situation and Outlook Report*, U.S. Department of Agriculture, ERS, August 1990.

Table 2.2. Fruit, per-capita utilization (pounds; 1 lb = 0.4536 Kg), United States, 1978–88

| Year | Grand Total* | Citrus Total* | Fresh† | Canned juice | Chilled juice | Frozen juice | Noncitrus Total* | Fresh‡ | Canned | Canned juice | Frozen fruit | Dried |
|------|-------------|---------------|--------|--------------|---------------|--------------|------------------|--------|--------|--------------|--------------|-------|
| 1978 | 191.53 | 107.46 | 26.56 | 11.10 | 12.16 | 57.64 | 84.07 | 57.19 | 12.00 | 3.90 | 3.61 | 7.37 |
| 1979 | 193.92 | 108.34 | 24.68 | 11.21 | 10.96 | 61.49 | 85.58 | 59.41 | 11.20 | 3.02 | 2.95 | 9.01 |
| 1980 | 200.11 | 112.52 | 28.85 | 10.23 | 11.75 | 61.68 | 87.59 | 61.08 | 11.21 | 3.27 | 3.37 | 8.66 |
| 1981 | 192.41 | 104.43 | 24.98 | 9.76 | 8.32 | 61.37 | 87.98 | 61.79 | 10.01 | 3.44 | 3.17 | 9.58 |
| 1982 | 198.48 | 109.30 | 24.73 | 7.88 | 7.04 | 69.65 | 89.19 | 62.15 | 10.32 | 2.90 | 3.22 | 10.60 |
| 1983 | 208.77 | 120.04 | 29.30 | 6.03 | 8.30 | 76.41 | 88.73 | 62.69 | 9.06 | 2.82 | 3.21 | 10.94 |
| 1984 | 196.65 | 102.75 | 23.99 | 5.70 | 7.39 | 65.66 | 93.90 | 67.56 | 9.17 | 2.52 | 3.32 | 11.32 |
| 1985 | 200.94 | 109.10 | 23.44 | 4.57 | 6.46 | 74.64 | 91.84 | 66.48 | 9.26 | 2.36 | 3.60 | 10.15 |
| 1986 | 213.64 | 117.25 | 26.67 | 4.14 | 7.63 | 78.81 | 96.39 | 69.18 | 9.59 | 2.24 | 3.91 | 11.48 |
| 1987 | 214.29 | 112.82 | 26.37 | 4.04 | 9.03 | 73.38 | 101.47 | 75.07 | 9.65 | 2.40 | 4.25 | 10.11 |
| 1988§ | 211.33 | 113.61 | 26.47 | 3.39 | 10.25 | 73.50 | 97.72 | 70.44 | 9.72 | 2.58 | 4.15 | 10.83 |

*Some figures may not add due to rounding.
†Includes only oranges, tangerines, tangelos, lemons, limes, and grapefruit.
‡Includes apples, apricots, avocados, bananas, cherries, cranberries, figs, grapes, kiwifruit, nectarines, peaches, pears, pineapples, papayas, plums and prunes, strawberries, mangos, olives, persimmons, and pomegranates.
§Preliminary.
Source: *Fruit and Tree Nuts Situation and Outlook Yearbook*, U.S. Department of Agriculture, ERS, August, 1989.

along with stronger health claims such as reduction of heart disease or prevention of cancer.

Broccoli and cauliflower are prime examples of vegetables that have gained dramatically in popularity due to evidence from the American Cancer Society that including them in a high-fiber, low-fat diet may reduce the risk of cancer. Per-capita consumption of broccoli and cauliflower increased from 1.1 pounds and 0.9 pound in 1978, respectively, to 4.2 and 2.9 pounds in 1988. *The Packer*'s "Fresh Trends '90" consumer survey indicated that consumer interest in broccoli and cauliflower remains strong. Asked to identify produce items they had included in salads for the first time in the past 12 months, consumers ranked broccoli first, cucumbers second, and cauliflower third.

Year-round availability of produce and improved postharvest handling and transportation have also contributed to greater consumption. For example,

fresh grape consumption grew from 3.1 pounds per capita in 1978 to 7.4 pounds in 1988, and strawberry consumption was up from 2.2 to 3.5 pounds per capita over the same period. Grapes have achieved year-round supply through contra-seasonal imports from Chile. Strawberries are now available year-round due to imports during the late fall and adoption of day-neutral varieties that have extended U.S. shipping seasons.

## Fruit and Vegetable Industry Profile

In 1988 farm-level cash receipts for citrus fruits were $2.5 billion, $5 billion for noncitrus fruits, and $9.8 billion for vegetables, making fruit and vegetable production a $17.3 billion industry.

Sales of fresh produce in grocery stores reached $26.5 billion in 1988 (Supermarket Business 1989). *Supermarket Business* also reported $763.2 million in retail floral sales in 1988, 14 percent through grocery stores. The Census Bureau reported an additional $1.8 billion in sales in 1987 through specialized produce stores (green grocers). Allowing for inflation and underreporting of Census Bureau data and produce sales through farmers' markets and roadside stands, it can be estimated that U.S. retail produce sales were approximately $32 billion in 1988. Estimates vary: *Supermarket Business* estimated that total retail produce sales were $39.2 billion in 1988. The difference reflects a significantly higher estimate of produce sold through specialized produce stores than that reported by the Census Bureau.

Produce sales through food service channels have grown since 1980, when 19 percent of total produce shipments moved through food service channels (McLaughlin 1983). Trade estimates for 1988 show that the food service share ranges from 38 to 45 percent of all fresh produce sales, or a possible $12.2 to $14.4 billion in sales. Food service now represents the dominant marketing channel for iceberg lettuce and potatoes, accounting for 55 and 65 percent, respectively, of all sales (Mayer 1988).

## Location of production

California, the largest producer of horticultural commodities in the U.S., contributes 35 percent of the nation's total vegetable production and 44 percent of its fruit and nut production. When the 15 major fresh market and processing vegetables are considered as a group, California accounts for more than half of national production. California is the nation's exclusive supplier of clingstone peaches, dates, figs, kiwifruit, olives, pomegranates, prunes, and raisins. California's share of U.S. production exceeds 60 percent for each of the following fruits and vegetables: lettuce, processing tomatoes, broccoli, cauliflower, carrots, celery, strawberries, grapes, nectarines, plums, apricots, avocados, lemons, and honeydew melons (California Department of Food and Agriculture 1989).

Florida, the second largest producer of horticultural crops, produces 13 percent of U.S. vegetables and 31 percent of U.S. fruit. Florida dominates in citrus, producing 69 percent of U.S. citrus and leading in oranges and grapefruit. Florida is also a major producer of fresh market tomatoes, snap beans, watermelons, and cucumbers, and accounts for more than half of the nation's production of escarole, endive, and eggplant.

The remainder of U.S. fruit and vegetable production is dispersed among all the other states, primarily Washington, Arizona, Texas, Michigan, New York, Idaho, Oregon, Wisconsin, Hawaii, and Georgia.

## Imports

Imports of fruits and vegetables to the United States expanded rapidly in the 1980s (table 2.3). Fresh produce shipments totaled 70.19 billion pounds in 1988, up 30 percent from 1980. The import market

**Table 2.3. Origin of fresh fruits and vegetables in U.S. markets (billion pounds; 1 lb = 0.4536 Kg)**

| Year | U.S. | Mexico | Canada | SA/CBI*† | Other* | Total |
|---|---|---|---|---|---|---|
| 1980 | 45.67 | 2.31 | 0.54 | 5.38 | 0.08 | 53.98 |
| 1981 | 48.31 | 1.88 | 0.69 | 5.68 | 0.08 | 56.64 |
| 1982 | 48.74 | 2.26 | 0.79 | 6.05 | 0.12 | 57.96 |
| 1983 | 50.88 | 2.54 | 0.73 | 5.83 | 0.12 | 60.10 |
| 1984 | 50.63 | 2.91 | 0.55 | 6.24 | 0.19 | 60.52 |
| 1985 | 50.73 | 2.94 | 0.87 | 7.34 | 0.28 | 62.16 |
| 1986 | 53.86 | 3.30 | 0.73 | 7.26 | 0.25 | 65.40 |
| 1987 | 55.16 | 3.77 | 0.90 | 7.45 | 0.24 | 67.52 |
| 1988 | 57.27 | 3.88 | 0.93 | 7.85 | 0.26 | 70.19 |

*Countries included change from year to year.
†SA/CBI = South America and Caribbean Basin Initiative countries.
Source: *Fresh Fruit and Vegetable Shipments by Commodities, States, and Months, Calendar Years 1980–1988*, U.S. Department of Agriculture, Agricultural Marketing Service, Fruit and Vegetable Division, various issues.

share increased from 15.4 percent in 1980 to 18.5 percent in 1988, the principal suppliers being Mexico, South America, and the Caribbean Basin Initiative (CBI) countries. Chile expanded horticultural exports to the U.S. market most rapidly, from $39.2 million in 1979 to $323.9 million in 1988 (USDA March 1989). The dollar value of fresh and processed U.S. vegetable imports from all sources was $1.62 billion in 1988. Fresh and processed fruit imports of which 44 percent was bananas were valued at $1.8 billion in 1988 (USDA July 1989).

## Supermarket Produce Department Profile

Today, 98 percent of American consumers report that the quality of produce has a major influence on where they shop for food (Food Marketing Institute 1990). Many retailers have repositioned their store formats and image around the produce department, and produce is a critical element in their competitive strategy. In 1988 the average household each week spent $5.67 on produce out of total weekly grocery store food expenditures of $48.60 (Supermarket Business 1989). According to the periodical *Supermarket Business*, 61 percent of customer transactions in 1988 included a purchase in the produce department, 10 percent more than the year before. High awareness of the health benefits of produce and improved produce quality and merchandising should continue to reinforce the produce department's role in attracting customers to the store.

## Profitability

The average produce department accounts for 8.7 percent of total store sales in conventional supermarkets and 9.5 percent in superstores. Supermarkets have at least $2 million in annual store sales. Superstores are operations with more than 30,000 square feet and expanded offerings. Its high turnover and average gross margin relative to other departments (32.1 percent for conventional supermarkets, 30.6 percent for superstores) make the produce department the largest contributor to net store profit—21 percent for conventional supermarkets and 24.6 percent for superstores. Storewide supermarket net profit after taxes averaged 1.64 percent of sales in 1987.

Due to their profitability produce departments are now generally placed first in store traffic patterns. Produce occupied 12.1 percent of total store space in 1988, compared with 3 to 4 percent in the smaller stores of the 1970s. According to *Supermarket Business*, in 1988 average square footage ranged from 1,940 (174.6 square meters) in conventional supermarkets to 4,850 (436.5 square meters) in superstores. However, this represents an average decline of 3.3 percent from 1987 and emphasizes a major issue in the produce department—the fight for shelf space.

## Product diversity

Demographic and lifestyle trends, expanded exports from other countries, and improvements in postharvest handling and international shipping have all promoted greater product diversity. The number of produce items handled continues to expand even though growth in shelf space has apparently peaked. The average produce department in 1988 handled 243 items in summer and 239 in winter (Produce Marketing Association 1989), compared with 141 and 126, respectively, in 1979 (McLaughlin 1983). Competition for shelf-space is likely to prompt shippers and grower commodity groups to commit more resources to promotion and advertising.

The rise in product diversity has greatly increased the volume of mixed (consolidated) load shipments from production regions. Mixed loads, due to temperature and ethylene incompatibilities, create notable postharvest handling challenges. The need for innovative handling methods presents a major marketing opportunity.

Despite the proliferation of new products, just six commodity groups make up 43 percent of total sales: bananas, apples, citrus fruits, potatoes, lettuce, and tomatoes. Fresh fruit/nuts/juices accounted for 44 percent of total produce sales in 1988, and fresh vegetables contributed 42 percent, salad bars 6 percent, floral products 5 percent, and other nonfresh items 3 percent (Produce Marketing Association 1989).

## Marketing Channels and Procurement Practices

The principal marketing channels in the U.S. fresh fruit and vegetable marketing system are shown in figure 2.1. The three primary sales outlets to consumers are: (1) food service establishments, hotels, restaurants, and institutions (schools, the military, hospitals, nursing homes, shelters, and prisons); (2) retail food stores; and (3) direct farmer-to-consumer sales via "u-pick" operations, farmers' markets, and

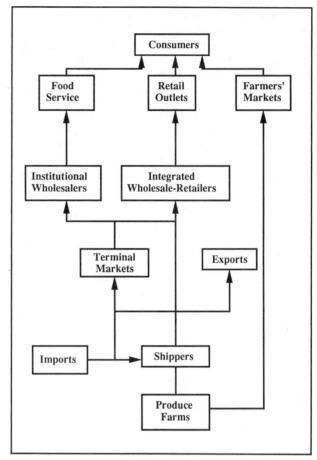

**Fig. 2.1 U.S. Fresh fruit and vegetable marketing system.***
*Brokers may assist in arranging sales transactions and transportation at any level of the system.

roadside stands. Although the majority of produce still moves through retail channels, food service may now account for 38 to 45 percent of total volume, and direct sales may account for 1 percent.

Produce sold in retail or food service outlets may be procured directly from shippers, brokers, or wholesalers operating in terminal (wholesale) markets or in independent warehouses in local communities. Since the 1950s, terminal markets have steadily declined in importance; today there are only 23 major terminal markets, and the volume of produce sales they handle is an estimated 35 percent of the national total. Product formerly moving through terminal markets now goes directly from shippers to final buyers.

## Integrated wholesale-retailers

The decline in terminal market share is partly a result of the increased buying power of integrated wholesale-retailers: the centralized buying operations of corporate chains (11 or more stores), and voluntary and member-owned wholesalers. Voluntary and member-owned wholesalers, referred to as affiliated groups, are composed of independent retailers (retailers operating less than 11 stores) which join together and affiliate with a central wholesale supply organization.

In the case of member-owned wholesalers, retailers own the central buying and warehousing facility. In the case of voluntary wholesalers, retailers affiliate with the established supply organization of a wholesaler, with no financial involvement by either. While the retailers remain independent, affiliation with a central wholesale organization brings them the benefits of joint buying, advertising, and merchandising programs, enabling them to compete with corporate chains. Well known examples of affiliated groups are Fleming, Super Valu Stores, Certified Grocers of California, and I.G.A. in the United States, and Spar in Europe.

In 1986, retailers in affiliated groups made 42.7 percent of all U.S. grocery sales, corporate chains 54.7 percent, and independent unaffiliated retailers 2.6 percent. Retailers in affiliated groups and corporate chains differ in their procurement practices. Because a corporate chain owns its stores, it controls the products it handles and essentially exercises forced distribution. With standardized store formats, chains have more consistent quality needs than do affiliated groups serving a wide diversity of independent retail members. Chains also typically have less ordering flexibility than affiliated groups, which can make more rapid store-level adjustments to accommodate sudden shipping-point changes in product availability and quality (McLaughlin 1983).

As the U.S. market has matured, competition has increased. The industry has undergone consolidation and larger operators have acquired smaller firms; thus, the number of integrated wholesale-retailer centralized buying operations has declined and sales per firm have increased. Tropicana estimated that there are 550 headquarters buying points in the U.S. food distribution system. These supply 17,065 chain supermarkets, 13,535 independent supermarkets, and 47,000 convenience stores (The Food Institute 1988).

Larger volume purchases are more efficiently handled by direct sales and distribution from the shipping point, rather than through terminal markets. Many chains put field personnel in the production regions to ensure product quality and availability. Therefore, integrated wholesale-retail buyers use terminal markets primarily to balance short orders and to procure small-volume exotic or specialty items, including highly perishable products.

## Wholesalers and brokers

Today, terminal markets and other wholesalers focus on independent, unaffiliated retailers and restaurant accounts. *Supermarket Business* reported 15,975 "mom and pop" grocery stores in 1988, mostly unaffiliated independents. In addition, in 1987 the Census Bureau identified 3,271 stores that specialized completely in fruits and vegetables.

Terminal market operators and local produce wholesalers do a substantial amount of inter-whole-saler buying. Primary market handlers (receivers, merchant wholesalers, and commission merchants) procure more than half of their product from the shipping point. Receivers and merchant wholesalers buy and resell products, and commission merchants operate on a consignment basis. Secondary market handlers (jobbers and purveyors) procure more than half of their product from other wholesalers, principally primary handlers. They service small-volume accounts such as independent retailers and restaurants, which require frequent deliveries of small lots. Purveyors focus almost exclusively on food service accounts.

While terminal markets in the Midwest and East are primarily destination markets, those located near the production regions on the West Coast and in Florida ship significant volumes to terminal market and other wholesalers in the destination markets. Wholesalers in all regions have expanded customer services to include such functions as ripening, sizing, repacking, consumer packaging, and suggested advertising for retail accounts.

Brokers are noteworthy players in fresh produce distribution. Brokers help negotiate sales on behalf of buyers or sellers for a percentage sales commission or a flat fee per unit. They do not physically handle or take title of the merchandise; thus, their fees are substantially lower than those charged by commission merchants. Usage of brokers varies greatly by type of buyer and commodity.

## Food service

The growth in fresh produce items handled on food service menus has affected distribution channels. "Fast-food" outlets have added salad bars and packaged salads, as well as other menu items which include produce. McDonald's reportedly used 2 percent of the total U.S. lettuce crop and 1 percent of the fresh tomato crop in 1987. Upscale, "white tablecloth" restaurants are expanding the demand for premium quality and exotic produce.

In the past, institutional wholesalers (food service distributors) supplying the food service industry with dry and packaged groceries did not handle produce; food service users procured their own produce, largely through terminal markets and local produce wholesalers. The rising volume of produce handled by food service establishments presented an opportunity for institutional wholesalers. During the 1980s the leading food service distributors (such as Sysco, Rykoff-Sexton, PYA/Monarch, and CFS Continental) formed entire divisions to procure and merchandise produce.

These changes in produce buying units have enabled a growing portion of food service produce to be procured directly from the shipping point. Some food service distributors invested in shipping-point firms or formed joint buying groups in production regions. This increased their negotiating power and enabled them to exert better control over product

**8**

quality, packaging, and consistency of supply. This is a significant development because before 1980, 80 percent of produce moved through retail channels, and retailers dictated standards in the produce trade.

Many shippers are developing special food service packs (smaller than retail packs) and lightly processed products (e.g., peeled garlic, broccoli florets, cut or cored lettuce) to cut waste and labor at the operator level. Limited availability of labor and high worker turnover are major problems for food service operators, and even fewer workers are projected to enter the labor force over the next decade. Furthermore, liability costs for restaurants and institutions are rising due to employee accidents from handling knives, and the cost of kitchen space is rising in urban areas. Consequently, demand for lightly processed products is expected to grow rapidly. A major industry debate exists over whether it is preferable to process at the shipping point, where product freshness is at its maximum level, or at the destination, where product reworking can occur. Both require optimal temperature management throughout the distribution system to maximize marketable yield.

The efforts of shippers to meet the special needs of food service users are complicated by the segmented nature of the food service industry. The commercial sector purchases 60 percent of the dollar volume of food sold through food service channels and the noncommercial sector accounts for 40 percent. Figure 2.2, showing the breakdown in purchases by user type, highlights the degree of segmentation. The restaurant category alone consisted of 256,600 establishments in 1988, including 105,800 fast-food establishments, 64,800 casual restaurants, 68,000 dinner houses, and 18,000 fine dining restaurants.

Food service establishments differ in their quality and packaging needs; some handle a much greater variety of produce than others. Figure 2.3, for example, shows that fine dining restaurants have the highest incidence of serving fresh fruits and vegetables. Over the next 20 years midscale (casual) restaurants can be expected to gain market share relative to fast-food restaurants, raising fresh produce sales through food service channels. This is because the U.S. population is aging, and midscale restaurants are patronized more by older consumers than by the population in general (National Restaurant Association 1989). Further, mature consumers have a greater preference than do most consumers for salads, vegetables, and nonfried potatoes.

## Shippers and new entrants

The change to fewer, larger integrated wholesale-retailer buyers and the rise in consolidated buying in food service channels have furthered the development of large-scale shippers based in production regions. Retailers and food service users demand more services today, including (1) information on product attributes, recipes, and merchandising, (2) ripening and other special handling and packaging, and (3) year-round availability of a wide line of consistent quality fruits and vegetables. Shippers have responded with improved communication programs and by becoming multiregional and multicommodity. Many California and Florida shippers obtain products from other countries during the off-season, sometimes via joint ventures. This enables shippers

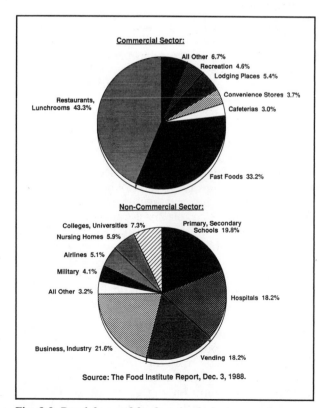

**Fig. 2.2. Breakdown of foodservice industry purchases: 1988 (percentages have been rounded).**

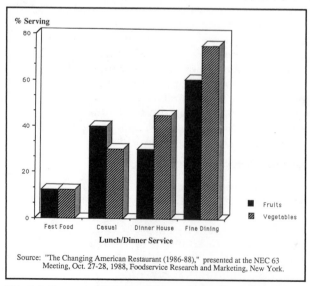

**Fig. 2.3. Percent of U.S. restaurants, by type, serving fresh fruits and vegetables.**

to extend shipping seasons and sell products produced in several locations via one marketing organization, maintaining a year-round presence in the marketplace. For example, shippers based in Salinas, California, also commonly ship out of the San Joaquin Valley, Imperial Valley, southwestern Arizona, and Mexico.

The rapid growth in multilocation firms has contributed to the integration of the Mexico-California-Arizona vegetable industries, in particular. Because most vegetable crops are not perennials, the location of production can shift readily, based on relative production and marketing costs and growing season.

New entrants to the produce industry are challenging independent shippers and grower cooperatives. Multinational food processors entered the fresh produce market during the 1980s as consumption of canned produce declined. These firms are applying their branded marketing strategies to produce and are contracting with producers here and in foreign regions to ensure a year-round market presence for their brands. They are acquiring produce wholesalers and shippers to broaden their base of commodities and distribution channels.

## Value-Added Products and Branding: Opportunities and Constraints

### Convenience products

The growing demand for convenience is influencing the product form in which fruits and vegetables are consumed. New, lightly processed products meet the demand for food that is both healthful and convenient. Pre-packaged salads, pre-cut fruit, broccoli and cauliflower florets, sliced mushrooms, cored pineapples, stir-fry vegetable mixes, packaged carrot and celery sticks, and pre-cut vegetables with cheese sauces in microwaveable trays are examples of the attempt to add value to produce without losing its fresh, natural image.

As vertical and horizontal integration increase in the fresh produce industry, investment in value-added products will stimulate new marketing and distribution strategies. For example, produce marketers are working with biotechnology firms to develop convenience-oriented products with unique flavor attributes. A case in point is VegiSnax, developed by FreshWorld Inc., a joint venture of DNA Plant Technology and DuPont. VegiSnax are pre-cut, packaged celery and carrot sticks, made from differentiated varieties bred for specific flavor and other attributes. The link between improved, proprietary varieties and branding is being explored by several other firms.

Modified-atmosphere packaging will increasingly be used with value-added products, already a common practice in Britain. If British investment in the U.S. food system continues to grow, some chains may introduce private labels of pre-cut produce in mod-ified-atmosphere packages (like those of Marks and Spencer), giving much greater retail control over product quality and packaging specifications. From the shipper's perspective, modified-atmosphere packaging will provide opportunities in new export markets but will make U.S. shippers more vulnerable to imports of certain commodities.

The still small but growing presence of value-added/convenience-oriented produce is evident from *The Packer*'s "Fresh Trends '90" survey. Fifty-two percent of consumers had self-service salad bars in the primary store they shop in, while 42 percent reported cut-and-cored pineapple and 18 percent had access to self-service or tended juice bars. The proportion of consumers who bought particular convenience items, (table 2.4), ranged from 75 percent for bagged grapes to 16 percent for overwrapped grapefruit halves. Although many consumers have tried these convenience items, frequency of purchase is still low. For example, only three items were purchased more than once a month by a majority of the consumers who had tried them: bagged grapes, broccoli and cauliflower florets, and the overwrapped mixed fruit trays.

Retailers report that although it is expanding, the market for convenience-oriented produce is still small, making it difficult to target heavy users. Inadequate targeting that contributes to low turnover and poor product appearance may further inhibit sales. Poor temperature management in retail displays also restricts expansion of this product category. Some sup-

**Table 2.4. Percent of consumers having purchased packaged-for-convenience/ready-to-eat produce items in past 12 months/1989–1990**

| Product | 1990 | 1989 |
|---|---|---|
| **Fruits** | | |
| Preportioned, bagged grapes | 75 | 74 |
| Tray-wrapped whole fruit items | 42 | NA |
| Tray-wrapped melon slices | 36 | 29 |
| Tray-wrapped individual melon halves (other than watermelon) filled with other fruit | 28 | 21 |
| Melon chunks with or without other fruit chunks in a lidded plastic container | 21 | 15 |
| Fresh fruit salad/"cocktail" in a single-serve lidded cut or tray | 21 | 16 |
| Tray-wrapped individual grapefruit halves with or without other fruit | 16 | 13 |
| None of the above | 10 | 13 |
| **Vegetables** | | |
| Shredded salad/slaw mix | 40 | 41 |
| Broccoli buds/florets | 32 | 27 |
| Cauliflower buds/florets | 29 | 25 |
| Asparagus tips/trimmed spears | 30 | 27 |
| Pre-cut carrot sticks | 24 | 21 |
| Pre-cut celery sticks | 21 | 20 |
| "Combination pack" of any pre-cut items | 20 | 19 |
| Sliced/pre-cut mushrooms | 23 | NA |
| None of the above | 31 | 34 |

Source: *Fresh Trends '90: A Profile of the Fresh Produce Consumer*, Vance Research Services and Market Facts, Inc. 1990.
NA = Not available

pliers of pre-cut packaged produce are providing special refrigerated cases along with frequent direct store delivery and stocking.

Successful development of convenience-oriented produce depends on whether enough consumers will pay for (and firms can achieve) the high level of management and coordination needed for efficient distribution of this kind of product.

### Concept of produce brands

Successful produce brands have been limited in the past because of the need for (1) year-round availability, (2) a consistent, high-quality supply, (3) a differentiated product, and (4) proper handling throughout the cold chain. Indeed, despite the efforts described above to market branded produce, "Fresh Trends '90" reported that most consumers viewed branded produce as about the same quality as nonbranded produce.

Furthermore, branding ranks last among 16 factors that influence produce purchases (table 2.5). Few "Fresh Trends '90" respondents preferred specific brands for items they had purchased in the last year. Those expressing no preference for a vegetable brand ranged from 54 percent for asparagus to 94 percent for iceberg lettuce. Potatoes were mentioned most as having a preferred vegetable brand (13 percent), but the responses were growing regions or varieties, not brand names. "Idaho," the principal response, was mentioned by 8 percent. The leading actual brand for a vegetable commodity was Campbell for mushrooms, identified by 2 percent. Fruits fared better: 17 percent each identified Chiquita and Dole as the preferred brand for bananas, 19 percent named Sunkist (29 percent expressed a preference for a brand of oranges), and 9 percent named Ocean Spray cranberries.

**Table 2.5. Factors indicated by consumers as influencing produce purchases, 1990**

| Factor | Rating of extremely or very important % |
|---|---|
| Taste/flavor | 96 |
| Freshness or ripeness | 96 |
| Appearance/condition | 94 |
| Nutritional value | 65 |
| Price | 63 |
| Storage/shelf life | 56 |
| Certified safe by residue testing | 52 |
| Convenient to eat/prepare | 47 |
| Size | 39 |
| In-season | 38 |
| Displayed loose (bulk) | 37 |
| Calorie content | 24 |
| Growing region/state/country of origin | 17 |
| Organically grown | 17 |
| Prepackaged | 11 |
| Brand name (of the grower or marketer) | 9 |

Source: *Fresh Trends '90: A Profile of the Fresh Produce Consumer,* Vance Research Services and Market Facts, Inc., 1990.

Hence, except for a few long-standing brands, branded messages have had limited impact. While obstacles to developing widely recognized consumer brands for produce are many, branding could deliver a positive jolt, in terms of advertising and merchandising, to categories where they are introduced. Fresh produce has been undermerchandised and underpromoted relative to packaged food products. Over the long run, successful brands could stimulate greater consumption of produce.

## Consumer Issues

### Postharvest handling

The importance of proper postharvest handling and temperature management in stimulating sales is highlighted by the results of "Fresh Trends '90" in table 2.5. Over 90 percent of consumers asked to rank the factors influencing their buying decisions almost unanimously put taste/flavor, freshness/ripeness, and appearance/condition in the top three. Consumers clearly base buying decisions on what looks good and appears likely to taste good. Least influential are the geographic origin of the product and whether it was organically grown, prepackaged, or branded.

The importance of in-store merchandising, appearance, and freshness is underscored by the extent of impulse purchases of produce. Only 32 percent of consumers report shopping with a written or mental list of the produce items they plan to buy, while 39 percent decide in the produce department and 29 percent know only the general category of produce they plan to buy. Shelf positioning will matter even more in the future as product diversity increases, and optimal postharvest handling will be critical to maximizing sales.

### Specialty produce

The niche market for unusual or exotic produce rapidly expanded in the 1980s. Larger ethnic populations and growth in their cultural expression have augmented the demand for product diversity as these consumers seek out traditional foods. Furthermore, foods once considered ethnic or regional are being consumed by a broader portion of the population; jicama, various chilies, cilantro, tomatillos, chayote, bok choy, napa cabbage, papayas, and mangos are now common in supermarkets. Demand for other specialty items grows as Americans are exposed to European specialties through international travel. Radicchio, arugula, endive, red oakleaf lettuce, escarole, and other Italian and French vegetables have become popular on the West and East coasts.

An expanding segment of specialty produce is varieties of traditional items grown primarily for their eating characteristics rather than for yield or shipping attributes. They often have superior taste and are generally distributed through upscale restaurants, farmers' markets, and specialized produce retailers. Common examples are Blenheim apricots, special varieties of vine-ripe tomatoes, tree-ripened peaches, lady apples, and sweet corn. Many of them, available years ago, are marketed as "heirloom" varieties. They are produced by "boutique growers," farmers who target restaurant chefs and upscale consumers that are willing to pay a premium.

Indeed, specialty products are generally introduced to the American palate first through upscale and ethnic restaurants, then through farmers' markets and exotic produce sections in supermarkets. Many items eventually are included in conventional produce displays.

Information from the U.S. Department of Agriculture (USDA) is available primarily for specialty vegetables and herbs rather than specialty fruit. Shipments of specialty vegetables reached 1.25 billion pounds in 1989, up 212 percent since 1980 (USDA March 1990). Twenty-six percent of the 1989 volume was imported, compared with 18.5 percent of all fresh produce shipments. The specialty vegetables imported in the largest volumes are chili peppers, dasheens, bamboo shoots, and jicama.

Reported shipments of specialty vegetables account for only about 5 percent of major fresh vegetable shipments. However, many exotic items do not appear in these USDA data due to their small individual volumes, meaning that the actual movement is underreported. Further, specialty shipments grew at an average annual rate of 13 percent during the 1980s, compared with 2 percent annual growth for the major vegetables, portending greater relative importance in the future. Specialty/leaf lettuces represent 57 percent of reported specialty vegetable shipments. California is the largest producer of specialties, with an estimated market share of 86 percent. In 1988 California harvested 155,069 acres of specialty and minor vegetables (USDA November 1989).

Figures from the 1987 Census of Agriculture demonstrate that despite its small size, the herb industry is one of the fastest growing segments of the produce industry. In 1987 production was 9.8 million pounds, up from 2.1 million pounds in 1978. Harvested acreage more than doubled over this period, and the number of farms more than tripled. California, the dominant producer, harvested 6 million pounds in 1987 from 1,675 acres. Florida and New Jersey rank second and third, respectively, in herb production. While most herbs are distributed through food service channels, retail distribution is expanding, and a variety of retail packs are now on the market.

## Food safety

The issue of food safety has been thrust into the national spotlight, affecting the fresh produce industry more than any other segment of the food industry. Several factors have combined to make food safety a high-profile issue: the advent of more sensitive residue-testing technology, the news media's fear-producing reporting style, inconsistencies in regulatory policies and the pronouncements of regulatory officials, exploitation of the issue by parts of the food industry, and the politicization of the issue as advocacy groups link it to their own agendas.

Despite extensive scientific evidence to the contrary, the public equates food safety hazards with pesticide residues. *The Packer*'s "Fresh Trends '90" survey found 86 percent of its respondents concerned about chemical residues on produce. A 1990 survey by the Food Marketing Institute (FMI) found that 80 percent of respondents were concerned about pesticide residues on food.

Yet the limitations of consumer survey methodology compel caution in interpreting these findings. The same FMI survey asked consumers whether they were confident that the food in their supermarket was safe, and 79 percent responded affirmatively. This would indicate that while consumers may be concerned in a general sense, this concern does not necessarily outweigh an underlying confidence in the system.

The "Fresh Trends '90" survey supports this conclusion. Sixty percent of respondents said that while they were concerned, they had not changed their buying habits. However, a notable 26 percent had modified their buying practices, up from 18 percent in 1988. Eleven percent had purchased organically grown fresh fruits and vegetables and 15 percent had sought out fresh produce merchandised as pesticide- or chemical-free. However, merely making one purchase of an item with a food safety-oriented marketing label in the past 12 months qualified respondents as having changed their buying behavior; hence, these figures cannot be translated into sales volumes. Furthermore, there was significant regional variation in the results, the incidence of pesticide-free or organic purchases being substantially higher in the West than in the South.

## Organic foods

Several surveys indicate that consumers prefer organic foods and in many cases say they are willing to pay more for them. For example, a national poll conducted in 1989 for *Organic Gardening* magazine found that 84 percent of respondents preferred organics, and 44 percent said they would pay more for organic produce. When consumers were asked in the "Fresh Trends '90" survey whether they prefer organic produce regardless of cost, 57 percent agreed to some extent (only 9 percent strongly agreed). However, it is important to recognize that these questions were asked in isolation, whereas consumers weigh numerous factors when making buying decisions. Recall from table 2.5 that when consumers were asked to *rank* how various produce attributes affect their buying decisions, only 17 percent felt that the classification, organic, was important. In other words, food safety labeling may only infrequently determine the actual buying decision of most consumers; greater weight is given other produce attributes.

Until recently organic products were primarily produced by small growers unable to meet chain store requirements for large volume, consistent quality packs. Organic products were mainly distributed through health food stores, consumer food cooperatives, and farmers' markets. It was hypothesized that if organic products were commonly available in supermarkets, organics would quickly develop into a mass market.

Developments have provided a 2-year experiment with organics through conventional supermarket channels. After the Alar publicity of February 1989, many U.S. retailers introduced organic sections, and several large-scale grower-shippers in California now produce organic products in large enough quantities to sell directly to chains. Organic distributors, too, have developed chain store distribution programs. Yet national interest in organic produce that surged during the Alar publicity quickly subsided. National marketplace research has shown that many chains have eliminated their organic sections or reduced the number of stores handling organics due to weak sales and high shrink. Just as with other niche markets, targeting heavy users can be difficult. It appears that most consumers motivated enough to change their buying behavior and pay a price premium on a regular weekly basis were already shopping in health food stores.

The nation's primary target market for products merchandised with food safety labels, especially organic, is the San Francisco Bay Area. Twelve major health food stores in the Bay Area sell high volumes ($15,000 to $30,000 weekly store sales) of organic, transitional organic, and "unsprayed" products. The most successful organic programs in conventional supermarkets are also in the Bay Area. Part of this success is likely due to high concentrations of target consumers in a small area, a large pool of organic growers in surrounding counties with direct marketing links to Bay Area retailers and distributors, and the efforts of distributors to extend shipping seasons and improve supply consistency by encouraging production in several California regions and developing joint ventures with producers in Mexico. Chain store distribution has been aided by decentralized ordering systems that place individual store produce managers and distributors in direct contact, so that the product does not lose freshness by passing through the chain's central distribution center.

Nationally, however, most chains report that organics contribute only 0.5 percent of produce department sales. Organic sections are generally unprofitable for conventional retailers because of slightly lower average gross margins, low product turnover, and higher shrink compared to conventional produce (25 percent shrink versus 6 percent for conventional produce). Total retail organic produce sales through both conventional and health food channels are an estimated $210 to $280 million.

On the supply side, many growers with large organic acreages report that most of their product must still be sold at conventional prices due to limited demand. Most organic growers remain small and target specialty markets. While the organic production industry in California has expanded dramatically, it is estimated to account for less than 2 percent of the state's production of horticultural commodities and $100 to $110 million in farmgate sales in 1989. There are an estimated 1,200 to 1,300 organic growers in California, most of whom are not certified as organic but rather market their product as complying with the California Organic Food Act. A private grower certification group, California Certified Organic Farmers (CCOF), had 585 members in 1989.

The second most important state in organic production, Washington, had $15 million in farmgate sales and 262 organic growers in 1989. Nationally, farmgate sales of all organically grown commodities are an estimated $200 million. Based on counts of organic producers obtained by the author from all state or private certification organizations and other information on uncertified producers, there are approximately 5,400 organic growers nationally. This includes organic producers of all agricultural commodities and is about 0.3 percent of all farmers.

High prices, substandard appearance of some commodities, and inconsistent supply are reported by retailers and consumers as the primary factors limiting consumer demand for organics. Price research by Consumer Alert in 1989 found that organic food shopping costs a typical family of four about $1,000 a year more (about 25 percent) than buying food from conventional retailers. Clearly, as organic production expands and prices decline the market will grow, but this will move the organic industry from a specialty toward a commodity orientation. As they cannot rely on price premiums, organic producers will probably strive for production and marketing costs that are competitive with those of conventional producers. As always, it is easy for growers to saturate a market niche.

This is particularly true for organically grown soft fruits, since use of most synthetic postharvest fungicides is prohibited. This constraint tends to limit these crops to a local market. Nonchemical postharvest treatments are critical to development of a large-volume organic soft fruit industry, capable of long-distance shipping. However, as more chemical registrations are lost over the next decade, conventional producers too will need to place greater reliance on nonchemical control methods. Hence, we can expect much more emphasis on temperature management at the carrier level, along with other efforts to streamline the distribution system.

## Conclusions

Fresh produce consumption will continue to expand in the 1990s. Improved postharvest handling and transportation technologies have led to the development of a modern distribution system for highly perishable fruit and vegetable commodities. Demands for even better performance will increase as product diversity grows, postharvest fungicides become less available, and world trade expands. Successful produce marketing firms will become more market driven, identifying and meeting the specific needs of each market segment for quality, packaging, product form, merchandising and information. More produce marketing firms will adopt a distribution system approach, emphasizing faster delivery, better temperature management, and improved packaging technologies, all based on better demand information.

## REFERENCES

1. California Department of Food and Agriculture. 1989. *California Agriculture: Statistical Review for 1988*. 28 pp.

2. Cook, R. L. 1990. Evolving vegetable trading relationships. *J. Food Distrib. Res.* 21(1):31-46.

3. ———. 1990. Challenges and opportunities in the U.S. fresh produce industry. *J. Food Distrib. Res.* 21(1):67-74.

4. *Food Chemical News.* 1989. Survey finds organic food shopping costs 25 percent more. *Food Chemical News* 31(9). (May 1)

5. Food Marketing Institute. 1990. *1990 Trends: Consumer Attitudes and the Supermarket.* Washington, DC: Food Mktg. Inst. 68 pp.

6. Foodservice Research and Marketing. 1988. The changing American restaurant (1986-1988). Paper presented at NEC-63 Meeting, Oct. 27-28, 1988.

7. Mayer, S. D. 1988. U.S. foodservice industry: responsive and growing. In *Marketing U.S. agriculture: 1988 yearbook of agriculture*. Washington, DC: U.S. Dept. Agric., pp. 86-90.

8. McCracken, V. 1990. The U.S. demand for vegetables. In *Vegetable markets in the western hemisphere*, eds. R. A. Lopez and L. C. Polopolus. Ames, IA: Iowa State Univ. Press.

9. McLaughlin, E. W. 1983. Buying and selling practices in the fresh fruit and vegetable industry: implications for vertical coordination. Unpub. Ph.D. Dissert., Michigan State Univ. 484 pp.

10. National Restaurant Association. 1989. A profile of the mature market. *Restaurants USA* 9(9). (October)

11. Produce Marketing Association. 1989. *Marketing insights '89: produce retailing performance and productivity 1988.* Produce Mktg. Assoc., pp. 9-17.

12. *Supermarket Business.* 1989. 42nd annual consumer expenditures study. *Supermarket Business* 44(9). (September)

13. The Food Institute. 1988. *Food retailing review.* Am. Inst. Food Distrib., Inc. 295 pp. (February)

14. ———. 1989. *Demographic directions.* Vol. 1. Am. Inst. Food Distrib., Inc. 207 pp. (July)

15. *The Food Institute Report.* 1988. Slow growth in foodservice. (December 3)

16. ———. 1990. New product introductions finished at record pace in 1989. (January 27)

17. U.S. Department of Agriculture, Agricultural Marketing Service. Various Issues. *Fresh fruit and vegetable shipments by commodities, states, and months. 1980-1988.*

18. U.S. Department of Agriculture, Economic Research Service. 1989. *Foreign agricultural trade of the United States (FATUS), calendar year 1988 supplement.* (July)

19. ———. 1989. *Fruit and tree nuts situation and outlook yearbook.* (August)

20. ———. 1989. *Vegetables and specialties situation and outlook yearbook.* (November)

21. ———. 1990. *Vegetables and specialties situation and outlook report.* (March)

22. ———. 1990. *Vegetables and specialties situation and outlook report.* (August)

23. U.S. Department of Agriculture, Foreign Agricultural Service. 1989. *Horticultural products review.* (March)

24. Vance Research Services and Market Facts, Inc. 1990. *Fresh trends '90: a profile of the fresh produce consumer.* Reports 1-4.

# 3

# Postharvest Biology and Technology: An Overview

Adel A. Kader

Losses in quantity and quality affect horticultural crops between harvest and consumption. The magnitude of postharvest losses in fresh fruits and vegetables is an estimated 5 to 25 percent in developed countries and 20 to 50 percent in developing countries, depending upon the commodity. To reduce these losses, producers and handlers must (1) understand the biological and environmental factors involved in deterioration and (2) use postharvest techniques that delay senescence and maintain the best possible quality. This chapter briefly discusses the first item and introduces the second, which is covered in detail in subsequent chapters.

Fresh fruits, vegetables, and ornamentals are living tissues subject to continuous change after harvest. While some changes are desirable, most—from the consumer's standpoint—are not. Postharvest changes in fresh produce cannot be stopped, but they can be slowed within certain limits. Senescence is the final stage in the development of plant organs during which a series of irreversible events leads to breakdown and death of the plant cells.

Fresh horticultural crops are diverse in morphological structure (roots, stems, leaves, flowers, fruits, and so on), in composition, and in general physiology. Thus, commodity requirements and recommendations for maximum postharvest life vary among the commodities. All fresh horticultural crops are high in water content and thus are subject to desiccation (wilting, shriveling) and to mechanical injury. They are also susceptible to attack by bacteria and fungi, with pathological breakdown the result.

## Biological Factors Involved in Deterioration

### Respiration

Respiration is the process by which stored organic materials (carbohydrates, proteins, fats) are broken down into simple end products with a release of energy. Oxygen ($O_2$) is used in this process and carbon dioxide ($CO_2$) is produced. The loss of stored food reserves in the commodity during respiration means (1) the hastening of senescence as the reserves that provide energy to maintain the commodity's living status are exhausted, (2) reduced food value (energy value) for the consumer, (3) loss of flavor quality, especially sweetness, and (4) loss of salable dry weight (especially important for commodities

destined for dehydration). The energy released as heat, known as vital heat, affects postharvest technology considerations, such as estimations of refrigeration and ventilation requirements.

The rate of deterioration (perishability) of harvested commodities is generally proportional to the respiration rate. Horticultural commodities are classified according to their respiration rates in table 3.1. Based on their respiration and ethylene production patterns during maturation and ripening, fruits are either climacteric or nonclimacteric (table 3.2). Climacteric fruits show a large increase in $CO_2$ and ethylene ($C_2H_4$) production rates coincident with ripening, while nonclimacteric fruits show no change in their generally low $CO_2$ and $C_2H_4$ production rates during ripening.

**Table 3.1. Horticultural commodities classified according to their respiration rates**

| Class | Range at 5°C (41°F) (mg $CO_2$/kg-hr)* | Commodities |
|---|---|---|
| Very low | <5 | Dates, dried fruits and vegetables, nuts |
| Low | 5–10 | Apple, beet, celery, citrus fruits, cranberry, garlic, grape, honeydew melon, kiwifruit, onion, papaya, persimmon, pineapple, potato (mature), sweet potato, watermelon |
| Moderate | 10–20 | Apricot, banana, blueberry, cabbage, cantaloupe, carrot (topped), celeriac, cherry, cucumber, fig, gooseberry, lettuce (head), mango, nectarine, olive, peach, pear, plum, potato (immature), radish (topped), summer squash, tomato |
| High | 20–40 | Avocado, blackberry, carrot (with tops), cauliflower, leeks, lettuce (leaf), lima bean, radish (with tops), raspberry |
| Very high | 40–60 | Artichoke, bean sprouts, broccoli, brussels sprouts, cut flowers, endive, green onions, kale, okra, snap bean, watercress |
| Extremely high | >60 | Asparagus, mushroom, parsley, peas, spinach, sweet corn |

*Vital heat (Btu/ton/24 hrs) = mg $CO_2$/kg-hr × 220.
Vital heat (kcal/1000 kg/24 hrs) = mg $CO_2$/kg-hr × 61.2.

## Table 3.2. Fruits classified according to respiratory behavior during ripening

| Climacteric fruits | | Nonclimacteric fruits | |
|---|---|---|---|
| Apple | Muskmelon | Blackberry | Lychee |
| Apricot | Nectarine | Cacao | Okra |
| Avocado | Papaya | Carambola | Olive |
| Banana | Passion fruit | Cashew apple | Orange |
| Biriba | Peach | Cherry | Peas |
| Blueberry | Pear | Cucumber | Pepper |
| Breadfruit | Persimmon | Date | Pineapple |
| Cherimoya | Plantain | Eggplant | Pomegranate |
| Durian | Plum | Grape | Prickly pear |
| Feijoa | Quince | Grapefruit | Raspberry |
| Fig | Rambutan | Jujube | Strawberry |
| Guava | Sapodilla | Lemon | Summer squash |
| Jackfruit | Sapote | Lime | Tamarillo |
| Kiwifruit | Soursop | Longan | Tangerine & |
| Mango | Tomato | Loquat | Mandarin |
| | | | Watermelon |

## Table 3.3 Horticultural commodities classified according to ethylene production rates

| Class | Range at 20°C (68°F) ($\mu$l $C_2H_4$/kg-hr) | Commodities |
|---|---|---|
| Very low | Less than 0.1 | Artichoke, asparagus, cauliflower, cherry, citrus fruits, grape, jujube, strawberry, pomegranate, leafy vegetables, root vegetables, potato, most cut flowers |
| Low | 0.1–1.0 | Blackberry, blueberry, casaba melon, cranberry, cucumber, eggplant, okra, olive, pepper (sweet and chili), persimmon, pineapple, pumpkin, raspberry, tamarillo, watermelon |
| Moderate | 1.0–10.0 | Banana, fig, guava, honeydew melon, lychee, mango, plantain, tomato |
| High | 10.0–100.0 | Apple, apricot, avocado, cantaloupe, feijoa, kiwifruit (ripe), nectarine, papaya, peach, pear, plum |
| Very high | More than 100.0 | Cherimoya, mammee apple, passion fruit, sapote |

## Ethylene production

Ethylene, the simplest of the organic compounds affecting the physiological processes of plants, is a natural product of plant metabolism and is produced by all tissues of higher plants and by some microorganisms. As a plant hormone, ethylene regulates many aspects of growth, development, and senescence and is physiologically active in trace amounts (less than 0.1 ppm). It also plays a major role in the abscission of plant organs.

The amino acid methionine is converted to S-adenosylmethionine (SAM) which is the precursor of 1-aminocyclopropane-1-carboxylic acid (ACC), the immediate precursor of ethylene. ACC synthase, which converts SAM to ACC, is the main site of control of ethylene biosynthesis. The conversion of ACC into ethylene is mediated by an enzyme (ethylene forming enzyme, EFE or ACC oxidase). This enzyme, as yet unidentified, is known to be very labile and is assumed to be membrane-bound.

Horticultural commodities are classified according to their ethylene production rates in table 3.3. There is no consistent relationship between the $C_2H_4$ production capacity of a given commodity and its perishability; however, exposure of most commodities to $C_2H_4$ accelerates their senescence.

Generally, $C_2H_4$ production rates increase with maturity at harvest, physical injuries, disease incidence, increased temperatures up to 30°C (86°F), and water stress. On the other hand, ethylene production rates by fresh horticultural crops are reduced by storage at low temperature, by reduced $O_2$ (less than 8 percent) levels, and elevated $CO_2$ (more than 2 percent) levels around the commodity.

## Compositional changes

Many changes in pigments take place during development and maturation of the commodity on the plant. Some may continue after harvest and can be desirable or undesirable.

**1.** Loss of chlorophyll (green color) is desirable in fruits but not in vegetables.

**2.** Development of carotenoids (yellow and orange colors) is desirable in fruits such as apricots, peaches, and citrus; red color development in tomatoes and pink grapefruit is due to a specific carotenoid (lycopene); beta-carotene is provitamin A and thus is important in nutritional quality.

**3.** Development of anthocyanins (red and blue colors) is desirable in fruits such as apples (red cultivars), cherries, strawberries, cane berries, and red-flesh oranges; these water-soluble pigments are much less stable than carotenoids.

**4.** Changes in anthocyanins and other phenolic compounds may result in tissue browning, which is undesirable for appearance quality.

Changes in carbohydrates include (1) starch-to-sugar conversion (undesirable in potatoes, desirable in apple, banana, and other fruits), (2) sugar-to-starch conversion (undesirable in peas and sweet corn; desirable in potatoes), and (3) conversion of starch and sugars to $CO_2$ and water through respiration. Breakdown of pectins and other polysaccharides results in softening of fruits and a consequent increase in susceptibility to mechanical injuries. Increased lignin content is responsible for toughening of asparagus spears and root vegetables.

Changes in organic acids, proteins, amino acids, and lipids can influence flavor quality of the commodity. Loss in vitamin content, especially ascorbic acid (vitamin C) is detrimental to nutritional quality. Production of flavor volatiles associated with ripening of fruits is very important to their eating quality.

## Growth and development

Sprouting of potatoes, onions, garlic, and root crops greatly reduces their utilization value and accelerates deterioration. Rooting of onions and root crops is also undesirable. Asparagus spears continue to grow after harvest; elongation and curvature (if the spears are held horizontally) are accompanied by increased toughness and decreased palatability. Similar geotropic responses occur in cut gladiolus and snapdragon flowers stored horizontally. Seed germination inside fruits such as tomatoes, peppers, and lemons is an undesirable change.

## Transpiration

Water loss is a main cause of deterioration because it results not only in direct quantitative losses (loss of salable weight), but also in losses in appearance (wilting and shriveling), textural quality (softening, flaccidity, limpness, loss of crispness and juiciness), and nutritional quality.

The dermal system (outer protective coverings) governs the regulation of water loss by the commodity. It includes the cuticle, epidermal cells, stomata, lenticles, and trichomes (hairs). The cuticle is composed of surface waxes, cutin embedded in wax, and a layer of mixtures of cutin, wax, and carbohydrate polymers. The thickness, structure, and chemical composition of the cuticle vary greatly among commodities and among developmental stages of a given commodity.

Transpiration rate is influenced by internal or commodity factors (morphological and anatomical characteristics, surface-to-volume ratio, surface injuries, and maturity stage) and external or environmental factors (temperature, relative humidity, air movement, and atmospheric pressure). Transpiration (evaporation of water from the plant tissues) is a physical process that can be controlled by applying treatments to the commodity (e.g., waxes and other surface coatings and wrapping with plastic films) or by manipulating the environment (e.g., maintenance of high relative humidity and control of air circulation).

## Physiological breakdown

Exposure of the commodity to undesirable temperatures can result in physiological disorders.

**1.** Freezing injury results when commodities are held below their freezing temperatures. The disruption caused by freezing usually results in immediate collapse of the tissues and total loss.

**2.** Chilling injury occurs in some commodities (mainly those of tropical and subtropical origin) held at temperatures above their freezing point and below 5° to 15°C (41° to 59°F) depending on the commodity. Chilling injury symptoms become more noticeable upon transfer to higher (nonchilling) temperatures. The most common symptoms are surface and internal discoloration (browning), pitting, watersoaked areas, uneven ripening or failure to ripen, off-flavor development, and accelerated incidence of surface molds and decay (especially organisms not usually found growing on healthy tissue).

**3.** Heat injury is induced by exposure to direct sunlight or to excessively high temperatures. Its symptoms include bleaching, surface burning or scalding, uneven ripening, excessive softening, and desiccation.

Certain types of physiological disorders originate from preharvest nutritional imbalances. For example, blossom-end rot of tomatoes and bitter pit of apples result from calcium deficiency. Increasing calcium content via preharvest or postharvest treatments can reduce the susceptibility to physiological disorders. Calcium content also influences the textural quality and senescence rate of fruits and vegetables; increased calcium content has been associated with improved firmness retention, reduced $CO_2$ and $C_2H_4$ production rates, and decreased decay incidence.

Very low oxygen (<1 percent) and high carbon dioxide (>20 percent) atmospheres can cause physiological breakdown of most fresh horticultural commodities. Ethylene can induce physiological disorders in certain commodities. The interactions among $O_2$, $CO_2$, and $C_2H_4$ concentrations, temperature, and duration of storage influence the incidence and severity of physiological disorders related to atmospheric composition.

## Physical damage

Various types of physical damage (surface injuries, impact bruising, vibration bruising, and so on) are major contributors to deterioration. Browning of damaged tissues results from membrane disruption, which exposes phenolic compounds to the polyphenol oxidase enzyme. Mechanical injuries not only are unsightly but also accelerate water loss, provide sites for fungal infection, and stimulate $CO_2$ and $C_2H_4$ production by the commodity.

## Pathological breakdown

One of the most common and obvious symptoms of deterioration results from the activity of bacteria and fungi. Attack by most organisms follows physical injury or physiological breakdown of the commodity. In a few cases, pathogens can infect apparently healthy tissues and become the primary cause of deterioration. In general, fruits and vegetables exhibit considerable resistance to potential pathogens during most of their postharvest life. The onset of ripening in fruits, and senescence in all commodities, renders them susceptible to infection by pathogens. Stresses, such as mechanical injuries, chilling, and sunscald, lower the resistance to pathogens.

# Environmental Factors Influencing Deterioration

**Temperature.** Temperature is the environmental factor that most influences the deterioration rate of harvested commodities. For each increase of 10°C (18°F) above optimum, the rate of deterioration increases by two- to threefold (table 3.4). Exposure to undesirable temperatures results in many physiological disorders, as mentioned above. Temperature also influences the effect of ethylene, reduced oxygen, and elevated carbon dioxide. Spore germination and growth rate of pathogens are greatly influenced by temperature; for instance, cooling commodities below 5°C (41°F) immediately after harvest can greatly reduce the incidence of Rhizopus rot. Temperature effects on postharvest responses of chilling-sensitive and nonchilling-sensitive horticultural crops are compared in table 3.5.

**Relative humidity.** The rate of water loss from fruits and vegetables depends upon the vapor pressure deficit between the commodity and the surrounding ambient air, which is influenced by temperature and relative humidity. At a given temperature and rate of air movement, the rate of water loss from the commodity depends on the relative humidity. At a given relative humidity, water loss increases with the increase in temperature.

**Atmospheric composition.** Reduction of oxygen and elevation of carbon dioxide, whether intentional (modified or controlled atmosphere storage) or unintentional (restricted ventilation within a shipping container and/or a transport vehicle), can either delay or accelerate deterioration of fresh horticultural crops. The magnitude of these effects depends upon commodity, cultivar, physiological age, $O_2$ and $CO_2$ level, temperature, and duration of holding.

**Ethylene.** The effects of ethylene on harvested horticultural commodities can be desirable or undesirable, thus it is of major concern to all produce handlers. Ethylene can be used to promote faster and more uniform ripening of fruits picked at the mature-green stage. On the other hand, exposure to ethylene can be detrimental to the quality of most nonfruit vegetables and ornamentals.

## Table 3.5. Fruits and vegetables classified according to sensitivity to chilling injury

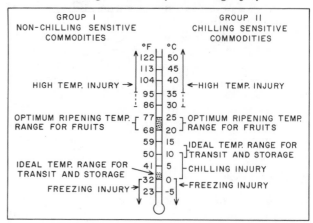

| GROUP I NON-CHILLING SENSITIVE COMMODITIES | | GROUP II CHILLING SENSITIVE COMMODITIES | |
|---|---|---|---|
| Apples* | Artichokes | Avocados | Beans, snap |
| Apricots | Asparagus | Bananas | Cassava |
| Blackberries | Beans, lima | Breadfruit | Cucumbers |
| Blueberries | Beets | Carambola | Eggplant |
| Cherries | Broccoli | Cherimoya | Ginger |
| Currants | Brussels sprouts | Citrus | Muskmelons |
| Dates | Cabbage | Cranberry | Okra |
| Figs | Carrots | Durian | Peppers |
| Grapes | Cauliflower | Feijoa | Potatoes |
| Kiwifruit | Celery | Guavas | Pumpkins |
| Loquats | Corn, sweet | Jackfruit | Squash |
| Nectarines* | Endive | Jujubes | Sweet potatoes |
| Peaches* | Garlic | Longan | Taro |
| Pears | Lettuce | Lychees | Tomatoes |
| Persimmons* | Mushrooms | Mangoes | Watermelons |
| Plums* | Onions | Mangosteen | Yams |
| Prunes | Parsley | Olives | |
| Raspberries | Parsnips | Papayas | |
| Strawberries | Peas | Passion fruit | |
| | Radishes | Pepinos | |
| | Spinach | Pineapples | |
| | Turnips | Plantain | |
| | | Pomegranates | |
| | | Prickly pear | |
| | | Rambutan | |
| | | Sapodilla | |
| | | Sapota | |
| | | Tamarillo | |

*Some cultivars are chilling sensitive.

**Light.** Exposure of potatoes to light should be avoided because it results in greening due to formation of chlorophyll and solanine (toxic to humans). Light-induced greening of Belgian endive is also undesirable.

**Other factors.** Various kinds of chemicals (e.g., fungicides, growth regulators) may be applied to the commodity to affect one or more of the biological deterioration factors.

# Postharvest Technology Procedures

## Temperature management procedures

Temperature management is the most effective tool for extending the shelf life of fresh horticultural commodities. It begins with the rapid removal of field heat by using one of the following cooling

## Table 3.4. Effect of temperature on deterioration rate of a nonchilling sensitive commodity

| Temperature (°F) | (°C) | Assumed $Q_{10}$* | Relative velocity of deterioration | Relative shelf life | Loss per day (%) |
|---|---|---|---|---|---|
| 32 | 0 | | 1.0 | 100 | 1 |
| 50 | 10 | 3.0 | 3.0 | 33 | 3 |
| 68 | 20 | 2.5 | 7.5 | 13 | 8 |
| 86 | 30 | 2.0 | 15.0 | 7 | 14 |
| 104 | 40 | 1.5 | 22.5 | 4 | 25 |

$$*Q_{10} = \frac{\text{Rate of deterioration at T} + 10°C}{\text{Rate of deterioration at T}}$$

methods: hydrocooling, in-package icing, top-icing, evaporative cooling, room cooling, forced-air cooling, serpentine forced-air cooling, vacuum cooling, and hydro-vacuum cooling.

Cold storage facilities should be well engineered and adequately equipped. They should have (1) good construction and insulation, including a complete vapor barrier on the warm side of the insulation, (2) strong floors, (3) adequate and well-positioned doors for loading and unloading, (4) effective distribution of refrigerated air, (5) sensitive and properly located controls, (6) enough refrigerated coil surface to minimize the difference between the coil and air temperatures, and (7) adequate capacity for expected needs. Commodities should be stacked in the cold room leaving air spaces between pallets and room walls to ensure good air circulation. Storage rooms should not be loaded beyond their limit for proper cooling. In monitoring temperatures, commodity temperature rather than air temperature should be used.

Transit vehicles must be cooled before loading the commodity. Delays between cooling after harvest and loading into transit vehicles should be avoided. Proper temperature maintenance should be ensured throughout the handling system.

### Control of relative humidity

Relative humidity can influence water loss, decay development, incidence of some physiological disorders, and uniformity of fruit ripening. Condensation of moisture on the commodity (sweating) over long periods of time is probably more important than is the relative humidity of ambient air in enhancing decay. Proper relative humidity is 85 to 95 percent for fruits and 90 to 98 percent for vegetables except dry onions and pumpkins (70 to 75 percent). Some root vegetables can best be held at 95 to 100 percent relative humidity.

Relative humidity can be controlled by one or more of the following procedures:

**1.** Adding of moisture (water mist or spray, steam) to air by humidifiers

**2.** Regulating of air movement and ventilation in relation to the produce load in the cold storage room

**3.** Maintaining the refrigeration coils within about 1°C (2°F) of the air temperature

**4.** Providing moisture barriers—insulating storage room and transit vehicle walls; polyethylene liners in containers and plastic films for packaging

**5.** Wetting floors in storage rooms

**6.** Adding crushed ice in shipping containers or in retail displays for commodities that are not injured by the practice

**7.** Sprinkling produce with water during retail marketing, use on leafy vegetables, cool-season root vegetables, and immature fruit vegetables (e.g., snap beans, peas, sweet corn, summer squash)

### Supplements to temperature and humidity management

Many technological procedures are used commercially as supplements to temperature management. None of these procedures, alone or in their various combinations, can substitute for maintenance of optimum temperature and relative humidity, but they can help extend the shelf life of harvested produce beyond what is possible using refrigeration alone (table 3.6).

Treatments applied to the commodity include (1) curing of certain root, bulb, and tuber vegetables, (2) cleaning followed by removal of excess surface moisture, (3) sorting to eliminate defects, (4) waxing and other surface coatings, film wrapping, (5) heat treatments (hot water or air, vapor heat), (6) treatment with postharvest fungicides, (7) sprout inhibitors, (8) special chemical treatments (scald inhibitors, calcium, growth regulators, anti-ethylene chemicals for ornamentals), (9) fumigation for insect control, and (10) ethylene treatment (degreening, ripening).

**Table 3.6. Fresh horticultural crops classified according to their relative perishability and potential storage life in air at near optimum temperature and relative humidity**

| Relative perishability | Potential storage life (*weeks*) | Commodities |
|---|---|---|
| Very high | <2 | Apricot, blackberry, blueberry, cherry, fig, raspberry, strawberry; asparagus, bean sprouts, broccoli, cauliflower, cantaloupe, green onion, leaf lettuce, mushroom, peas, spinach, sweet corn, tomato (ripe); most cut flowers and foliage; minimally processed fruits and vegetables |
| High | 2 to 4 | Avocado, banana, grape (without $SO_2$ treatment), guava, loquat, mandarin, mango, melons (honeydew, crenshaw, Persian), nectarine, papaya, peach, pepino, plum; artichoke, green beans, Brussels sprouts, cabbage, celery, eggplant, head lettuce, okra, pepper, summer squash, tomato (partially ripe) |
| Moderate | 4 to 8 | Apple and pear (some cultivars), grape ($SO_2$ treated), orange, grapefruit, lime, kiwifruit, persimmon, pomegranate, pummelo; table beet, carrot, radish, potato (immature) |
| Low | 8 to 16 | Apple and pear (some cultivars), lemon, potato (mature), dry onion, garlic, pumpkin, winter squash, sweet potato, taro, yam; bulbs and other propagules of ornamental plants |
| Very low | >16 | Tree nuts, dried fruits, and vegetables |

Treatments to manipulate the environment include (1) packaging, (2) control of air movement and circulation, (3) control of air exchange or ventilation, (4) exclusion or removal of ethylene, (5) controlled or modified atmospheres (CA or MA), and (6) sanitation.

## Future Trends in Perishables Handling

Research and development efforts are continually aimed at improving existing technology and testing new ideas for possible alternatives to current procedures. Some trends follow:

**1.** Development of more effective and economical methods of controlling and monitoring temperature and relative humidity in storage and transport environments.

**2.** Reduction of quantitative and qualitative losses due to chilling injury (via avoidance or amelioration) of chilling-sensitive commodities.

**3.** Replacements of chemicals used for control of some physiological disorders, decay-causing pathogens, and insects.

**4.** Expanded use of film wrapping and other treatments that replace waxing to minimize water loss.

**5.** Expedited handling—rapid transportation and more efficient distribution systems at the local, national, and international levels.

**6.** Increased mechanization—in harvesting, bulk handling from field to packinghouse, quality grading at the packinghouse, and during transport to destination markets.

**7.** Continued reduction of the variety of shipping containers to a few sizes, each adequate for numerous commodities, and increased recycling of packaging materials used for fresh produce.

**8.** Increased use of textural and flavor quality (in addition to appearance quality) in the grade standards for quality and maturity.

**9.** Development of MA or CA technology for use during transit, storage, and marketing of appropriate commodities. Some goals: (a) Improved systems for generating, maintaining, and monitoring MA and CA atmospheres; (b) more effective removal of $C_2H_4$ and other volatiles when needed; (c) improved safety procedures that would permit expanded use of carbon monoxide as a fungistat on $CO_2$-sensitive commodities; (d) innovations to allow greater use of MA during transport and distribution, and (e) expanded use of MA packaging for intact and lightly processed fruits and vegetables.

**10.** Research to develop new procedures for maintaining the quality and safety of lightly processed produce (such as fruit salads, peeled citrus, cut lettuce, carrots and celery sticks, and florets of broccoli and cauliflower) for food service and consumer use.

**11.** Modifications in handling procedures to economize in labor, materials, and energy use, and to protect the environment.

## REFERENCES

1. Brady, C. J. 1987. Fruit ripening. *Annu. Rev. Plant Physiol.* 38:155-78.
2. Cappellini, R. A., and M. J. Ceponis. 1984. Postharvest losses in fresh fruits and vegetables. In *Postharvest pathology of fruits and vegetables: postharvest losses in perishable crops*, ed. H. E. Moline, 24-30. Univ. Calif. Bull. 1914.
3. Grierson, D. 1987. Senescence in fruits. *HortScience* 22:859-862.
4. Grierson, W., and W. F. Wardowski. 1978. Relative humidity effects on the postharvest life of fruits and vegetables. *HortScience* 13:570-74.
5. Harvey, J. M. 1978. Reduction of losses in fresh market fruits and vegetables. *Annu. Rev. Phytopathol.* 16:321-41.
6. International Institute of Refrigeration. 1979. *Recommended conditions for cold storage of perishable produce*. Internat. Inst. Refrig. Bull. Supp. 148 pp.
7. Kader, A. A. 1983. Postharvest quality maintenance of fruits and vegetables in developing countries. In *Postharvest physiology and crop preservation*, ed. M. Lieberman, 520-36. New York: Plenum.
8. Kader, A. A., J. M. Lyons, and L. L. Morris. 1974. Quality and postharvest responses of vegetables to preharvest field temperature. *HortScience* 9:523-27.
9. Lidster, P. D., P. D. Hilderbrand, L. S. Brard, and S. W. Porritt. 1988. *Commercial storage of fruits and vegetables*. Can. Dept. Agric. Publ. 1532. 88 pp.
10. Lipton, W. J. 1987. Senescence in leafy vegetables. *HortScience* 22:854-59.
11. Mayak, S. 1987. Senescence in cut flowers. *HortScience* 22:863-65.
12. National Academy of Sciences. 1978. *Postharvest food losses in developing countries* (Science & Technology for International Development). Washington, DC: Natl. Acad. Sci. 202 pp.
13. Poovaiah, B. W. 1986. Role of calcium in prolonging storage life of fruits and vegetables. *Food Technol.* 40(5):86-89.
14. Rhodes, M. J. C. 1980a. The maturation and ripening of fruits. In *Senescence in plants*, ed. K. V. Thimann, 157-205. Boca Raton, FL: CRC Press.
15. ———. 1980b. The physiological basis for the conservation of food crops. *Prog. Food Nutr. Sci.* 4(3-4):11-20.
16. Romani, R. J. 1987. Senescence and homeostasis in postharvest research. *HortScience* 22:865-68.
17. Shewfelt, R. L. 1986. Postharvest treatment for extending the shelf-life of fruits and vegetables. *Food Technol.* 40(5):70-89.
18. Tindall, H. D., and F. J. Proctor. 1980. Loss prevention of horticultural crops in the tropics. *Prog. Food Nutr. Sci.* 4(3-4):25-40.
19. United Nations Food and Agriculture Organization. 1981. *Food loss prevention in perishable crops*. FAO Agric. Serv. Bull. 43. 72 pp.
20. Wang, C. Y., ed. 1990. *Chilling injury of horticultural crops*. Boca Raton, FL: CRC Press. 313 pp.

# 4

# Maturation and Maturity Indices

Michael S. Reid

We come here to the first step in the postharvest life of the product, the moment of harvest. For most fresh perishables, harvest is manual, so the picker is responsible for deciding whether or not the product has reached the correct maturity for harvest. The maturity of harvested perishable commodities has an important bearing on their storage life and quality, and may affect the way they are handled, transported, and marketed. An understanding of the meaning and measurement of maturity is, therefore, central to postharvest technology. The meaning of the term *mature*, the importance of maturity determination, and some examples of approaches to determining and applying a satisfactory index of maturity, are discussed here.

## Definition of Maturity

To most people "mature" and "ripe" mean the same thing when describing fruit. For example, *mature* is defined in Webster's dictionary as:

mature (fr. L *maturus* ripe): **1**:Based on slow and careful consideration; **2 a** (1): having completed natural growth and development: RIPE (2): having undergone maturation **b**: having attained a final or desired state; **3 a**: of or relating to a condition of full development . . . .

In postharvest physiology we consider "mature" and "ripe" as distinct terms for different stages of fruit development (fig. 4.1). *Mature* is best defined by 2 a (1) above (having completed natural growth and development), and for fruits, is defined in the U.S. Grade standards as "that stage which will ensure proper completion of the ripening process." This latter definition lacks precision in that it fails to define "proper completion of the ripening process." Most postharvest technologists consider that the definition should be "that stage at which a commodity has reached a sufficient stage of development that after harvesting and postharvest handling (including ripening, where required), its quality will be at least the minimum acceptable to the ultimate consumer."

Horticultural maturity is the stage of development when a plant or plant part possesses the prerequisites for utilization by consumers for a particular purpose. A given commodity may be horticulturally mature at any stage of development (fig. 4.1). For example, sprouts or seedlings are horticulturally mature in the early stage of development, whereas most vegetative tissues, flowers, fruits, and underground storage organs become horticulturally mature in the midstage, and seeds and nuts in the late stage of development. For some commodities, horticultural maturity is reached at more than one stage of development, depending on the desired use of the product. In zucchini squash, for example, the mature product can be the fully open flower, the young fruit, or the fully developed fruit.

A qualitative difference in the relationship between maturity and edibility distinguishes many fruits from the vegetables. In many fruits, for example mature (but green) bananas, the eating quality at maturity will be far from optimal. The fruit becomes edible only after proper ripening has taken place. In contrast, in most vegetables optimal maturity coincides with optimal eating quality.

## Indices of Maturity

The definition of maturity as the stage of development giving minimum acceptable quality to the ultimate consumer implies a measurable point in the commodity's development, and the need for techniques to measure maturity. The maturity index for a commodity is a measurement or measurements that can be used to determine whether a particular example of the commodity is mature. These indices are important to the trade in fresh fruits and vegetables for several reasons.

**Trade regulations.** Regulations published by grower groups, marketing orders, or legally appointed authorities (such as the state departments of agriculture and the USDA) frequently include a statement as to the minimum (and sometimes maximum) maturity acceptable for a given commodity. Objective maturity standards are available for relatively few commodities and most regulations rely on subjective judgments related to the broad definitions quoted above.

**Marketing strategy.** In most markets the laws of supply and demand mean price incentives for the earliest (or sometimes the latest) shipments of any particular commodity. This encourages growers and shippers to expedite (or delay) the harvesting of their crop to take advantage of premium prices. The minimum maturity statements in the grade standards are placed there to prevent sale of immature or overmature product and consequent loss of consumer confidence. Objective maturity indices enable grow-

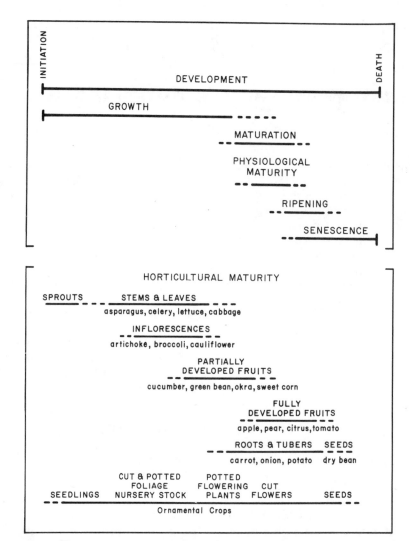

**Fig. 4.1 Horticultural maturity in relation to developmental stages of the plant (Watada et al. 1984).**

ers to know whether their commodity can be harvested when the market is buoyant.

**Efficient use of labor, resources.** With many crops the need for labor and equipment for harvesting and handling is seasonal. In order to plan operations efficiently, growers need to *predict* the likely starting and finishing dates for harvest of each commodity. Objective maturity indices are vital for accurate prediction of harvest dates.

## Characteristics of a maturity index

Maturity measures to be made by producers, handlers, and quality control personnel must be simple, readily performed in the field or orchard, and should require relatively inexpensive equipment. The index should preferably be objective (a measurement) rather than subjective (an evaluation). The index must consistently relate to the quality and postharvest life of the commodity for all growers, districts, and years. If possible, the index should be non-destructive.

The search for an objective determination of maturity has occupied the attention of many horticulturists, working with a wide range of commodities, for years. The number of satisfactory indices that have been suggested is nevertheless rather small, and for most commodities the search for a satisfactory maturity index continues.

Two rather different problems will be addressed here. The first is to measure maturity at harvest or at a subsequent inspection point. The second, more complex problem is to find some way of predicting the time at which a commodity will mature. For both problems, similar techniques may be appropriate, but the ways in which they are applied differ.

## Developing maturity indices

Many features of fruits and vegetables have been used in attempting to provide adequate estimates of maturity. Examples of those that have been proposed, or that are presently in use, are shown in table 4.1. The wide range of methods that have been

## Table 4.1. Maturity indices for selected fruits and vegetables

| Index | Examples |
|---|---|
| Elapsed days from full bloom to harvest | Apples, pears |
| Mean heat units during development | Peas, apples, sweet corn |
| Development of abscission layer | Some melons, apples, feijoas |
| Surface morphology and structure | Cuticle formation on grapes, tomatoes<br>Netting of some melons<br>Gloss of some fruits (development of wax) |
| Size | All fruits and many vegetables |
| Specific gravity | Cherries, watermelons, potatoes |
| Shape | Angularity of banana fingers<br>Full cheeks of mangoes<br>Compactness of broccoli and cauliflower |
| Solidity | Lettuce, cabbage, brussels sprouts |
| Textural properties<br>  Firmness<br>  Tenderness | <br>Apples, pears, stone fruits<br>Peas |
| Color, external | All fruits and most vegetables |
| Internal color and structure | Formation of jellylike material in tomato fruits<br>Flesh color of some fruits |
| Compositional factors<br>  Total solids<br>  Starch content<br>  Sugar content<br><br>  Acid content, sugar/acid ratio<br>  Juice content<br>  Oil content<br>  Astringency (tannin content)<br>  Internal ethylene concentration | <br>Avocados, kiwifruit<br>Apples, pears<br>Apples, pears, stone fruits, grapes<br>Pomegrantates, citrus, papaya, melons, kiwifruit<br>Citrus fruits<br>Avocados<br>Persimmons, dates<br><br>Apples, pears |

## Table 4.2. Methods of maturity determination

| Index | Methods of determination | Subjective | Objective | Destructive | Nondest. |
|---|---|:---:|:---:|:---:|:---:|
| Elapsed days from full bloom | Computation | | × | | × |
| Mean heat units | Computation from weather data | | × | | × |
| Development of abscission layer | Visual or force of separation | × | × | | × |
| Surface structure | Visual | × | | | × |
| Size | Various measuring devices, weight | | × | | × |
| Specific gravity | Use of density gradient solutions, flotation techniques vol/wt | | × | | × |
| Shape | Dimensions ratio charts | × | × | | × |
| Solidity | Feel, bulk, density, γ-rays, X-rays | × | × | | × |
| Textural properties: | | | | | |
| Firmness | Firmness testers, deformation | | × | × | |
| Tenderness | Tenderometer | | × | × | |
| Toughness | Texturometer, fibrometer (also: chemical methods for determination of polysaccharides) | | × | × | |
| Color, external | Light reflectance | | × | | × |
| | Visual color charts | × | | | × |
| Color, internal | Light transmittance, delayed light emission | | × | | × |
| Internal structure | Light transmittance, delayed light emission | | × | | × |
| | Visual examination | × | | × | |

(*continued on next page*)

devised to measure these characteristics are summarized in table 4.2.

The strategy for developing a maturity index is:

**1.** To determine changes in the commodity throughout its development.

**2.** To look for a feature that correlates well with development, and

**3.** To use storage trials and organoleptic assays (taste panels) to determine the value of the index which defines minimum acceptable maturity. When the relationship between changes in the index quantity and the quality and storage life of the commodity has been determined, an index value can be assigned for the minimal acceptable maturity.

**4.** To test the index system in several years and in several growing locations to ensure that it consistently reflects the quality of the harvested product.

## Estimating maturity

Among the characters that have been used to establish maturity are the following:

**Chronological.** With certain crops (fast rotation vegetables, such as radish, and perennial tree crops growing in short summer environments), maturity

**Table 4.2. Methods of maturity determination** (*cont'd*)

| Index | Methods of determination | Sub-jective | Objec-tive | Destruc-tive | Non-dest. |
|---|---|---|---|---|---|
| Compositional factors: | | | | | |
| Total solids | Dry weight | | × | × | |
| Starch content | KI test, other chemical tests | | × | × | |
| Sugar content | Hand refrac-tometer, chemical tests | | × | × | |
| Acid content | Titration, chemical tests | | × | × | |
| Juice content | Extraction, chemical tests | | × | × | |
| Oil content | Extraction, chemical tests | | × | × | |
| Tannin content | Ferric chloride test, chemi-cal tests | | × | × | |
| Internal ethylene | Gas chroma-tography | | × | × | |

can be defined chronologically, for example as days from planting, or days from flowering. Chronological indices are seldom perfect, but they do permit a degree of planning, and are widely used. For some crops, the chronological method is refined by calculating accumulated heat units during the growing period, which modulates the chronological index according to the weather pattern during the growing season.

**Physical.** A wide range of physical characteristics of commodities are used to assess their maturity. Some of the most important are discussed briefly below:

*Size, shape, and surface characteristics.* Changes in size, shape, or surface characteristics of fruits and vegetables are a common maturity index. Vegetables, particularly, are harvested when they have reached a marketable size, and before they become too large. Banana maturity is determined as a change in diameter of the fingers, and changes in surface gloss, or feel (waxiness), are used as a practical tool in harvesting of some melons such as honeydew melons (see Chapter 29, table 29.2).

*Abscission.* During the later stages of maturation and start of ripening in many fruits, a special band of cells, the abscission zone, develops on the stalk (pedicel) attaching the fruit to the plant. Measurement of the development of this zone, whose purpose is to permit the fruit to separate from the plant, is possibly the oldest of all maturity indices. Children know that the fruit that are easiest to pull from the tree are the sweetest. Abscission force is not used as a formal maturity index, but the development of the abscission zone, or "slip" in the netted muskmelons

(see Chapter 29, fig. 29.3) is used to determine their maturity.

*Color.* The color change which accompanies maturation in many fruits is widely used as a maturity index. Objective measurement of color requires expensive equipment (fig 4.2) and although the human eye is unable to give a good evaluation of a single color, it is extremely sensitive to color differences. Color comparison techniques are therefore commonly used to assess fruit maturity (fig 4.3). Color swatches may be used to determine external or internal color. Accurate devices employing state of the art electronics and optics now permit objective color measurements (fig 4.2). As the price of such devices

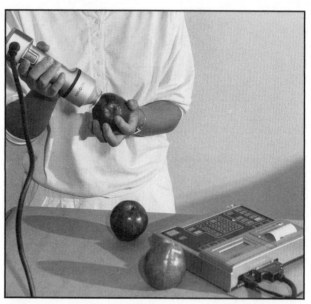

**Fig. 4.2 Colorimeter used to measure surface color of apples.**

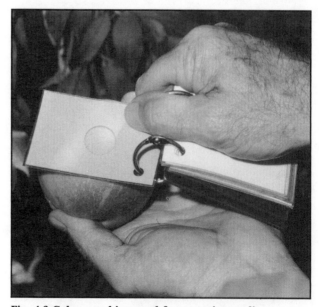

**Fig. 4.3 Color matching used for maturity grading.**

falls, they may start to replace comparison techniques for color evaluation.

*Texture.* Maturation of fruits is often accompanied by softening. Overmature vegetables frequently become fibrous or tough. These properties can be used to indicate maturity, and are measured with instruments that determine the force required to push a probe of known diameter through the flesh of the fruit or vegetable (fig 4.4). The solidity of lettuce, cabbage and brussels sprouts is an important quality and maturity characteristic. In the case of lettuce, X-ray equipment has been devised to measure head firmness, but the technique has not been adopted commercially.

**Chemical changes.** The maturation of fruits and vegetables is often accompanied by profound changes in their chemical composition. Many of these changes have been used in studies of maturation, but relatively few have provided satisfactory maturity indices. The requirement that the maturity index be easy to measure mitigates against measurement of changes in chemical composition, although enzymatic techniques devised for medical diagnoses may, in the future, be adopted for maturity estimation in perishables. Chemical changes that are used for maturity estimation include the change in total soluble solids, measured using a refractometer (fig 4.5), changes in the distribution of starch in the flesh of the commodity, measured using a starch/iodine reaction (fig 4.6), and the sugar/acid ratio, which is used as the legal maturity index for citrus. The unsatisfactory nature of chemical tests for maturity is exemplified by the old oil content measurement for avocados, which has been replaced by determination of percent dry weight (see below) because of the time-consuming and complex nature of oil determination.

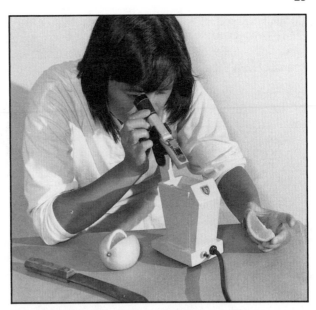

Fig. 4.5 Measuring soluble solids content with a refractometer.

Fig. 4.6 Treatment of cut apples with an iodine solution reveals the disappearance of starch (which stains dark) as apples mature.

**Physiological changes.** The maturation of commodities is associated with changes in their physiology, as measured by changing patterns of respiration and ethylene production. The problem with using these parameters in assessing maturity is the variability in absolute rates of ethylene production and respiration amongst similar individuals of the same commodity. Furthermore the techniques are complex and expensive to implement on a commercial scale. Nevertheless, the rate of ethylene production of a sample of apples is used by some producers to establish the maturity of apples, and particularly to identify those

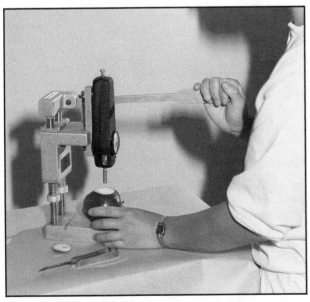

Fig. 4.4 Using the UC Firmness tester to measure the flesh firmness of apples.

that will be suited to long-term controlled atmosphere storage.

## Predicting Maturity

Prediction of maturity is more complex than assessing the maturity at or after harvest. The basic requirement is for a measurement whose change through development can be modelled mathematically so that the time at which the measurement will reach the maturity index (minimum acceptable) can be predicted. Theoretically then, once the pattern of change is established for the chosen index quantity, measurements made early in the season can be used to predict the date at which the commodity will reach minimum acceptable maturity.

The way in which this strategy has been applied can be illustrated best by the following three examples.

### Apples

Literature relating to the prediction of maturity in apples is voluminous, yet no really satisfactory method has been proposed. The use of climatic data to predict the date of harvest by a modification of the "days from full bloom" index noted in table 4.1, even when adapted by using "days from the 'T' stage" has provided only general predictions of the harvest date.

In an attempt to provide a more satisfactory prediction of the maturation date, workers have examined a number of changes during fruit development. Measurement of respiration, ethylene production, sugar content, starch content, and changing firmness of the fruit each failed to meet some of the criteria outlined above for a satisfactory maturity index, and were too variable to permit prediction of maturation date. The "starch pattern," an old method of determining maturity, refined by assigning scores to a range of patterns, has proved to be a good index. Changes in the mean starch index score before harvest are readily analyzed as a linear regression, and the date of minimal acceptable maturity can be predicted several weeks in advance (fig 4.7).

### Avocados

The State of California for many years promulgated a minimum oil content as the maturity standard for avocados. This index has been unsatisfactory, being difficult to apply, and because some avocados having more than the minimum oil requirement, may be lacking in organoleptic qualities. However, raising the minimum oil content might eliminate from the market particular avocado varieties or crops whose organoleptic quality is adequate. Using taste panel evaluations to determine quality, researchers have shown that the patterns of dry weight accumulation, or growth of avocado fruit can be used, not only to determine the date at which minimum acceptable maturity is achieved, but also to predict that date (fig 4.8). Taste panel scores increased as oil content increased. Oil content was found to be closely corre-

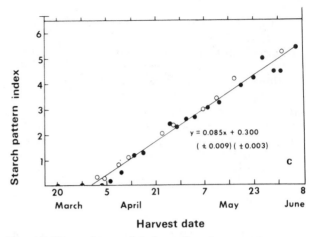

Fig. 4.7 Change in starch index values for maturing Granny Smith apples in New Zealand.

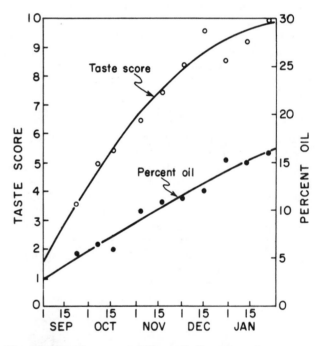

Fig. 4.8 Changing acceptability and oil content of maturing avocado (Lee and Young 1983).

lated with the percent dry weight (fig 4.9). Consequently, the California minimum maturity index was changed from oil content to percent dry weight.

### Kiwifruit

As a prelude to developing a maturity index for kiwifruit, researchers measured changes in a wide range of chemical and physical parameters during growth and development (fig 4.10). This information was compared to storage and taste panel results to decide on possible methods for determining, and if possible predicting, the time of minimal acceptable maturity for this crop. It seemed possible that soluble solids content and fruit firmness might provide a suitable maturity index, but in repeating the experi-

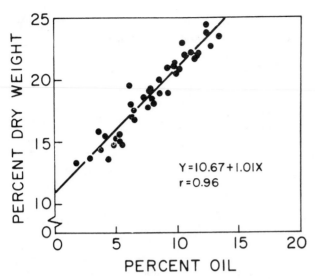

Fig 4.9 Relationship between dry weight and oil content of avocado (Lee et al. 1983).

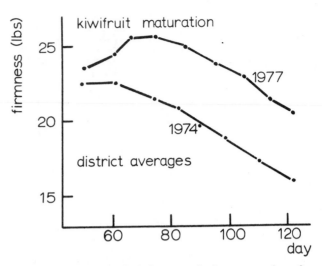

Fig. 4.11 Changes in fruit firmness during maturation of kiwifruit.

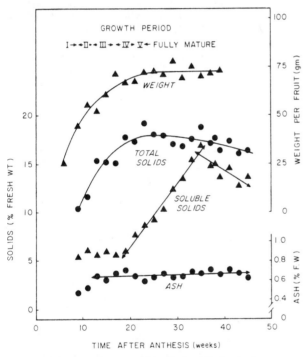

Fig 4.10 Changes in some chemical and physical parameters of kiwifruit during growth and development.

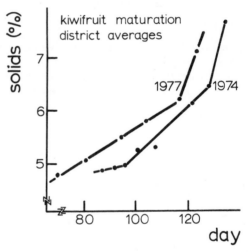

Fig. 4.12 Changes in soluble solids content during the 6 weeks before harvest of kiwifruit can be used to predict harvest date.

### Honeydew melons

Although a number of changes take place during the maturation of honeydews (see Chapter 29, table 29.2), visual separation of the earlier stages (stages 1 and 2) is difficult. The legal minimum maturity standard for honeydew melons is 10 percent soluble solids. In a study of different honeydew seed lines changes in percent soluble solids were compared with a physical parameter (firmness) and a physiological measure (internal ethylene concentration) (table 4.3). The NK seed line showed a close correlation between the onset of ripening (shown by marked increase of internal ethylene content) and achievement of the statutory 10 percent soluble solids. In other seed lines (Peto, HM, Asgrow), however, solids were already above the legal minimum before ripening had commenced. Changes in ethylene content correlated well with reduced firmness, and these two factors might

ments over several seasons, changes in firmness of the fruit were found to be very variable, and unrelated to fruit quality (fig 4.11). In New Zealand, a minimum maturity index of 6.25 percent soluble solids has now been used for a number of years. The change in soluble solids in the 6 weeks prior to the normal harvest date can be used, with regression analysis, to predict the date of harvest for different orchards, seasons, and growing districts (fig 4.12).

**Table 4.3. Changes in maturity parameters during ripening of honeydew melons**

| Seed Source | Maturity Class | Internal $C_2H_4$ (*ppm*) | Flesh Firmness (*kg-f*) | Soluble Solids (%) |
|---|---|---|---|---|
| Peto | 1 | 1.3 | 2.9 | 12.5 |
| | 2 | 7.2 | 2.1 | 12.9 |
| | 3 | 24.8 | 1.6 | 14.6 |
| | 4 | 30.8 | 1.0 | 14.6 |
| NK | 1 | 0.3 | 3.0 | 8.9 |
| | 2 | 3.3 | 2.0 | 11.4 |
| | 3 | 31.0 | 1.2 | 14.9 |
| | 4 | 31.4 | 1.4 | 13.5 |
| HM | 1 | 1.5 | 2.8 | 12.9 |
| | 2 | 4.5 | 1.8 | 12.1 |
| | 3 | 29.4 | 0.6 | 13.2 |
| | 4 | 31.2 | 1.0 | 13.5 |
| Asgrow | 1 | 0.7 | 3.7 | 11.3 |
| | 2 | 5.6 | 2.4 | 13.1 |
| | 3 | 23.4 | 1.5 | 13.1 |
| | 4 | 24.1 | 1.0 | 13.6 |

reasonably be used to define the minimum sugar content for the different seed lines. In this case, a complex physiological measurement (internal ethylene content) would then be a maturity "bench mark" used to determine proper levels of the legal maturity index (sugar content). These preliminary results must, of course, be repeated in different growing areas and seasons to ensure that the relationships suggested here are consistent. The relationship of the chosen index to minimum eating quality must also be determined using taste panel evaluations.

## REFERENCES

1. Arthey, V. D. 1975. *Quality of horticultural products.* New York: Halstead Press, John Wiley & Sons. 228 pp.

2. Eskin, N. A. M., ed. 1989. *Quality and preservation of vegetables.* Boca Raton, FL: CRC Press. 313 pp.

3. Hulme, A. C., ed. 1971. *The biochemistry of fruits and their products.* Vol. 2. New York: Academic Press. 788 pp.

4. Lee, S. K., and R. E. Young. 1983. Growth measurement as an indication of avocado maturity. *J. Am. Soc. Hort. Sci.* 108:395-97.

5. Lee, S. K., R. E. Young, P. M. Schiffman, and C. W. Coggins, Jr. 1983. Maturity studies of avocado fruit based on picking dates and dry weight. *J. Am. Soc. Hort. Sci.* 108:390-94.

6. Pattee, H. E., ed. 1985. *Evaluation of quality of fruits and vegetables.* Westport, CT: AVI Publ. Co. 410 pp.

7. Ryall, A. L., and W. J. Lipton. 1979. *Handling, transportation and storage of fruits and vegetables.* Vol. 1, *Vegetables and melons.* 2nd ed. Westport, CT: AVI Publ. Co. 588 pp.

8. Ryall, A. L., and W. T. Pentzer. 1982. *Handling, transportation and storage of fruits and vegetables.* Vol. 2, *Fruits and tree nuts.* 2nd ed. Westport, CT: AVI Publ. Co. 610 pp.

9. Watada, A. E., R. C. Herner, A. A. Kader, R. J. Romani, and G. L. Staby. 1984. Terminology for the description of developmental stages of horticultural crops. *HortScience* 19:20-21.

# 5

# Harvesting Systems

James F. Thompson

The goals of harvesting are to gather a commodity from the field at the proper level of maturity, with a minimum of damage and loss, as rapidly as possible, and at a minimum cost. Today, as in the past, these goals are best achieved through hand harvesting in most fruit, vegetable, and flower crops. Tables 5.1 and 5.2 list the level of hand harvesting in various vegetable, fruit, nut, and berry crops. All flower crops are hand harvested.

## Hand Harvesting

### Primary advantages

**1.** Humans can accurately select for maturity, allowing accurate grading and multiple harvest.

**2.** Humans can handle fruit with a minimum of damage.

**3.** Rate of harvest can easily be increased by hiring more workers.

**4.** Hand harvesting requires a minimum of capital investment (although some farmers provide housing for their employees).

The main problems with hand harvesting center around labor management. Labor supply is a prob-

**Table 5.2. Level of hand harvesting for selected U.S. fruit, nut, and berry crops**

| Acreage hand harvested (%) | Commodity | | | |
|---|---|---|---|---|
| 76–100 | Apple | Apricot | Avocado | Banana |
| | Breadfruit | Sweet cherry* | Coffee* | Grape* |
| | Guava* | Kiwi | Kumquat* | Loquat |
| | Lychee | Mango | Nectarine | Peach* |
| | Pear* | Persimmon | Pineapple* | Pomegranate |
| | Quince* | Rosehips* | Wild blueberry* | Currant* |
| | Gooseberry* | Strawberry | Grapefruit* | Lemon* |
| | Lime | Orange* | Olive* | Papaya |
| | Passion fruit* | Tangelo* | Tangerine | Cashew |
| | Coconut* | Chestnut* | Jojoba* | |
| 51–75 | Red raspberry* | Macadamia | | |
| 26–50 | Prune* | Blackberry* | Highbush blueberry* | Black raspberry* |
| | Pecan | | | |
| 0–25 | Tart cherry* | Date | Fig | Cranberry* |
| | Almond | Filbert | Peanut* | Pistachio |
| | Walnut | | | |

*More than 50% of crop is processed.

lem for farmers who cannot offer a long employment season. Labor strikes during the harvest period can be costly. In recent years, costs associated with complying with government labor regulations have increased significantly.

In spite of these problems, quality is so important to marketing fresh market commodities successfully that hand harvesting remains the dominant method of harvest (fig. 5.1). In fact, tables 5.1 and 5.2 indicate that the crops that are largely machine harvested are used for processing or are crops that are less easily damaged, such as nuts, roots, and tubers.

Effective use of hand labor requires careful management. New employees must be trained to harvest the product quality needed at an acceptable rate. Employees must know what level of performance is expected of them, and must be encouraged and trained to reach that level. Well-managed employees enjoy their jobs more and can be more productive than those that are poorly managed. Benefits such as paid vacations, insurance, and so on, help ensure the return of already trained employees.

Machines are used to aid hand harvest with some commodities. Belt conveyors are used in such vegetable crops as lettuce and melons to move them to a central loading or in-field handling device. Scoops with rods protruding from the end are used by workers to comb through some berry crops. Platforms or moveable worker positioners have been used in place

**Table 5.1. Level of hand harvesting for selected U.S. vegetable crops**

| Acreage hand harvested (%) | Commodity | | | |
|---|---|---|---|---|
| 76–100 | Artichoke | Asparagus | Broccoli* | Cabbage |
| | Cantaloupe | Cauliflower | Celery | Cucumber* |
| | Lettuce | Green onions | Collards | Cress |
| | Dandelion | Eggplant | Endive | Escarole |
| | Fennel | Kale | Kohlrabi | Mushrooms* |
| | Okra | Peppers | Rapini | Rhubarb* |
| | Romaine | Sorrel | Squash | Watercress |
| | Cassava | Celeriac | Ginger | Parsley root |
| | Parsnip | Rutabaga | Salsify | Turnip |
| | Taro* | Jerusalem artichoke | | |
| 51–75 | Sweet potatoes | Mustard greens | Parsley | Swiss chard |
| | Turnip greens | | | |
| 26–50 | Dry onion | Pumpkin* | Tomato* | |
| 0–25 | Carrots | White potato* | Lima beans* | Snap beans* |
| | Sweet corn* | | Spinach* | Horseradish* |
| | Red beet* | Peas* | Garlic | Brussels sprouts* |
| | Malanga | Boniato | | |
| | | Radish | | |

*More than 50% of crop is processed.

**Fig. 5.1. Hand harvesting, shown here for strawberries, is the primary method for harvesting fresh market horticultural commodities in the U.S.**

of ladders in such crops as dates, papayas, and bananas. Lights are used in California for night harvest of some crops, when temperatures are cool and when worker effectiveness and product quality are at their best. Numerous other mechanical aids have been tried, but few increase productivity enough to warrant their expense.

## Mechanical Harvesting

### Primary advantages

**1.** The potential for rapid harvest is available.

**2.** Working conditions are improved.

**3.** Problems associated with hiring and managing hand labor are reduced.

Effective use of mechanical harvesters requires operation by dependable, well-trained people. Improper operation results in costly damage to expensive machinery and can quickly cause great crop damage. Regular and emergency maintenance must both be available. The commodity must be grown to accept mechanical harvest. For example, trees must be pruned for strength and to minimize fruit damage caused by fruit falling through the tree canopy. Maximum and uniform stand establishment is necessary for vegetable crops. Cropping patterns also must be

set up to utilize the expensive equipment as long as possible to pay for the high capital investment. (This can severely limit the production choices of some farmers.)

Mechanical harvest is not presently used for most fresh market crops because machines are rarely capable of selective harvest, they often cause excessive product damage, and they are expensive. Commodities that can be harvested at one time and are less sensitive to mechanical injury (roots, tubers, and nuts) are amenable to mechanical harvesting. Rapid processing after harvest will minimize the effects of mechanical injury.

### Mechanical harvest problems

**1.** Damage can occur to perennial crops (e.g., bark damage from a tree shaker).

**2.** Processing and handling capacity may not be able to handle the high rate of harvest.

**3.** Equipment may become obsolescent before it is paid for.

**4.** There are social impacts to lower labor requirements.

Mechanical harvesting of crops now hand harvested will probably require breeding new varieties that are more suited to mechanical harvest. This lengthy process has been done for only a few commodities. Equipment has been developed for the easiest crops to mechanically harvest. Other crops will be mechanized at a slow rate compared with the rate of mechanization over the past 40 years.

## REFERENCES

1. ASAE. 1983. *Status of harvest mechanization of horticultural crops.* St. Joseph, MI: Am. Soc. Agric. Eng. 78 pp.

2. Grierson, W., and W. C. Wilson. 1983. Influence of mechanical harvesting on citrus quality: cannery vs. fresh fruit crops. *HortScience* 18:407-09.

3. Kader, A. A. 1983. Influence of harvesting methods on quality of deciduous tree fruits. *HortScience* 18:409-11.

4. Kasmire, R. F. 1983. Influence of mechanical harvesting on quality of nonfruit vegetables. *HortScience* 18:421-23.

5. Morris, J. R. 1983. Influence of mechanical harvesting on quality of small fruits and grapes. *HortScience* 18:412-17.

6. ———. 1990. Fruit and vegetable harvest mechanization. *Food Technol.* 44(2):97-101.

7. O'Brien, M., B. F. Cargill, and R. B. Fridley. 1983. *Principles and practices for harvesting and handling of fruits and nuts.* Westport, CT: AVI Publ. Co. 636 pp.

8. Studer, H. E. 1983. Influence of mechanical harvesting on the quality of fruit vegetables. *HortScience* 18:417-21.

# 6

# Preparation for Fresh Market

## I. Fruits

F. GORDON MITCHELL

Protection of fresh fruit must begin with cultural practices in the field and continue until the fruit are consumed. Deterioration can result during production from improper pruning, thinning, fertilization, disease control, and so on. Many problems result from cumulative insults to the fruit during the postharvest handling period. Thus, protection is vital, both in the field and the packinghouse, to avoid immediate causes of deterioration and to delay its onset later in the distribution channels.

Table 6.1 presents data on cumulative impact bruising throughout a Bartlett pear handling operation in California. Although these results represent an extreme of what might be experienced commercially, they do show the serious consequences of repeated insults to the fruit at all stages of handling.

## Components of Harvesting

### Field containers

Most fresh market fruits are now harvested by hand into buckets or bags, which are then emptied into field bins for subsequent handling. Metal or plastic picking buckets are typically used for the softer fruits, and bottom-dump picking bags are used for fruits with a lesser potential for compression bruises. Certain delicate fruits are still transferred from buckets to field lugs (many sweet cherries), picked directly into field lugs (table grapes), or picked into buckets and packed into packages directly from the bucket (some stone fruits). Very soft, delicate fruits may be harvested, sorted, and packed directly into the package by the picker (strawberries, bushberries). Despite extensive past research on mechanical harvesting, presently there is no mechanical harvest of fresh market fruits in California.

In California, most field bins are standardized at 47 inches by 47 inches outside length and width and 24 inches inside depth (119 × 119 × 61 cm). Some kiwifruit and sweet cherries are handled in 12-inch (30-cm) deep bins to avoid bruising. Bins are typically of ¾-inch (19-mm) plywood, smooth or coated on the inside, and vented. To avoid cutting the fruit, ventilation slots are normally routed so that inside edges are tapered. Ventilation requirements for cooling are discussed in chapter 8, *Cooling Horticultural Commodities*.

Bin surfaces should be clean and smooth. Frequent washing, water dumping, or hydrocooling can cause surface roughness of the plywood bins and increase fruit abrasion problems. Coatings (paint or varnish-type) are available to reduce this problem; plastic-coated plywood, while expensive, is sometimes used. Separate plastic liners within bins effectively reduce abrasion injury, but special care is needed to maintain side venting, and liners need care where submersion water dumping is practiced.

Careful field supervision is critical in protecting fruits from injury. Physical injuries can result from dropping fruit into buckets or bags, overfilling these containers, striking the containers (especially soft-sided bags) against limbs and ladders, lack of care in transferring fruits into field bins or lugs, and overfilling these field containers. Avoiding these problems requires constant, visible, strict supervision. Even short drops can cause substantial impact bruising, as shown by the data for Bartlett pears in table 6.2.

### Transport from the field

Many opportunities for fruit bruising occur during field transport. Impact bruises occur when bins or lugs are dropped or bounced. Compression bruises

**Table 6.1. Cumulative levels of impact bruising on 'Bartlett' pears during postharvest handling**

| Location | Bruised fruits (%) |
|---|---|
| Tree | 0 |
| Picker bag | 14 |
| Field bin | 26 |
| After dump | 38 |
| After size | 82 |

**Table 6.2. Effect of drop height on incidence and severity of impact bruising on 'Bartlett' pears**

| Drop height (in) | (cm) | Fruit bruised (%) | Bruise severity (score)* |
|---|---|---|---|
| 0 | 0 | 0 | 0 |
| 4 | 10 | 40 | 0.6 |
| 6 | 15 | 44 | 0.6 |
| 9 | 23 | 56 | 1.0 |
| 12 | 30 | 78 | 1.2 |
| 16 | 41 | 100 | 2.3 |

*Scored on 0–5 scale: 0 = no damage; 5 = unmarketable.

result from stacking of overfilled field containers. Depth restrictions, even underfilling of bins, may be needed to avoid compressing the bottom layers of some soft fruits. Abrasion or vibration bruises may occur when fruits move or vibrate against rough surfaces or other fruits during transport.

Supervision is needed at all stages of field transport to minimize the accumulation of physical injuries. Despite the best supervision to avoid injuries during loading, considerable damage can occur during truck transport. Steps to take to avoid such problems include:

**1.** Avoid extended forklift movement of bins through the field from point of harvest to loading site.

**2.** Supervise truck or trailer loading to avoid rough handling or dropping of bins or lugs.

**3.** Grade farm roads to eliminate ruts, potholes, and bumps.

**4.** Where necessary, route truck movement to avoid public roads that are in poor condition.

**5.** Restrict transport speeds to a level that will avoid free movement of fruit. This may require different speed limits on different roads.

**6.** Use suspension systems on all transport equipment. Consider installing air suspension systems on all axles of transport equipment—tests have shown that these reduce damaging motion and fruit damage levels more than 50 percent.

**7.** Reduce tire air pressure on transport vehicles to reduce motion transmittal to the fruit.

**8.** Install plastic liners inside bin sides. Bottom liners are not needed. Plastic bubble liner material, fastened to bin sides with bubbles facing the fruit, performs well. Side vents can be cut in the liners to match those on the bin sides.

**9.** In difficult situations, such as long-distance transport, top bin pads can further reduce damage. These are pieces of light (⅜ inch or 10 mm) plywood cut to fit inside the bin, faced with a double layer of ½-inch (13-mm) thick bubble liner, and held against the fruit by short rubber trucker straps.

### Temperature protection

The effects of high temperatures on field deterioration are fully discussed elsewhere. Temperature protection in the field involves shading the harvested fruit to minimize warming (and sunscald), and prompt handling and cooling to minimize exposure to heat. Various studies have shown that even a mild breeze causes harvested fruit in the shade to quickly warm to near the ambient air temperature. Fruit in the sun can warm many degrees above air temperature (fig. 6.1). Surface color may also affect the extent of warming.

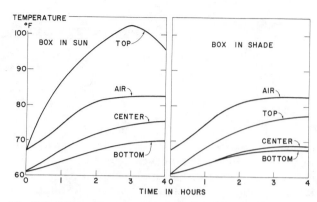

**Fig. 6.1. Effect of sun exposure and position in box on field warming of sweet cherries.**

Protection from high temperatures in the field should start with shading fruit after harvest. Whenever possible, harvested fruit should be moved to the shade of trees or vines until transport. If natural shade is not available, use portable shading. Placing empty packages or lugs over the top of stacks of packages can provide some protection. During periods of high field temperatures it may be desirable to avoid harvesting at midday.

During transport, speedy handling is usually the only protection that is provided. Frequent transfers to the cooler or packinghouse minimize the opportunity for fruit heating and deterioration. Where transport time is extended because of distance or delays, covering the load may be helpful. Fruit exposed to the sun warm quickly whether in the field or on top of a loaded truck. Further, airflow through the load during transport can quickly warm the fruit to the ambient air temperature. If transport of cool, early harvested fruit is delayed until temperatures are high, the resulting warming speeds fruit deterioration and increases the cost of subsequent cooling.

Any tarpaulin used to cover loads during transport should be a light color (white or silver are best) and should be kept clean to maintain good heat reflectance. Use only a tarpaulin that is supported to maintain an air space over the load. The only exception is when insulated tarpaulins are used to delay heating the fruit. If the tarpaulin extends down all sides of the load, it blocks airflow through the load and limits fruit warming when ambient air temperatures are high. Wetting fabric tarpaulins can further reduce warming by providing an evaporative cooling surface. Take care to ensure that any load cover does not confine heat within the load. Under difficult transport conditions, with water-tolerant fruits, place top ice over fruit in field bins before tarping and transport.

## Preparation for Packing

### Delivery to the packer

Most fruits are dumped directly onto a grading and packing belt. This was once done by hand dumping of small field lugs, which is still in limited use. In large-scale operations lug dumping is mechanized to provide uniform fruit flow and to reduce supervision problems. Commodities that are easily injured do not tolerate dumping. Table grapes are packed directly into the shipping package from field lugs.

Most fruits are now in field bins and dumped by either dry dumps or water dumps. In dry dumps (fig. 6.2) the bin is covered with a padded lid, then slowly inverted, and the fruit is delivered through a controlled opening in the lid. Electronic control of delivery belts allows adjustment of the flow of fruit to the sorting line. Properly designed dry dumps deliver a uniform product flow with minimum fruit injury.

Water dumps are of several types. A few still dump fruit from the field bin directly into water. Others submerge the bins and the fruit floats free (fig. 6.3). The most common flotation dumps submerge the entire bin as it travels along a conveyor. Pumps circulate the dump-tank water to move the free-floating fruit to an elevator, where it is rinsed and transferred to the sorting line. For flotation of fruits that are heavier than water, a salt (often sodium sulfate) must be added to the dump water to make it denser.

Sanitation is important in water dumps. Dump water quickly accumulates a high concentration of fungal spores, which can infect harvesting and handling wounds on the fruit. Dump tanks should be designed for rapid draining and filling and for easy cleaning. Chlorine at a concentration of 50 to 200 ppm is often maintained in the dump tank water as

Fig. 6.3. Water-submersion bin dumps are sometimes used, especially for apples and pears. Filled bins are carried under water to allow fruit to float free. The empty bin is then elevated out of the tank. For fruit denser than water, such as pears, water density is increased by adding an appropriate salt.

a fungistat, but the tolerance of the fruit species and variety to chlorine must be known.

### Sorting line

Efficient sorting demands careful attention to a series of specific equipment and supervision requirements (fig. 6.4). Adequate sorting space is probably the most limiting factor in new fruit-packing facilities. This has become especially critical as volume-fill packing techniques have been widely adopted. Although limited use of electronic color sorting has occurred, most separations must still be done by hand. Sorting requirements include the following:

**1. Adequate space for sorting.** This cannot be designated only in terms of packages per hour. Space requirements also depend on the percent of diverted fruit (subgrade, alternate grade, and so on), the number of decisions or separations required (by color, shape, and various defects), and the relative size of fruit being sorted (number of decisions per package).

**2. Ability to adjust flow of fruit.** A supervisor must have instant control of the sorting belt speed to adjust for variations in fruit quality, defects, size, and so on.

**3. Assignment of responsibility.** Each worker must have a specific task or responsibility. This usually involves a specific area, zone, or lane of the sorting belt. When multiple decisions are made, it should also involve separating responsibilities for these decisions. Periodically rotating worker positions on the sorting line may reduce monotony and fatigue.

**4. Ability to view the product.** Workers must have a complete view of the entire surface of the product for efficient sorting. Avoid designs in which part of the belt is hidden from the worker's view. Consider various systems to alter fruit position as it progresses

6.2. Inversion-type dry bin dumps are widely used in fruit packing lines. The padded lid is clamped over the bin before inversion; the gate in the lid is then opened to allow fruit flow during dumping.

**Fig. 6.4. Adequate facilities for fruit sorting are essential. This sorting table limits worker reach. The primary separation is at sorting table height; fruit are rotated horizontally as they pass down the belt to allow good sorter viewing.**

along the sorting belt. Slow rotation or periodic movement is desirable. Systems that both turn and rotate the fruit provide the greatest surface visibility. Adequate lighting is essential.

**5. Worker comfort.** Remember that workers are expected to operate at peak efficiency during long working hours. Platforms and stools adjusted to the proper work height for each worker are essential. Sorting and removal belts and chutes should be designed to eliminate unnecessary reaching and stretching. Worker fatigue can also be influenced by proper lighting and noise control.

**6. Avoidance of product injury.** A sorting line that causes fruit injury is self-defeating. Thus, the delivery system, the sorting belt, and the distribution system must be designed to avoid injuries. The height and number of drops and shears (direction changes) must be minimized. Activated sorting belts, which may aid in worker viewing of the fruit, must be carefully adjusted to avoid fruit injury. Thorough periodic cleaning of the facility to eliminate accumulated dirt will help reduce fruit injury. Fruit should flow along the belts only one layer deep.

**7. Worker training and supervision.** No sorting system will be effective if workers are not well trained and supervised. Workers must have clearly specified responsibilities and must be familiar with defects and with segregation categories and limits. Posting of visual aids may help in worker training. Supervisors must be familiar with worker performance limits and must be able to identify "under sorting" and "over sorting." Supervisors must also be able to adjust product flow to stay within the limits of worker performance.

## Fruit sizing

Fruit sizers come in a wide range of designs, but they all segregate sizes by either weight or dimension (figs. 6.5 and 6.6). Traditional dimension sizers measure the fruit by two-, three-, or four-point contact. Newer sizers monitor the image of each fruit from at least two positions and calculate the volume from the image. These image sizers and newer weight sizers use computers to estimate fruit size and to signal fruit separation into the correct size category. Sizers can work efficiently only if the delivery system is properly designed and adjusted to deliver a uniform flow of properly separated fruit across the full width of the sizer.

**Fig. 6.5. Weight sizers segregate fruit by mass. Fruit are distributed into individual cups that carry them forward until fruit mass exceeds the mass adjustment for the cup position. At that point the fruit drops onto a cross-flow belt for transport to the packing station.**

**Fig. 6.6. Dimension sizers are of many designs—this one is a volumetric sizer. Fruit are in contact with active rolls at four points. As fruit move forward, the rolls separate to provide a continuously expanding area for the turning fruit to pass through and drop onto the cross-flow belt.**

There are three requirements to consider in selecting a sizer:

**1. Capacity.** The sizer must meet the volume requirements of the packing operation. The practical commercial capacity of a sizer is estimated to be about two-thirds of the theoretical rated capacity. Because sizer capacity is related to fruit count, determine the capacity requirement based upon the smallest average fruit size anticipated.

**2. Accuracy.** The sizer must segregate fruits with the required accuracy to meet uniformity requirements for a place pack, for size segregation to meet marketing requirements, or simply to meet legal sizing requirements.

**3. Injury.** Sizers must meet the performance requirement without injuring the fruit, including all varieties and maturities that are commercially encountered.

Other special features may be desirable in selecting a sizer, provided the three basic requirements are met. These special features may include (1) ease of adjustment as incoming fruit sizes change, (2) ability

to adjust fruit diversion patterns as peak sizes change, and (3) ease of cleaning and maintenance.

### Special treatments

Depending upon the commodity to be packed, other special treatments may be required.

**Presizing.** Presizers, usually located immediately after the dump, are designed to eliminate all fruit below a minimum size. This reduces the volume flowing over the balance of the packing line, and thus increases overall equipment capacities. Presizers are of many styles and have the same basic requirements as sizers, except that only a single separation is required.

**Cleaning and washing.** Fruit may need cleaning and washing to remove soil or contamination, or to remove natural waxes in preparation for wax applications. Detergent washes are sometimes used with soft brushes or sponges followed by clear water rinsing. Many peaches receive wet brushing to remove the trichomes (peach fuzz).

**Waxing.** Some fruits are waxed (fig. 6.7) as part of the packing operation. Waxes may be used to reduce water loss, to replace natural waxes removed during washing, to cover injuries such as those caused by peach defuzzing, to act as carriers for fungicides, or to improve the fruit's cosmetic appearance. Waxes must be approved "food grade" materials. Studies indicate that waxes should not reduce the rate of water loss by more than about one-third lest they interfere with the normal aerobic respiratory activity of the fruit.

**Disease control.** Some postharvest disease treatments may be applied during packing. Heat treatment, especially hot water treatment, has been studied for many fruits. It is widely used for papayas, typically before or at the start of packing. Fungicide applications, when needed, are commonly applied while the fruit is spread on the conveyor belts, often immedi-

**Fig. 6.7. Some fruits are washed and waxed during preparation for market. Here a cold-water-emulsion wax, often containing fungicides, is applied to the fruit after washing, and allowed to dry without added heat.**

ately after washing. Fungicides are often incorporated into fruit waxes to aid in achieving a uniform surface application. All chemical applications must be made in strict conformity to label provisions.

## Packing the Fruit

### Purposes of packing

Packing can be viewed as a convenience in achieving orderly marketing of fresh fruit. The package is a convenient unit for transferring the product from the point of production to the point of final sale or consumption. If it is to function well, the package must be designed and used in a manner that protects individual fruit.

There are three requirements in packing fresh fruit to protect them from deterioration during handling and distribution:

**1. Fruit must be immobilized within the package.**Fruit can become injured by movement during transit. Immobilization can be ensured by wrapping and place-packing of carefully sized fruit, by use of various types of trays, or by certain volume-fill techniques (described below). Proper design of packages and padding is essential.

**2. Fruit must be cushioned against impacts.**Impact bruises can occur during packing when fruit are dropped. After packing, impact bruises can occur when packages are dropped during handling. Unitized handling (usually pallet-size units) greatly reduces rehandling of individual packages and thus helps reduce impact bruising. Various cushion pad materials, used as bottom pads or between layers of fruit, are effective in absorbing impacts and reducing bruising.

**3. Fruit must be protected from compression.**Compression bruising can result when overfilled packages are compressed, during lidding or after stacking. The packer may also cause compression bruising by pattern-packing oversized fruit. Much compression bruising results from package failure, a situation that causes the fruit, rather than the package, to assume the stacking stresses. Thus, proper package design and specifications, selected for the particular application, are vital in preventing fruit compression bruising (see chapter 7, *Packages for Horticultural Crops*).

### Packing line

The packing line must be designed to minimize opportunities for fruit injury. Shears and drops should be eliminated whenever possible, and live shears (often moving belts or air jets) and padding should be incorporated where diversion is needed. Careful control of fruit flow, whether to packing tubs or stations or on return-flow packing belts, is needed to avoid unnecessary fruit accumulation.

For certain packing procedures, it may be desirable to have the packing line accommodate limited fruit resorting before packing. This can be especially important in volume-fill packing systems, where delivery of a large volume of fruit to machine fillers taxes the system's sorting capacity. Because the fruit have been previously sized and graded, final sorting may be more properly considered a quality control procedure—assurance that fruit quality standards have been uniformly met.

Delivery of packaging materials to the packing stations and removal of filled packages are as important to packing efficiency as is adequate fruit delivery. Various mechanical and hand supply systems are possible, depending on the type and volume of the operation. These facilities must be carefully designed for efficiency so that they do not become bottlenecks in packing.

Fruit leaving the packing stations are usually inspected before final padding and lidding. This inspection, a final quality control procedure before completion of the pack, is to assure that both fruit and packing procedures meet grade and quality requirements. In some larger operations, continuous third-party grading is used, with the inspector sampling at this point to assure compliance with legal grades and standards. Here also, packages are usually stamped with necessary information on variety, size, grade, and so on.

### Hand-packing operations

Fruit are hand packed to create an attractive pack, often to pack a fixed count, sometimes to select for size, and always to immobilize the fruit within the package (fig. 6.8). This requires fairly precise sizing of fruits, at least within single layers of the package. Immobilization usually requires packing to lateral tightness. In tray packs, the presence of an oversized

**Fig. 6.8. Plastic tray pack used here for peaches illustrates one type of hand packing. Fruit are selected and placed into cups in trays to provide uniform lateral tightness within package. After packing, a pad is placed over top layer, and package flaps are folded and glued closed.**

fruit may prevent top pads or trays from contacting surrounding smaller fruit, which may then be subject to motion injury during transport. Similarly, an undersized fruit may negate lateral tightness, allowing surrounding fruit to turn and be injured.

Packaging materials, serving to isolate or immobilize the fruit, are often as important as pattern packing in preventing fruit damage within the package. These materials may include trays, cups, wraps, shims, liners, or pads. They often add significantly to packaging material costs.

Hand-packing facilities range from return-flow belts to a variety of tubs, bins, and chutes. The return-flow belt is normally used without mechanical sizing, with the packer selecting fruit of a desired size from the range of sizes on the belt. Tubs and bins are filled with presized fruit from which packers can draw at random.

Older hand-pack systems simply positioned the packer and packing station within easy reach of the fruit. A modification for nonwrapped packs is the rapid-pack system where the work area is designed so the worker faces the fruit for easy two-hand access. One innovation is an automated tray-pack where the tray passes on a belt just below the fruit delivery chute. By carefully regulating the speed of the tray belt, the operator can cause fruit to fill most cups in the tray. Other workers then orient fruits, fill voids, remove excessive fruits, and place the trays into packages.

### Mechanical packs

Mechanical fruit packing systems (volume-fill/tight-fill) are in increasing use in California. These systems deliver carefully sorted and sized fruit, along with empty packages, to automatic fillers. After filling, the packages pass through standard inspection, marking and closing operations and may receive special top padding, vibration settling, and lid fastening (tight-fill). A mechanical place-packing system available for citrus packing accumulates sized fruit into the desired packing pattern and uses rubber suction cups to transfer each layer of fruit into the package.

Mechanical packing systems normally handle large volumes of fruit at high speed, and afford no opportunity for further grading at the filling chute. It is therefore mandatory for the arriving fruit to have been adequately sorted and sized to assure meeting the desired grade.

Mechanical packers should be adjusted to properly fill the volume of the package. Most fillers are designed to use weight as an estimate of volume (fig. 6.9). Many fill to within a small amount of the desired weight, but final adjustment must be done by hand as the packages pass over scales. Some fillers are designed to adjust the fill weight to the nearest fruit, so that only check weighing is then necessary, but maintaining this adjustment has been difficult. Lack of accuracy in fill weight adjustment is a major cause of poor performance of volume-fill packs.

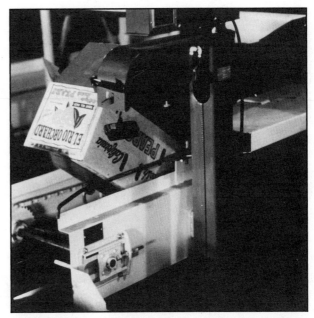

Fig. 6.9. Automatic volume-filling is widely used for fruit packing. Here the empty package is tilted up to the filling belt and is settled to horizontal position during filling with presized fruit. The scale is adjusted to the desired fill weight.

The most frequent problem in new volume-filling installations is that the flow of fruit of certain peak sizes exceeds filler capacity while some other sizes flow at very small volumes. To compensate, one or two extra fillers may be installed to expand the filling capacity of peak-size fruit. Because peak sizes vary with the fruit, variety, and orchard, the delivery system to the fillers must be easily adjustable.

A major problem with mechanical fillers is the height of drop of fruit into the package. This can be minimized by properly designed equipment. Modifications such as tilting packages during filling to reduce fruit drop heights are common. Padding of filler chutes and use of decelerator curtains may reduce impacts. Cushion pads in the bottom of the package can further reduce fruit impact injury during filling.

Tight-fill packing (fig. 6.10) is a special modification of volume-filling designed to ensure fruit immobilization during transport. It involves volume-filling fruit into the package to an exact weight, settling with a few seconds of carefully controlled vibration to eliminate voids, top padding with a special pad that nests around the top fruit, and tightly fastening the lid. When all of these steps are properly performed, and the package is designed for the packing system, the fruit are held tightly in place without compression bruising. There are specific requirements for package and pad design, fill density, and vibration characteristics that must be met for successful tight-fill packing. The method is described in University of California Circular 548 (Mitchell et al. 1968).

38

**Fig. 6.10. Tight-fill packing unit. The machine vibrates fruit into place, seals package flaps, and fastens a lid to the package.**

## REFERENCES

1. Gentry, J. P., F. G. Mitchell, and N. F. Sommer. 1965. Engineering and quality aspects of deciduous fruit packed by volume filling and hand placing methods. *Trans. Am. Soc. Agric. Eng.* 8:584-89.

2. Grierson, W., W. M. Miller, and W. F. Wardowski. 1978. *Packingline machinery for Florida citrus packinghouses.* Univ. Fla. Bull. 803. 30 pp.

3. Hardenburg, R. E. 1967. *Wax and related coatings for horticultural products; a bibliography.* U.S. Dept. Agric. ARS 51-15.

4. LaRue, J. H., and F. G. Mitchell. 1964. Bulk handling of shipping fruits. *Calif. Agric.* 18(6):6-7.

5. Mitchell, F. G., J. H. LaRue, J. P. Gentry, and M. H. Gerdts. 1963. Packing nectarines to reduce shrivel. *Calif. Agric.* 17(5):10-11.

6. Mitchell, F. G., N. F. Sommer, J. P. Gentry, R. Guillou, and G. Mayer. 1968. *Tight-fill fruit packing.* Univ. Calif. Agric. Exp. Sta. Ext. Ser. Circ. 548. 24 pp.

7. O'Brien, M., B. F. Cargill, and R. B. Fridley, eds. 1983. *Principles and practices for harvesting and handling fruits and nuts.* Westport, CT: AVI Publ. Co. 636 pp.

8. Smith, R. J. 1963. *The rapid pack method of packing fruit.* Univ. Calif. Agric. Exp. Sta. Circ. 521. 20 pp.

# II. Vegetables

ROBERT F. KASMIRE

Preparation for marketing vegetables begins with harvesting (maturity selection and some quality control) and, depending upon the commodity, may include some or all of the following: product assembly, receiving, cleaning, trimming, sorting, grading, sizing, waxing, packaging, packing, cooling, ripening initiation, curing, storage, unitizing, and shipping. Preparation may be done mostly in the field (lettuce, celery, cauliflower) or in packing sheds (most vegetables), or both (broccoli, cantaloupes). Vegetables may be prepared entirely at the shipping point (most commodities), partly at the shipping point and partly at the destination market (tomatoes, potatoes), or mostly at the destination market (commodities shipped to prepackers in destination markets).

Modern equipment and facilities used are expensive; therefore, the shipping season is extended for as long as possible. Equipment for some commodities (lettuce, celery) is portable, or mobile, and may be used in two or more shipping districts each year, either by the same grower-shipper or leased to others, or by custom operators. This amortizes the costs of equipment over a longer period each year.

## Harvesting

**Maturity selection (at harvest).** Select products of desired maturity for the intended markets. Harvest maturity influences susceptibility to handling dam-

age, ripening required (if any), shelf life, size, length of cooling cycles required, and market availability.

**Product assembly.** Assemble for field packing or for hauling to packinghouses or coolers. May include limited sorting and grading (for products to be hauled to packinghouses) or final grading (for field-packed vegetables). Includes assembling packages of field-packed vegetables or assembling loads of field containers (bulk bins, gondolas, field boxes) at packinghouses for subsequent operations.

## Packinghouse Operations

**Receiving.** Unload products from loaded field containers onto a conveyor or into a water dump for conveying into a packinghouse. May include a small-size eliminator and some sorting for removal of decayed products. Dry receiving (dumping) operations often cause considerable product damage.

**Cleaning.** Removes soil and other foreign material from product surface by washing, brushing, or both. The wash water may or may not be chlorinated. Recycled water should always be chlorinated.

**Trimming.** Remove unwanted leaves, stems, or roots before grading, packaging, and packing (lettuce, celery, cauliflower, asparagus, dry onions). Trim asparagus spears to predetermined lengths by cutting off excess length at the basal end.

**Grading.** Separate products by market quality (grades) either before or after sorting.

**Sorting.** Select product by maturity, shape, color, or some other physical parameter. Some commodities are machine sorted. Culling is part of both sorting and grading.

**Curing.** Some products are cured (garlic, sweet potatoes, dry onions, and new crop potatoes) after harvesting and before storage or marketing. Onions and garlic are cured to dry the necks and outer scales. Potatoes and sweet potatoes are cured to develop wound periderms over cut, broken, or skinned surfaces. Curing helps heal harvesting injuries, reduces water loss, and prevents entry of decay-causing organisms during storage. Curing may be done in the field (garlic, onions), in curing rooms (sweet potatoes), or during transit (new crop potatoes).

**Sizing.** Separate product units into physical sizes (weight, volume, length, diameter, or other parameters). Most commodities are sized by mechanical or electronic sizers, but many products are still visually sized.

**Waxing.** Cover surfaces of product with food-grade wax to reduce water loss through epidermal openings. Waxes are generally applied only to fruit-type vegetables but may also be applied to roots (rutabagas). A fungicide may be incorporated into the wax.

**Packaging.** Enclose consumer units of a product in individual packages (wraps, bags, sleeves, trays, or other units) that are then packed in master packages. Most materials used for consumer unit packaging are plastic films composed of two or more types of film material combined into a single film. They may be used separately or as wraps over molded polystyrene or pulp trays. Paper bags are also used. Product units in a consumer unit package should be of comparable weight, size, maturity, and grade. Some packaging involves enclosing a single product unit (a head of lettuce or cauliflower), while in other packaging several product units are enclosed in a single consumer unit (potatoes, radishes, brussels sprouts, carrots). Packaging is done both automatically and manually, at the shipping point or destination market.

**Packing.** Assemble a given quantity (count or weight) of comparably sized product units or consumer units in packages. When counts are used, products are often packed in specific arrangements within packages. Counts, arrangements, and weights are often specified in, and regulated by, various government and industry codes or tariffs. Packages may be bags, cartons, crates, lugs, or bulk bins. Some products are shipped unpacked to markets in bulk trucks or railroad cars.

**Unitizing.** Assemble completed packages of products into larger units (pallet loads or slip-sheet loads) for better, less expensive handling, for load stabilization, and improved transit temperature management.

**Ripening initiation.** Apply ethylene or ethylene-producing materials to stimulate ripening of tomatoes and honeydew melons.

**Cooling.** Remove heat from product before shipping or long-term storage by cooling. Lowering product temperatures extends storage and shelf life.

**Loading.** Load into transit vehicles for shipment.

**Shipment.** Transport to markets in commercial or privately owned conveyances (trucks, rail cars, container vans, ships, and airplanes).

## Storage

Some products are held for long periods between harvesting and marketing so as to fulfill the market demand. Primarily potatoes, garlic, dry onions, winter squashes, and sweet potatoes are stored in California. Celery, cabbage, and carrots and some other root crops are also stored in other producing regions. Storage is generally in specially constructed facilities; however, storing may also be in the ground (undisturbed plant beds) before harvesting but after the crop is matured (dry onions). Products may be stored in bulk (usually) and before packing-shed operations, or in packages. Some fall-season muskmelons are stored for a few weeks to enable shippers to capitalize on higher-priced markets later in the season. Storage involves economic risks but also provides a way to postpone shipping products to already glutted markets.

## Problems

Most problems involved in preparing fresh vegetables for market involve communications among and with workers or are related to equipment or materials used. The more important types of problems confronted are the following: (1) inadequate training and supervision of workers; (2) poor understanding by handlers of their roles in maintaining product quality and of how their decisions affect product quality loss; (3) inadequate communication among supervisors and among handlers in marketing and distribution; (4) rough handling when operations are conducted too fast in order to reduce labor costs; (5) unsatisfactory, improper, or inadequate equipment or materials; and (6) inadequate product cooling.

### REFERENCES

1. Ryall, A. L., and W. J. Lipton. 1979. *Handling, transportation, and storage of fruits and vegetables*. Vol. 1, *Vegetables and melons*. 2nd ed., 43-117. Westport, CT: AVI Publ. Co.

# III. Automation Trends in Packinghouse Operations

PICTIAW CHEN

The 1980s markedly increased automation of packinghouse operations. Reasons for the change are listed below.

## Packinghouse Automation Advantages

Reasons for the introduction of automation and its advantages include:

1. **Increased productivity**

    Higher product flow rate

    Better coordination of different operations to permit a system to operate at nearly full capacity most of the time

2. **Improved quality of products**

    More accurate sizing and quality grading

    Product units can be packed into more categories

    Less handling of products, reducing mechanical damage

3. **Reduced cost**

    More efficient use of hand labor

    Fewer conveying lines, resulting in a more compact system

    Fewer culls as a result of more accurate sizing and/or sorting

4. **Accountability facilitated**

    Accountability of incoming fruits by producer's identification

    Accountability of packouts—numbers of fruits, weight, size, maturity, grade, number, and type of packages

    Accountability of stored fruits

    Accountability of buying and selling prices, hours and wages of workers, and packing efficiency

5. **Easy control of operations facilitated**

    Control of packing rate

    Control of packout categories—size, grade, container

    Control of operations—receiving, fruit flow, water temperature, chemical concentration, and palletizing

    Instant feedback and instant change in operations to optimize operations and to match demand-supply conditions

## Factors facilitating automation

**Advancement of computer technology.** In the past 15 years computer technology has advanced at an incredible rate. The size and price of computer components have decreased to less than 1 percent of what they were 15 years ago while their capacity and reliability have increased tremendously. As a result a large selection of small, low-cost, high-capacity computers are readily available for use in packinghouses.

**Improvement of electronic and sensing components.** Electronic components are also getting better, cheaper, smaller, and more reliable. The availability of low-cost, high-quality sensing elements, such as transducers and photoelectric cells (electric eyes), makes it possible to measure and detect such parameters as temperature, weight, color, and surface blemishes at an extremely high speed and to transform the parameter values into electric signals that can be fed into the computer for further processing.

**Increased knowledge of physical properties of agricultural products.** The responses of agricultural products to different kinds of inputs are important to the design of automated systems. The following are examples of some physical properties and their applications to packinghouse operations.

1. *Physical*

    Parameters: size, shape, volume, weight, density, and surface area

    Application: singulation, orientation, sizing, packing, quality evaluation, waxing, and coloring

2. *Mechanical*

    Properties: force-deformation, firmness, response to impact, vibration, and stress-strain

    Application: fruit handling, cause and prevention of damage, quality evaluation

3. *Optical and radiation*

    Related properties: light reflectance, transmittance, absorption and emission, X-ray and gamma-ray absorption

    Application: color sorting, blemish detection, maturity evaluation, composition analysis, and other quality evaluations, sizing, counting

4. *Vibrational and sonic*

    Properties: vibration of individual fruit, vibration of fruits in packages, natural frequency, sonic and ultrasonic transmission

    Application: quality evaluation, material sorting, and prevention of damage

## Present automated systems

Automated fruit packing systems presently available consist of an electronic sizing unit, a quality evaluating unit, and a computer, which analyzes the size and quality data and determines the destination of fruit as dictated by a program, which can be easily changed at any instant. The computer also keeps track of each fruit and accounts for its size, weight, grade, and destination, and finally prints out the packout sheet. In addition to the computer, some systems also have mechanical devices for other automated operations, such as bin dumping, chemical mixing, tray packing, bagging, box filling, and palletizing.

## Future Trends

1. **Broader application of computer and electronic devices**

   Integration of all operations via the computer

   Control of each operation and optimization of entire system

   Analysis of supply-demand conditions and optimization of packout categories

2. **Increased efficiency**

   Higher speed—increased production

   Elimination of singulation requirement—fruits can go through grading devices in batches instead of one at a time

3. **Improved techniques for quality evaluation**

   Improved accuracy in evaluation of quality factors

   Inclusion of more quality factors—bruise, external and internal defects, sugar-acid ratio, and overall quality

   Increased use of scanning devices, multi-spots viewing equipment, and machine-vision systems to improve sorting accuracy

4. **Improved fruit handling system to reduce damage**

## REFERENCES

1. Birth, G. S. 1979. Radiometric measurement of food quality—a review. *J. Food Sci.* 44:949-57.

2. Chen, P. 1978. Use of optical properties of food materials in quality evaluation and materials sorting. *J. Food Process Eng.* 2(4):307-22.

3. Dull, G. G. 1986. Nondestructive evaluation of quality of stored fruits and vegetables. *Food Technol.* 40(5):106-10.

4. Finney, E. E., Jr. 1973. *Measurement techniques for quality control of agricultural products.* St. Joseph, MI: Am. Soc. Agric. Eng. 53 pp.

5. ———. 1978. Engineering techniques for nondestructive quality evaluation of agricultural products. *J. Food Protec.* 41(1):57-62.

6. Foraker, J. D. 1980. Computer age apple packing. *Fruit Grower*, March.

7. Gaffney, J. J., comp. 1976. *Quality detection in food.* St. Joseph, MI: Am. Soc. Agric. Eng. Publ. 1-76. 241 pp.

8. Johnson, M. 1985. *Automation in citrus sorting and packing.* In *Proc. Agri-Mation 1 Conference and Exposition*, Chicago, Feb. 25-28, 1985, 63-68.

9. McClure, W. F., and R. P. Rohrbach. 1978. Asynchronous sensing for sorting small fruit. *Agric. Eng.* 59(6):13-14.

10. Thomason, R. L. 1986. High speed, machine vision inspection for surface flaws, textures and contours. Presented at Vision '88 Conference, Detroit, MI, June 3-5.

11. Werth, L. 1986. *Machine vision implementation in agri-applications: the general purpose approach.* In *Proc. Agri-Mation 2 Conference and Exposition*, Chicago, March 3-5, 63-77.

# IV.  Cull Utilization

JAMES F. THOMPSON

About 10 to 15 percent of the fruit and vegetables delivered to California packinghouses are rejected as culls. They are culled because of scars, split pits, deformities, mechanical injuries, sunburn or sunscald, mold or insect damage, immaturity, overripeness, softness, or small size. These defects result in a huge quantity of material that must be utilized or disposed of.

## Disposition of Culls

The most obvious approach to reducing culls is to reduce the number that reach the packinghouse. The farmer should use the best cultural practices to produce a well-sized, unblemished commodity. It should be harvested and handled carefully to minimize injury, and harvesting personnel should be encouraged to discard poor-quality fruit and vegetables in the field.

Culls can also be reduced by lowering the quality standards; however, this often results in the poorer quality fruit and vegetables being culled by the retail distributer or the consumer. The added costs of handling and shipping poor-quality fruit and vegetables can increase the cost of good-quality produce. A packer shipping high-quality fruits and vegetables has a competitive advantage.

However, even careful attention in the field does not eliminate the need for management of culls at the packinghouse. The cost liability to the packinghouse operator can be minimized if culls are sold as a by-product rather than returned to the field and dumped.

In California, the largest use of fruit that is not suitable for the fresh market is as processed fruit product or as a by-product source. Fruit that is too ripe for long-distance shipment can be sent to local markets, if they are available. Some culls can be processed into juice or a canned or frozen product. A large quantity of cull apples, pears, oranges, and papayas are used for juice extraction. Cull avocados are made into guacamole, and undersized artichokes are marinated and canned. Some culls can be turned into dried fruit for human consumption. However, good-quality dried fruit is made only from good-quality fresh fruit. Only undersized or slightly over-ripe fruit should be considered for drying. Citrus culls are used as a source for flavorings (lemon and orange oil), pectin, and juice. Other culls can be used as a source of natural food colors.

### Cattle feed

Cull fruit is palatable and a good source of energy for animals, but it is low in protein and has other characteristics that make it different from other feed sources. For example, stone fruit (peaches, plums, nectarines) contain 85 percent water, 9 percent digestible dry matter, 4 percent pits, and 2 percent indigestible dry matter. The high water content diminishes the real value as feed because it makes culls expensive to transport, requires large trough volumes, and allows the feed to spoil quickly. If fed in large proportions, cull fruit causes almost continuous urination, and consequently the animals have a high salt requirement. The only potential advantage to the high water content is that animals in a remote, dry location will not need extra water hauled to them.

Low protein levels in cull fruit limit the quantity that can be fed. Where rapid weight gain is important—in feed lots, for example—only about 20 percent of the ration can be composed of cull fruit. As a maintenance ration, up to 80 percent of the feed can be culls.

Stone fruit pits rarely cause internal injuries or choking. Cattle spit out some pits while eating, and many of the remaining pits are regurgitated with the cud and spit out. In fact, the main problem with pits is disposing of them as they tend to fill feed troughs.

Cull fruit typically is bought for $2 to $5 per ton. In terms of feed value, this is equivalent to buying barley for $20 to $50 per ton. However, the costs of handling and transporting culls must be added to this cost. Also, some cost must be added to account for the uncertain effects of using a feed that has not been thoroughly tested for nutrient levels and trace chemicals.

Cull potatoes are good source of feed for animals. Like stone fruit they are high in water content (about 77 percent), high in energy value, and low in protein. Beef steers can be fed up to 50 percent potato waste in finishing rations and still have acceptable weight gain. However, the cattle must be carefully adapted to a potato ration, and the ration should not be changed rapidly.

To a limited extent cattle are also fed culled cantaloupes and other muskmelons. Cool-season vegetable culls have also been used as feed. But these culls all have the same general limitations already discussed.

### Alcohol production

Most fruit and some cull vegetables (especially roots and tubers) can be used for alcohol production. Alcohol for human consumption has a much higher value than alcohol for motor fuel. Some cull pears, kiwifruits, and apples are used for fruit wine production in California, and some of the apple wine is converted to cider vinegar.

The use of culls for fuel alcohol production is limited mainly by the low sugar content of most fruits and vegetables. The 8 percent to 12 percent sugar content of most cull fruits results in an alcohol yield of about 10 gallons per ton of fruit. Potatoes have one of the best yields of alcohol for culls at 20 to 25 gallons per ton, but this is still low compared to better feedstocks such as corn, with a yield of 90 gallons per ton. The low yield makes it uneconomical to haul culls any significant distance. If fuel alcohol production from culls is to be economical, it must be done near the packinghouse. Low sugar content also results in 4 percent to 5 percent alcohol "wines," which require considerable energy per gallon of alcohol to process and distill.

### Waste Management

Stillage waste left after distillation of alcohol is a waste management problem. It has very little protein, so it is not suitable as an animal feed, but it has a high pollution potential as measured by BOD (biological oxygen demand). Stillage waste may be usable as a feedstock for methane generation. The effluent from a methane generator is low in pollution potential.

In general, there are several key constraints to any method of cull utilization. First, transportation is expensive as 80 percent to 90 percent of the weight of culls is water, which usually has no value to the one using the culls. Second, culls are produced in large quantities during a short time period. A year-round operation must be able to store the culls, usually by drying, which adds a cost to the product, or by utilizing other products during the off-season. Operations that run only during the season cannot make large capital investments in equipment that will be utilized a few months a year. Finally, there must be a market for the by-product. For example, in California dried fruit has a specialized and limited market which cannot absorb extra amounts of product made from culls.

### Pollution potential

Unfortunately, the limits to utilization often result in large portions of the culls being discarded. Improper disposal results in fly, odor, health, or pollution problems.

Flies and odor problems are prevented by ensuring rapid drying. Fly maggots hatch into adults within 7 to 10 days, and odor problems can develop before fly problems. The culls should be crushed and spread no more than one or two layers deep; sometimes this is done on orchard roads or fallow fields. Culls can be disced into the soil, although this tends to cover the fruit with soil and slows drying. Also, insects or diseases that may have caused the fruit to be culled in the first place may infect a future crop. Disposal sites should be as far away from neighbors as possible.

Flies can travel up to 5 miles from the place where they hatch.

Culls should not be dumped near streambeds. Fruit dump sites can attract the dumping of all kinds of refuse. If culls are deposited away from the point of production, use municipal solid waste disposal sites if available.

# 7

# Packages for Horticultural Crops

F. GORDON MITCHELL

Packages are convenient units for marketing and distribution of horticultural products, and they have many special requirements. Packages must protect the contents against damage during distribution and must maintain their shape and strength, often for long periods at a relative humidity near saturation and sometimes after water drenching. Many must be designed to facilitate rapid cooling of the contents from warm field temperatures to low storage or transport temperatures, and must allow continual removal of heat produced by the contents. Because of the fragile nature of many horticultural products, packages usually must assume all stacking stresses throughout distribution and be adaptable to high-volume packing operations. When used for display, packages must be attractive to the consumer.

## Today's Horticultural Packages

Many materials, sizes, and shapes are represented in packages for horticultural products. More than 500 different packages are used for produce in the United States. Past efforts at standardization have had limited success. Major changes have been in response to economic considerations—the use of less expensive materials, the need to adapt to less expensive packing and handling procedures, or ability to increase load density during transport. In the United States, major shifts have been from wood to corrugated packages (with limited use of plastic packages), from hand packing to mechanical volume-fill packing operations (many fruits and some vegetables), and from single package handling to unitized handling on pallets. These changes have forced a general review of package requirements for use with horticultural products.

## Product Requirements

Success in developing a package for horticultural products is based upon emphasizing the requirements of the product being packaged. These requirements vary widely with the commodity, marketing program, packing method, and so on. However, there are many generalities that transcend most commodities.

## Protection from injuries

All physical injuries must be avoided wherever possible during handling and distribution. Some of the more obvious open wounds (e.g., cuts, punctures) often occur before packaging and can be eliminated by good supervision and sorting. Certain bruises, however, may accumulate throughout all stages of handling, including packaging and distribution.

**Impact bruises** (fig. 7.1). Impact bruising results from dropping the product onto a hard surface, either individually or in packages. Impact injury may not be visible from the surface, so careful quality control is needed to protect against it. Dropping the product into the package is a common cause of impact injury during packing. Careful padding at drop points, use of decelerator strips at filling chutes, and designing fillers to raise empty packages to reduce drop heights during volume-filling can all reduce incidence and severity of impact bruising. Cushion pads in the bottom of packages may provide further protection.

Packaged products can receive impact bruising from being dropped during palletization, loading, unloading, commodity segregation, and so on. Bottom cushion pads in the packages may reduce impacts. Unit handling reduces package rehandling and thus the number of impacts. However, even rough handling by forklifts can cause impact bruis-

**Fig. 7.1. Impact bruise on Anjou pears. Bruising extends into the flesh and may or may not be visible on the surface.**

ing. Careful supervision is essential to reduce exposure to impacts.

**Compression bruises.** Compression bruising (fig. 7.2) results from improper packing and from inadequate package performance. The package dimensions must be carefully adjusted to accommodate the volume of product being packed. For pattern packs, product size must be carefully selected to avoid compression injuries from overpacking. Probably the greatest single cause of compression bruising is intentional overpacking to create a bulge pack. Whatever significance it may once have had, a bulge pack will not convince modern buyers that they are getting something for nothing.

Compression of the product after it has been packed into the package is a major cause of bruising. The most obvious cause is when overfilled packages are stacked; because of package distortion, the commodity absorbs much of the stacking force. Compression bruising also occurs if a package is not strong enough to support packages stacked on top of it. Packages must be able to withstand the expected weight of additional packages.

Avoid stacking packages beyond their design limits. Most produce is stored at high relative humidity, some of it for long periods after packaging. It is not economically feasible to design corrugated packages to withstand stacking three or four pallets high during storage. Even two-high stacking should be considered a temporary requirement—shippers should recognize that temporary stacking often occurs during distribution, and allow for this in package design. For multipallet stacking heights and for storage, pallet racks or "crutches" (discussed below) are less expensive than extra-strength packages.

**Vibration/abrasion bruises.** Less well recognized is the damage that results when products move within the package during transit (fig. 7.3). The resulting vibration bruising is usually restricted to the product surface, but it still reduces saleability. On soft com-

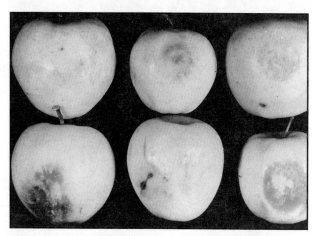

**Fig. 7.2. Compression bruise on Golden Delicious apples. Bruise damage occurs on surface and extends into flesh of fruit.**

**Fig. 7.3. Vibration bruise on Bartlett pears. Bruise damage appears on fruit surface and usually does not extend into flesh.**

modities, deep flesh bruising can result. To prevent vibration injury the product must be immobilized within the package by properly sizing the package for the product and its density.

In hand packing, immobilization begins by achieving lateral tightness within the package. Supplemental materials, such as wraps, trays, cups, shims, liners, and pads, may be useful. Volume-fill packs can be immobilized by proper filling, settling, padding, and lidding. A special procedure called tight-fill packing is designed specifically to immobilize the product after volume-filling.

After filling, the packages must perform properly if the product is to remain immobilized. Any bulging increases the volume of the package and allows more space between contents, thus making transit injury more likely. To immobilize the product the package must be strong enough to resist bulging throughout distribution, even during high-humidity storage and transport.

## Temperature management

The horticultural package must accommodate the special temperature requirements of the product. Temperature management depends upon good contact between the product in the package and the external environment. For some products, providing for airflow past package surfaces may be sufficient. Usually ventilation is needed for airflow through the package to remove heat rapidly. Within limits, increasing the size of ventilation openings speeds heat exchange. For corrugated packages, 5 percent venting of side or end panels allows rapid cooling without overly weakening the package. A few large vents perform better than many small vents. Vertical slots, kept at least 2 inches (5 cm) from the package edges, perform well. The effect of these vents during transport depends on a loading pattern that lets cold air reach the vent openings.

Certain fruits require ripening before retail marketing. They must be uniformly warmed to ripening temperature, and they often need treatment with ethylene. A package that is properly vented for cooling performs well for both warming and gassing.

Package vents should be unobstructed by internal packaging materials. Liners, wraps, trays, shims, or pads should minimize blockage of airflow through the package. If these restrictive materials are essential, then airflow must be appropriately increased to compensate for their effects.

Some packages are designed to restrict heat flow into the package. Packages used for air transport may be designed without ventilation (and sometimes with insulation) to delay product warming during transit without refrigeration. Flower boxes are designed so vents can be sealed after forced-air cooling. It must be recognized that heat flow restriction will be equal in both directions; thus, the heat of respiration of the product will be retained within the package to the same degree that outside heat is excluded. Package-icing may be useful in such a system if the commodity is packaged cold and both the product and package can tolerate prolonged water contact, and ice-gel packets can be used with high value commodities.

## Protection from water loss

Many horticultural products suffer wilting, shriveling, or drying as a result of water loss during handling and marketing. Water loss occurs because of a water vapor pressure gradient between the product, which is normally near saturation (100 percent RH), and the surrounding, drier environment. During storage it is desirable to hold most products at a high relative humidity to minimize water loss. Because there is usually little or no environmental humidity control available during transport and marketing, the package may be designed to provide a partial barrier to movement of water vapor from the product.

Many package moisture barriers are available. Plastic (poly) liners, usually with small perforations to allow some gas exchange, maintain an essentially saturated atmosphere within the package, but can result in surface cracking problems with some commodities. Poly curtains, which may be open at the package ends and folded around the product sides and top, provide a partial moisture barrier and are a successful compromise for some fruits. Corrugated packages can be treated with various coatings to act as moisture barriers. Because most package materials absorb moisture, surface coatings slow water vapor uptake by the packages and delay their deterioration. The most widely used, called curtain-coating, is a poly-wax emulsion that is coated on the surface of the corrugated board.

Moisture barriers inside a package must not impede essential airflow through the vents. Poly liners are particularly troublesome because they completely block all vents. With poly curtains, tests with tray-packed peaches show that increased airflow during cooling and storage could compensate for this blockage. In the case of vacuum cooling, heat removal from the wrapped product can occur if the wrap is perforated.

Packages with surface coatings can be vented normally, so that the moisture barrier does not affect air movement. In most conditions, package venting causes only a limited increase in water loss from the product. For long-term storage, water loss can be reduced in such packages by partially restricting ventilation, with compensating upgrading of airflow requirements for cooling.

## Facilitating special treatments

Certain commodities have special treatments that must be considered in packaging selection and design. Examples are sulfur dioxide fumigation of grapes for disease control and methyl bromide fumigation of various commodities for insect control. These treatments require well-vented packages through which the fumigant can readily flow. Venting sufficient for rapid cooling is considered adequate for fumigation. Some grapes are packed with sulfite pads that slowly release sulfur dioxide; such packing requires a package or plastic liner with restricted ventilation.

Ethylene may benefit or degrade various horticultural products. The need for package venting to achieve uniform warming and ethylene treatment during fruit ripening has been mentioned. Conversely, some commodities must be protected from ethylene. Certain in-package ethylene scrubbing procedures, which are in limited use, perform best if package ventilation is restricted. Room scrubbers that circulate air from the storage room through an ethylene scrubbing unit may require good package venting to be effective.

Modified atmosphere storage with some commodities, especially apples and pears, calls for the use of partially sealed poly package liners. Accumulation of 2 percent to 3 percent carbon dioxide inside the liners reportedly improves fruit storage life. Because such poly liners inhibit heat exchange and slow product cooling, the real benefits under modern handling conditions are questionable. Use of this technique has been declining in recent years. There is some interest in the use of film wraps for individual fruit within the package.

Other special packaging requirements may exist for certain commodities. For example, gladiolus and asparagus must be packed upright to avoid curvature caused by geotropism. Orchids must be packed with a water supply to maintain freshness, and asparagus with a moist pad to maintain growth and to slow toughening. Asparagus must also be packed with some headspace above the spear tips to allow for growth and elongation.

## Compatibility with Handling Systems

Most handling systems still require some hand lifting, thus limiting package weight. A few systems are designed for mechanical lifts and can accommodate pallet bins (e.g., watermelons). The package may need special design features to make it compatible with packing equipment and handling procedures. Package top flaps may be a detriment in hand packing, depending on the packing procedure. The package must be sized to facilitate unitization and mixed load handling. Possible problems with weather and contamination must be considered in advance. Packages may be constructed of various materials to meet special requirements. Wood packages (fig. 7.4), long standard in the horticultural industry, are still used for long-term storage or high-moisture conditions, and foam packages are now sometimes used.

Some products, especially root, bulb, and tuber crops, are often handled in large plastic or burlap bags without other packaging. These are compatible with high-volume production and dense loading of transport vehicles. However, handling systems must be compatible with the necessary product protection if excessive deterioration is to be avoided. This can include such factors as temperature management and protection from injury.

### Packing facilities

In designing a package, its compatibility with conveyors and other packing equipment must be considered. Modifying equipment to accommodate a new package can be expensive. In volume-fill operations, package top flaps (which can pose problems with hand packing) can be important in containing the product during filling. This advantage may justify

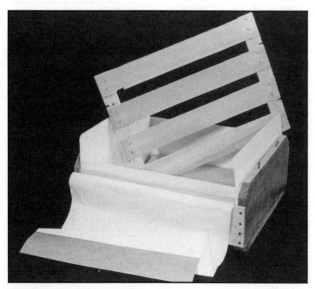

**Fig. 7.4. Wood lug used for shipping grapes. A package similar in style and design has been widely used for many horticultural crops. These packages often have wood ends, a wood/paper veneer side, and bottom wrap.**

the cost of redesigning the conveyor system to accommodate the height of the flaps.

A new package for field use must be compatible with field conditions. Some package treatments, such as wax coatings, may deteriorate at high temperatures, discoloring the packages and destroying much of their effectiveness. Some packages (e.g., polystyrene) are lightweight when empty, and precautions may be needed to stabilize them during strong winds. Other problems such as rains, heavy dews, or package soilage may require special attention.

As a new package is introduced, inventory problems must be considered. New packages should simplify operations, so early elimination of existing packages should be scheduled. Careful planning can minimize the inventory problem during package changeover.

### Unitized handling

Almost all horticultural packages in the United States must be suited for unitized handling. They must be designed for secure palletization, with packages either stacked in-register or cross-stacked. This is difficult to do with bulge packs, so attention to the package mass/volume requirements must be part of designing a new package. The package must withstand expected stresses in the stacking column. If packages are cross-stacked, their vents must be located on both ends and sides to allow air circulation through all packages on the pallet. This will require a geometric relationship between horizontal package dimensions and vent locations on package sides and ends.

Package dimensions must be compatible with pallet dimensions. Receivers have long voiced a desire to standardize on 120 cm × 100 cm (or 48 × 40 inch) pallets. The trend toward such a goal has been slow partly because of poor space utilization in the older 240-cm (96 inch) transport vehicles, and partly because of many vested interests in existing packages, pallets, and handling equipment. Conversion should now be speeded as most U.S. refrigerated vans are 260 cm (102 inch). Although standard pallet dimensions are desired, the paramount need is for the package to be compatible with the pallet size likely to be used for that commodity. In selecting package dimensions, check their legality in the production and marketing areas.

Some packages are designed to stabilize the palletized load. An example is the use of stacking tabs on certain styles of packages (fig. 7.5). Many packages can be glue-bonded between layers on the pallet with special breakaway palletizing glues. Normally, at least one horizontal nylon or steel pallet strap is required to assure stability of the glue-bonded pallet unit. Where gluing is not done, two or three vertical straps may also be needed to stabilize the pallet load.

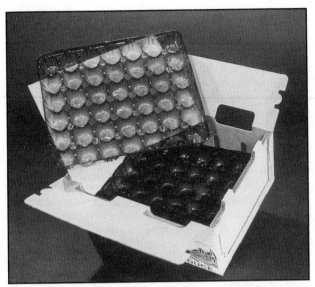

**Fig. 7.5. Corrugated "Bliss type" lug used for shipping many fruits. Package shown is equipped with plastic trays often used for hand or hand-assisted packs of fruit. End-stacking tabs are used to stabilize loaded pallets.**

## Package standardization

Someone who observes produce handling in distribution warehouses may be appalled at the large number of package types, shapes, and dimensions in use. These packages are not designed to be compatible in loading, and unnecessary, often serious product injury occurs as commodities are loaded together in mixed shipments or for retail distribution. Most products would fit into packages of just a few horizontal dimensions. This need has triggered a serious effort to establish a list of standard-sized packages, and there has been interest in making them meet the international metric dimensions.

There are many considerations in developing new packages for the so-called "metric pallet" (1200 × 1000 mm). Most facilities designed for metric pallets could also handle 48 × 40-inch (1219 × 1016 mm) pallets. Outside horizontal dimensions being suggested for U.S. horticultural products include 600 × 400 mm, 500 × 400 mm, 500 × 300 mm, and 400 × 300 mm. Equivalents on 48 × 40 inch pallets would be 24 × 16 inch, 20 × 16 inch, 20 × 12 inch and 16 × 12 inch. A few more size combinations that afford stability and good coverage of the pallet area may be needed to accommodate some commodities. For example, a 400 × 333 mm (or 16 × 13¼ inch) package would approximate the current widely used "Los Angeles lug" dimensions and may allow conversion with a minimal effect on packing and handling equipment.

Because of side bulge, packages of full dimensions will not fit a 1200 × 1000 mm pallet. How much bulge to allow depends on the package specifications, including board weight, package height and design, and relative package size. If packages are too wide

or long, with bulge, to fit the pallet, then the pallet dimensions would be exceeded and some package walls would extend over the edges, unsupported by the pallet. The resulting tilting and shifting of the load could cause severe damage to the product. Shortening and narrowing of package dimensions by as much as 10 mm (0.4 inch) may be needed to compensate for bulging.

There is also some effort to standardize package depths for all commodities. This poses serious problems in meeting packing and marketing requirements and in protecting the product. Many volume-fill packs have minimum fill depth requirements to achieve stability within the package.

Product requirements must be carefully considered in selecting a new package design. If a telescope-style package must be used (a two-piece package with a lid that fits over the body) (fig. 7.6), then the inside dimensions are less than in a single-wall package. The density of the product within the package must be carefully calculated based upon the usable inside space. Fill density information is available for many fruits in volume-fill and tight-fill packs. Some of these packs have very specific package depth requirements to perform satisfactorily. Tight-fill packages must allow a filling depth of three to four times the diameter of the largest size fruit to be packed. Soft commodities, such as berries, require shallow packing depths to avoid damage.

Trays for tray-packed commodities must be redesigned to fit the new package dimension. Many commodities are packed in 8 to 10 different sizes of trays. Because of the expense of retooling for new trays, it is essential that their exact inside horizontal dimensions be well established before changeover. Further, the tray packs may fit poorly into packages of certain depths, resulting in poor packing density. All of these

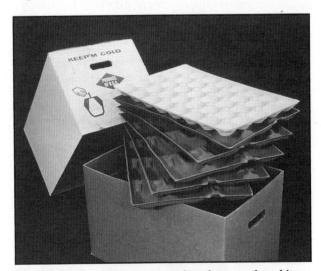

**Fig. 7.6. Full telescope corrugated package used to ship apples. Similar packages are used for many other horticultural commodities. Apples are layer-packed into trays shown.**

requirements must be carefully considered for each commodity before conversion commences.

## Adaptability to Handling Requirements

The package must perform efficiently under the handling conditions that will be encountered. This includes not only product marketing and distribution, but also inventory and handling of packaging materials.

### Moisture

Some commodities, especially certain vegetables, are in contact with water during and after packing. Packages used for these commodities must be able to withstand water contact, often for long periods. Water contact may occur when a packed product is hydrocooled, when ice is placed into the package (package-icing), or when top ice is placed over a load during transport. Corrugated packages made to withstand such conditions are usually more expensive than packages suited only for dry conditions.

Most horticultural packages must tolerate exposure to high relative humidity. Even if storage facilities operate at a moderate relative humidity (often 85 to 95 percent), water released from the product creates near 100 percent RH within the package. In terminal markets, atmospheric water may condense onto the cold surfaces of packages removed from transport vehicles or storage rooms. Special treatments or heavy corrugated board construction are needed for packages to tolerate long exposure to these conditions.

### High temperature

The problems of package exposure to sun and heat in the field have already been mentioned. Packages for field use must be designed and used to avoid deterioration. Light-colored board should reflect more radiant heat than natural kraft board. Flower boxes may be insulated to protect flowers from heat or cold when temperature protection is not available (e.g., delivery of a single package at a rural bus stop). High ambient temperatures affecting carry-over inventories of corrugated packages reportedly can speed delamination of the corrugated board, creating potential problems during the next season.

### Product storage

The package must withstand environmental conditions during the entire storage and marketing season of the commodity. To minimize problems during storage, stacking aids such as pallet frames, racks, and "crutches" (rigid steel or wood corners fitted around the loaded pallet) are used. Even with these aids, the package must be capable of withstanding handling abuses at the end of the storage and distribution period.

### Inspection

The package should facilitate easy inspection of the contents. Telescope-style corrugated packages are easily inspected by removing the lid. Snap-on lids and inspection ports serve the same function. Whatever the procedure, the package must securely reclose and protect the product during the balance of the distribution period.

### Retail display

Some packaging systems are designed for use in retail displays. Most notable examples are berries that are packed in small baskets, apples, oranges, carrots, and other commodities that are bagged in consumer size units, and some shallow tray or place-packed items. Packages to be used in retail display must fit the needs of the retailer, and package cosmetics become much more important than for other types of packaging.

Many commodities are most satisfactorily packed into small consumer units after arrival in distribution markets because product deterioration can require reconditioning if packing occurs earlier. The design of packages used for retail display must not compromise the other requirements of protecting the product throughout the marketing system.

### Package disposal and reuse

In recent years problems of package disposal have become increasingly difficult. With incineration now restricted or eliminated in most metropolitan areas, disposal choices include reuse, recycling, and landfill. Some wood packages are reconditioned and reused. Corrugated fiberboard packages can be recycled, especially if they are not heavily treated with wax or resin against movement of water vapor. Very limited recycling of the material in plastic foam packages and sale of ground foam for insulation have been attempted. The greatest problems are with the disposal of plastic or foam packages and corrugated fiberboard packages that are plastic-coated or heavily waxed.

Many proposals have been made for the use of collapsible, returnable packages that are manufactured to withstand many reuses. This would require a national network to assemble, recondition, clean, and return packages after each use. To date, implementing such a system has been defeated by the complicated logistics.

## Economic Considerations

In considering the cost of any new package, all of the costs of adapting it into the marketing system must be considered. These involve costs of packaging material, labor, modifications in packing and handling operations, and potential changes in product condition.

1. **Packaging costs**

   Package component costs

   Transportation costs

   Package make-up costs—labor and materials

   Needed internal packaging materials—liners, shims, pads, trays, wraps

   Storage costs of package components

2. **Packing costs**

   Adaptability to mechanized package distribution

   Effect on packing operation

   Effect on packing labor efficiency

   Number of packing steps required

   Costs of modifying or converting packing facilities

3. **Palletizing and handling costs**

   Effect on pallet stacking efficiency

   Effect on costs of strapping labor and materials

   Adaptability to pallet gluing

   Adaptability to various pallet materials and substitutes

4. **Marketing costs**

   Effect on load density in storage and in transport vehicles

   Special labor or equipment needed for handling

   Adaptability of package as a display unit

5. **Product value costs**

   Effect of package in modifying product deterioration

   Value of adjustments due to package failures

   Value of brand "reputation" related to package performance

## Simulated Transit Testing

Testing is an important aspect of any package development program. Objective laboratory testing is the first step in evaluating new packages and packing procedures; it can reduce costly and time-consuming trial shipments and allow more precise evaluation of a large number of variables. After laboratory testing, a promising package or treatment can be compared with the standard in a trial shipment. Use of laboratory tests involves a minimum time expenditure and provides confidence that major problems will not occur in subsequent usage.

### Basis for test procedures

To be effective, laboratory test procedures must meet certain conditions.

1. The types of abuses to products that occur in actual transport and handling must be duplicated.

2. Laboratory treatment should equal a severe transport and/or handling condition.

3. Laboratory procedure should allow rapid testing of a large number of variables.

4. Tests should emphasize transport and handling effects on the product as well as on packages and packaging materials.

### Test procedures

A number of standard laboratory testing procedures are designed to evaluate the package. In the Pomology Postharvest Laboratory at Davis a test procedure has been developed that emphasizes performance of the product. It includes a series of impact and vibration treatments.

**Vertical impact test.** Number and height are adjustable, but 30 impacts from a 5-cm (2 inch) height are standard.

**Horizontal impact test.** Number and intensity are also adjustable but 30 impacts at 3.2 km (2 miles) per hour are standard.

**Transit vibration test.** Stroke, frequency, and time are adjustable, but 30 minutes vibration at 1.1 g acceleration, using a 0.6-cm (¼ inch) stroke at 550 cycles per minute, is standard.

**Actual transit test.** Normally the best components of all variables tested in the laboratory are combined in designing the new package or packing method. This is first compared with the industry standard in a laboratory test. If promising, it is then tested in a well-replicated trial shipment in comparison with the industry standard.

### REFERENCES

1. Anon. 1976. *Fibre box handbook.* Chicago: Fibre Box Assn.
2. Ben Yehoshua, S. 1985. Individual seal-packaging of fruit and vegetables in plastic film—a new postharvest technique. *HortScience* 20:32-38.
3. Boustead, P. J., and J. H. New. 1986. *Packaging of fruit and vegetables: a study of models for the manufacture of corrugated fiberboard boxes in developing countries.* London: Tropical Development and Research Institute Publication G199. 44 pp.
4. Gentry, J. P., F. G. Mitchell, and N. F. Sommer. 1965. Engineering and quality aspects of deciduous fruit packed by volume filling and hand placing methods. *Trans. ASAE* 8:584-89.
5. Guillou, R. 1964. *Orderly development of produce containers.* p. 20-25. *Proc. Fruit Veg. Perish. Handling Conf.*, Univ. of Calif., Davis: Div. of Agric. Sciences, Univ. of Calif.
6. Guillou, R., N. F. Sommer, and F. G. Mitchell. 1962. Simulated transit testing for produce containers. *TAPPI* 45(1): 176-79A.

7. Hardenburg, R. E. 1966. Packaging and protection. In *Protecting our food supply*. U.S. Dept. Agric. Yearb. (1966):102-17.

8. Heiss, R., ed. 1970. *Principles of food packaging, an international guide*. West Germany: P. Keppler Verlag KG. 332 pp.

9. Hochart, B. 1972. *Wood as a packaging material in developing countries*. United Nations Publication E.72. II. B. 12. 111 pp.

10. Mitchell, F. G., N. F. Sommer, J. P. Gentry, R. Guillou, and G. Mayer. 1968. *Tight-fill fruit packing*. Univ. Calif. Agric. Exp. Sta. Circ. 548. 24 pp.

11. O'Brien, M., J. E. Gentry, and R. C. Gibson. 1965. Vibrating characteristics of fruits as related to in-transit injury. *Trans. ASAE* 8:241-43.

12. O'Brien, M., and R. Guillou. 1969. An in-transit vibration simulator for fruit handling studies. *Trans. ASAE* 12:94-97.

13. Smith, R. J. 1963. *The rapid pack method of packing fruit*. Univ. Calif. Agric. Exp. Sta. Circ. 521. 20 pp.

14. Sommer, N. F., and D. A. Luvisi. 1960. Choosing the right package for fresh fruit. *Pack. Eng.* 5:37-43.

15. Stokes, D. R., and G. W. Woodley. 1974. *Standardization of shipping containers for fruits and vegetables*. U.S. Dept. Agric. Mktg. Res. Rept. 991. 118 pp.

# 8

# Cooling Horticultural Commodities

## I. The Need for Cooling

F. Gordon Mitchell

Many changes have occurred in packing, handling, transporting, marketing, and distributing horticultural crops that directly influence cooling requirements and results. New packages and packing materials, packing methods, and palletization procedures often make produce cooling more difficult. The current desire to reach more distant markets, prolong storage life, and market a product that better satisfies consumers often requires more uniform or faster cooling. Discussed here are some of these changes, their implications for product cooling requirements, and some possible responses that may improve the cooling system.

## The Need for Cooling

An understanding of the cooling requirements of horticultural commodities begins with adequate knowledge of their biological responses. All fresh horticultural crops are living organisms, carrying on the many biological processes that are essential to the maintenance of life. They must remain alive and healthy until processed or consumed. The energy needed for these life processes comes from the food reserves that are accumulated while the commodities are still attached to the plant (with some exceptions among flowers).

The process by which food reserves are converted into energy is called respiration. In a complex series of steps, the stored food reserves (starches and sugars) are converted first to organic acids, then to more simple carbon compounds. Oxygen from the surrounding air is consumed in the process, and carbon dioxide is released. If oxygen is severely limited, anaerobic respiration occurs, aldehydes, alcohols, and other undesirable materials are produced, and the tissue ultimately dies.

Horticultural commodities are not the tight, dense, closed tissues that they appear to be. Rather, they have an amazingly open structure, with a complete network of interconnecting air spaces (called intercellular spaces) throughout. This can be illustrated with many commodities by injecting air while the intact tissue is under water: Air bubbles appear almost instantly and simultaneously over the entire surface. Due to this structure, the oxygen concentration in the intercellular spaces in the center of many commodities is almost as high as in the surrounding atmosphere.

Some of the energy produced through respiratory activity goes to maintain the life processes. Excess energy is released in the form of heat, called vital heat. The amount of vital heat varies with the type of product, variety, maturity or stage of ripeness, injuries, temperature, and other stress-related factors. This heat must be considered in any temperature management program.

Product temperature is a major determinant of respiratory rate. Because the final result of respiratory activity is product deterioration and senescence, it is desirable to achieve as low a respiratory rate as possible without risking tissue injury or death. An exception is during controlled ripening of fruits. Each 10°C (18°F) temperature reduction reduces respiratory activity by a factor of 2 to 4. For example, the respiratory rate of a product at 5°C (41°F) would be only one-fourth to one-sixteenth of what it would be at 25°C (77°F). Good cooling and temperature management practices are, therefore, critical to slowing physiological deterioration.

### Ethylene's effects

Temperature affects both the rate of ethylene production and the sensitivity of products to ethylene. Ethylene gas is naturally produced in most if not all plant tissue. This simple chemical compound is generally recognized as a fruit-ripening hormone. It can have beneficial or detrimental effects on fresh commodities, depending on management needs. For these effects to occur a minimum concentration must accumulate within the internal atmosphere of the product, and the temperature must be above a minimum. These minimums are not well defined. However, since the production and action of ethylene are both temperature-dependent, rapid cooling and good temperature management are vital if fruit ripening and other deterioration processes are to be delayed.

In terms of respiratory activity, fruits—including fruit-type vegetables—have been grouped into two classifications, "climacteric" and "nonclimacteric." Climacteric fruits normally ripen after harvesting, during which time sugars increase and volatile constituents (flavors and odors) develop. Flesh softening accompanies ripening, but if fruits are picked while immature they may soften without development of sweetness and flavor. Ethylene initiates ripening, accompanied by a rapid rise in the respiratory activity in the fruit called the climacteric rise. Included are apples, peaches, papayas, cantaloupes, tomatoes, and many fruits.

The nonclimacteric fruits, including citrus, grapes, strawberries, and others, do not ripen after harvest and exhibit no rise in respiratory activity. Even under optimum ripening conditions, dessert quality does not improve noticeably after harvest. Ethylene may affect color changes among these fruits; for example, the breakdown of the green chlorophyll pigment causes orange fruits to color. However, the sugar, acid, and flavor of the fruits are not influenced by the treatment.

Ethylene is also implicated in a number of product injury and deterioration problems, and ethylene synthesis and action are temperature related. Among these are russet spotting of lettuce, sleepiness in carnations, leaf abscission in many commodities, rind pitting in citrus, bitter flavor in carrots, yellowing of cucumbers, softening of kiwifruit, and lignification in asparagus.

## Moisture loss

Fresh commodities constantly lose water to the surrounding environment. After harvest this lost water cannot be replaced by the plant (with the exception of flowers and some leafy green vegetables), and weight loss occurs. Many products show visible shriveling or wilting after losing 3 to 5 percent of their initial weight. They lose water as a result of a water vapor gradient between their saturated internal atmosphere (within the intercellular spaces) and the less saturated external atmosphere. Water vapor migrates in the direction of lower concentration, primarily through natural openings on the fruit surface, but also through surface injuries. The rate of migration is a function of the resistance (usually of the skin) of the particular product to water vapor movement. It is also controlled by the vapor-pressure difference between the product and its environment, which is governed by temperature and relative humidity. Warm air can hold much more water vapor than cold air. Relative humidity indicates the amount of water vapor in the air as a percent of the maximum amount the air can hold at that temperature. A product in the field at 25°C (77°F) and 30 percent relative humidity loses water 36 times faster than it does when cooled and in storage at 0°C (32°F) and 90 percent relative humidity. Thus, maintaining low product temperature is essential to reduce water loss and subsequent shriveling and wilting. These relationships vary with some commodities, such as leafy vegetables, depending on whether stomates on the product surface are open or closed.

## Decay organisms

Postharvest management of fresh horticultural commodities actually involves two living systems: the product and the microorganisms that attack it. Of thousands of potential microbial pathogens, only a few cause problems. These few, however, cause extensive direct loss of fresh fruits, vegetables, flowers, and ornamentals.

Temperature affects the rate of growth and spread of these microorganisms the same way that it affects the commodity—the lower the temperature, the slower their life processes progress. Certain organisms that can cause severe losses do not grow at low storage temperatures. For example, Rhizopus rot ceases growth at about 5°C (41°F), and germinating spores have been found to be killed after about 2 days at 0°C (32°F). Other organisms continue growth but at a very slow rate near 0°C. In studies with Botrytis rot of strawberries during a 7-day marketing period, at 2°C (36°F) or below germinating spores would not penetrate into the fruit, and at 0°C mycelium (the vegetative fungal growth) would not penetrate a healthy fruit from an adjacent invaded fruit. Thus good temperature management plays a vital role in reducing microbial loss problems.

## Injuries

Physical injuries can result from abuses to fresh commodities at any temperature, but temperature affects the severity of the product response to those injuries. Bruises and other wounds cause increased ethylene production, which may accelerate respiration, cause deterioration problems, or initiate fruit ripening. Bruising usually damages the natural barriers on the product surface, increasing the opportunity for water loss and for entry of rot organisms. Prompt cooling and maintenance of a low temperature reduces the effects of injuries by influencing all of these processes.

## Low temperatures

Good temperature management, it should be clear by now, is the single most important factor in delaying product deterioration; prompt cooling and maintenance of proper temperatures are both essential. For many products, this means maintaining as low a temperature as possible without danger of freezing. The freezing point varies with soluble solids content; freezing-point guides are available for some products. How closely the freezing point can be approached depends on the accuracy and sensitivity of the temperature controls.

Many horticultural crops suffer chilling injury at temperatures considerably above their freezing point and must be held at warmer temperatures to avoid injury. Chilling injury symptoms include surface or internal browning, surface pitting, failure to degreen, failure to ripen, increased susceptibility to microorganisms, texture changes (mealy or wooly texture), and loss of flavor. Most crops of tropical and subtropical origin and even some deciduous fruits are subject to chilling injury. Guides are available for the lowest safe temperature for various commodities. Though these products must be stored above their threshold chilling temperature, most can withstand a few hours below it without injury.

## Speed of cooling

Most products benefit from prompt, thorough cooling. For example, strawberries have increasing losses to deterioration as delays between harvesting and cooling exceed 1 hour. The same happens to sweet cherries when delays exceed about 4 hours. These fruits benefit from cooling even when rewarming will occur during subsequent handling; deterioration is proportional to the total exposure time to warm temperature, not to the pattern of cooling and warming.

Some Bartlett pears should be cooled to −0.5°C (31°F) within 24 hours of harvest to reduce losses from an enzymatic problem called watery breakdown. This problem is especially serious among fruits harvested in mid and late season. Rapid cooling to low temperature also limits decay and ensures maximum storage life, and thus is generally recommended for all Bartlett pears that are to be stored.

There are exceptions to the need for prompt cooling after harvest. For many years freshly harvested freestone peaches in South Africa have been held at ambient temperature for about 36 hours before cooling. This fruit is subject to chilling injury (dry or brown tissue), and the holding period was found to delay onset of the problem. Research suggests that a similar treatment may be useful for honeydew melons if they are gassed with ethylene to initiate ripening before shipment at chilling temperatures.

Although most fruits are not expected to be harmed by moisture condensation during warming, some table grapes reportedly develop berry cracking (and subsequent rot) when condensation moisture remains in the cluster for prolonged periods. For that reason, prompt cooling might be desirable only if low temperatures can be maintained. Thus, it is important to know the product and its temperature and marketing requirements before selecting a cooling program.

## Changes Affecting Cooling

Cooling is affected by many different factors throughout the handling system—from the field to the consumer. Some handling changes have increased the need for rapid, thorough cooling, such as the desire to market more mature (or even ripe) products, to extend the storage period, to supply more distant markets, and to meet new consumer quality standards. Other changes have made fast, thorough cooling more difficult to attain and have prompted the development of new cooling procedures: the change from field lugs to large bins, new package designs, unitized handling, and changing transport equipment. The effects of these changes must be understood if improved product performance is to be achieved.

## Extended market life

Handlers want to extend postharvest life for a variety of reasons. Processors may store to accumulate product for processing, to protect quality when deliveries exceed processing capacity, to hold the product over weekends to avoid labor complications, or simply to extend the processing season. Storage may be for a few days to several months.

Packers may want to store the product to extend the marketing season, to reach distant markets, to accumulate a supply for holiday periods, to facilitate orderly marketing, or to avoid price declines during periods of oversupply. Thus, longer shelf life may be required. As deterioration is a function of time and temperature, faster cooling can significantly extend shelf life.

## Consumer demands

There is increasing evidence that consumer satisfaction with fresh commodities is related to appearance, flavor, and maturity or ripeness, or ability to quickly ripen after purchase. With consumers purchasing larger quantities of fresh commodities, handlers are eager to make the changes needed to sustain consumer demand. This requires them to ship slightly more mature products that have developed more of the characteristic aromas and flavors. Fast cooling and good temperature management are vital to protecting such commodities.

## Transportation

Refrigerated transport equipment has changed during the past two decades from ice in bunkers to mechanical refrigeration systems. The correct use of ice bunker equipment, which included proper loading patterns, frequent reicing, and adequate air circulation capacity, generally allowed reasonable cooling of most products during transport. Most mechanical refrigeration equipment in current use can maintain temperature, but lacks the airflow capacity needed for rapid cooling. Furthermore, packages and loading patterns have changed. High-density loads are used to minimize transportation costs. These factors all inhibit cooling during transit. Therefore, thorough product cooling before loading and protection against warming are more important than ever before.

## Pallet bins

In harvest operations, small hand-carried field lugs have generally been replaced by large pallet bins that are moved by lift truck, often holding about one-half ton of product. The field lugs were separated by cleats, which facilitated airflow when cooling was needed. In contrast, the size of pallet bins results in much of the product being far from the bin top or sides. This allows less opportunity for cold air to penetrate into the bin in normal room cooling operations.

### Shipping containers

Corrugated fiberboard has replaced wood in shipping containers for many commodities. Wood can be used with forced-air "wet" cooling systems, but fiberboard must be especially treated for use with "wet" systems.

Corrugated containers normally stack more tightly than wooden containers and lack the cleats of wooden containers that allow air circulation during storage and transport. They also tend to have much less vent area than the wood containers they replaced. Internal packaging materials such as wraps, trays, liners, pads, and plastic bags further restrict heat removal. Despite these adverse properties, proper stacking and adequate spacing of corrugated shipping containers on storage pallets, together with provision for high-speed airflow, can provide reasonable cooling speed.

### Unitization

Many horticultural products are now handled and shipped unitized on pallets. When corrugated containers are tightly stacked on a pallet for unitized handling, cooling problems become more severe. Container vents, even if adequately sized, are often blocked when containers are cross-stacked on pallets, and much of the product has little access to cold air. Cooling of such units in conventional cold rooms is slow, irregular, and often inadequate without some modification in container design and venting, pallet stacking patterns, fan operation, and cooling systems.

## Meeting Changing Needs

Conflict between the need to achieve more rapid cooling and the increasing difficulty of cooling appears almost impossible to resolve through the use of room cooling procedures. However, rapid cooling systems are available that can achieve good results if properly used.

### Cooling as part of handling

The entire product handling system must be considered when planning cooling facilities because any change may affect cooling rate, uniformity, and requirements. The packing method dictates how, and sometimes when, the product is presented for cooling. Packaging materials and design affect access of the cooling medium to the product, and pallet stacking patterns influence coolant flow through and around containers. Loading patterns, transport equipment, and marketing procedures all greatly affect cooling requirements. The maximum market life of many commodities may be approached when they are shipped by refrigerated ocean transport for export marketing, thus thorough cooling is essential.

### Cooling and storage as separate operations

Refrigeration capacity needed for fast cooling and for cold storage are quite different. For example, it takes about 100 times more refrigeration capacity to cool pears in 24 hours than to cold store them for 24 hours. Even when fruit is cooled over a 6-day period, the daily refrigeration capacity during cooling is almost 25 times that of maintaining a properly cooled product. This ignores the refrigeration required to remove heat entering the facility through walls, doors, fans, and forklifts.

Other differences between cooling and cold storage must be considered. During fast cooling, high air speed does not significantly increase total water loss from the product if it occurs only while the product is being cooled. However, during subsequent cold storage, high air speed desiccates horticultural commodities over the long period of exposure to rapidly moving air. High relative humidity is essential to prevent excessive water loss during cold storage, but is less important during the short cooling period. Ethylene control is generally not needed during cooling, but for some commodities may be essential during storage.

Thus, cooling and storage are two separate operations that have vastly different requirements. The specific requirements for achieving fast, uniform cooling must be considered independently of the cold storage requirements.

# II.  Cooling Methods

F. Gordon Mitchell

Several cooling methods are used with horticultural commodities—room cooling, forced-air cooling, hydrocooling, package icing, and vacuum cooling—are used before storage or loading for shipment. Top-icing, channel-icing, and mechanical refrigeration in transport vehicles are used for cooling during transit. A few cooling methods (e.g., room cooling, forced-air cooling, and hydrocooling) are used with a wide range of commodities. Some commodities can be cooled by several methods, but most commodities respond best to one or two cooling methods.

Most users are concerned with the time to "complete cooling," which usually means the time to reach a desired temperature before transfer to storage or transport. In cooling comparisons we may report times as "half-cooling" or "⅞-cooling." Half-cooling time is the time to cool the product halfway from its initial temperature to the temperature of the cooling medium. This is a constant value for a given system, and thus the speed of cooling may appear to slow as cooling continues. For example, if a load of peaches in a cooler with 0°C (32°F) air takes 4 hours to cool from a pulp temperature of 20°C (68°F) to 10°C (50°F) (half-cooling), it will then take an additional 4 hours to cool to 5°C (41°F), and 4 hours more to reach about 2.5°C (36.5°F), and so on. We often use ⅞-cooling (three half-cooling times—in this example, 12 hours) as a reference point. Seven-eighths-cooling would be defined as the time required to cool the product ⅞ of the way from its initial pulp temperature to the temperature of the cooling medium.

Both initial pulp temperature of the product and temperature of the coolant influence the ⅞-cooling temperature of the product. The peaches in the example might be harvested in early morning, but by late afternoon in California they could have pulp temperatures near 32°C (90°F), in which case, with 0°C (32°F) coolant, the ⅞-cooling temperature would be about 4°C (39°F) instead of 2.5°C (36.5°F), and an additional 3 hours of cooling would be required to reach the same pulp temperature as the early morning fruit. If cold air at −1°C (about 30°F) could be used without freezing the peaches, then at ⅞-cooling the peaches with 20°C initial temperature would be at about 1.6°C (35°F) rather than the 2.5°C (36.5°F) achieved using 0°C air. Thus, once you know the half-cooling time of a particular cooling system you can project the effect of variables in product and coolant temperature on cooling times to reach a given product temperature.

The cooling rate is influenced by the mass flow rate of the cooling medium. At higher flow rates heat is promptly removed from the surfaces of the package or of the product, creating a greater gradient for heat to move to the surfaces. The higher flow rates also create more turbulence, which can help to release heat from all surfaces.

Package position and access to the cooling medium also affect the cooling rate. Packages on the outside corner of a pallet have more surface contact with the cooling medium than inside packages. Product in the center of a package may cool more slowly unless vent alignment is maintained and the cooling medium moves actively through the vents. Placement of packages and pallets should assure positive contact with the cooling medium and prevent bypass or "short circuiting" of the medium.

## Room Cooling

This widely used cooling method involves placing field or shipping containers of produce into a cold room. Its commonest use is for products with relatively long storage life that are stored in the same room they are cooled in. Examples include potatoes, sweet potatoes, citrus, and controlled-atmosphere stored apples.

Cold air from the evaporator enters the room near the ceiling, moves horizontally under the ceiling, and then sweeps past the produce containers in returning to the evaporator (fig. 8.1). The main advantage of room cooling is that produce can be cooled and stored in the same room without the need for transfer. It is suited for crops that are marketed soon after harvest, crops that are stored unpacked, and crops that require only mild cooling after harvest. Its disadvantages are that (1) it is too slow for most commodities, (2) it requires more space than is needed for good storage, and (3) it can result in excessive water loss. Cooling requires days for packed products, but can be fast for unpacked products with good exposure to the cold air. For example, bunched flowers in buckets can cool in 15 minutes, but the same flowers packed in boxes and loaded on a pallet takes days to cool.

For best results, containers should be stacked so the moving cold air can contact all container surfaces. An airflow of at least 60 to 120 m (200 to 400 feet) per minute is needed for adequate heat removal. Well-vented containers can greatly speed room cooling by allowing air movement through the containers.

Room cooling allows produce to be cooled in the same location where it will be stored, thus requiring less rehandling. Cooling, however, requires more space than is needed for good storage management, thus rehandling may be needed to better use storage space.

Products for room cooling must be tolerant of slow heat removal, for much of the cooling is by heat conduction through the container walls. Because the air speed needed for cooling is greater than that needed for storage, products stored in the cooling

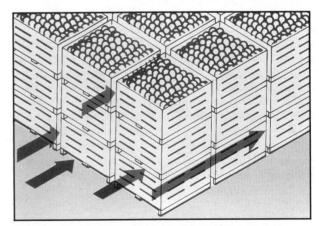

Fig. 8.1. Diagrammatic view of air path during room cooling of produce in bins. Air circulating through the room passes over surfaces and through forklift openings in returning to the cooling coils. In this system the air takes the path of least resistance in moving past the product. Cooling from the surface to the center of bins is largely by conduction.

58

room lose water faster than under more ideal storage conditions.

Room cooling has become increasingly difficult as more commodities are being handled in large field bins or in containers that are tightly unitized on pallets. The difficulty results from the longer path for conduction cooling and the inadequate air movement within the large product units.

### Ceiling jets

A modification of room cooling directs airflow past container surfaces to improve heat removal. Cold air is injected between a false ceiling and the roof. The resulting positive pressure forces the air into the room through cones in the ceiling. The floor of the cooling room is marked for spaced stacking of pallets or bins so that air from the cones sweeps down the corners of the stacked units and spreads into the channels between them. An air-return plenum then draws the air in one direction.

### Cooling bays

For both cooling and storage, a single large room is divided into bays by installing partitions part way into the room from each side. Air-supply channels direct the air into the back of each bay. When a single bay is filled with warm product, supply ducts are opened to direct a large volume of cold air behind the product. Air return occurs down the center forklift aisle. When cooling is completed, the air supply is reduced in that one bay to create desired storage conditions. With this system, a cold product in one bay is not warmed by warm product in other bays.

## Forced-Air Cooling (Pressure Cooling)

Forced-air cooling is adaptable to the widest range of commodities. It can solve many difficult cooling problems because it provides for cold air movement through, rather than around, containers. The system creates a slight pressure gradient to force air through container vents, achieving rapid cooling through the intimate contact between cold air and warm product. With proper design, fast, uniform cooling can be achieved through stacks of pallet bins or unitized pallet loads of containers. This is the most widely adaptable cooling method for small-scale operations.

The speed of forced-air cooling is controlled by adjusting the volume of cold air passing over the product. However, as the volume of air passing through containers increases, the static pressure required also increases, raising the energy consumption of the fan. Similarly, increasing the number of packages through which the air must pass (lengthening the path of air flow) increases the total volume of air that must pass through the container vents, and thus increases the static pressure requirement. While water loss and shrivel are less than for room cooling, refrigeration systems may need to provide high relative humidity for products that are sensitive

to surface drying (e.g., strawberries, cherries, grape stems, and mushrooms).

Various cooler designs can be used, depending on specific needs. Converting existing cooling facilities to forced-air cooling is often simple and inexpensive if enough refrigeration capacity and cooling surfaces (e.g., evaporator coil surfaces) are available. A well-designed forced-air cooler is separate from cold storage rooms. Some of the variations in forced-air cooler design are described here.

### Forced-air tunnel

In this, the most used forced-air cooling system, a row of palletized containers or bins is placed on either side of an exhaust fan, leaving an aisle between the rows. The aisle and the open end are then covered to create an air plenum tunnel (fig. 8.2). The exhaust fan creates negative air pressure within the tunnel. Cold air from the room then moves through the openings in or between containers toward the low-pressure zone, sweeping heat from the product as it moves past. The exhaust fan can be a portable unit that is placed to direct the warm exhaust air toward the air return of the cold room. It is more commonly a permanent unit that also circulates the air over the refrigeration coils and returns it to the cold room (fig. 8.3).

### Cold wall

This forced-air cooling system uses a permanent air plenum equipped with exhaust fans (fig. 8.4). It is often located at one end or side of a cold room, with the exhaust fans designed to move air over the refrigeration coils. Openings are located along the room side of the plenum against which stacks or pallet

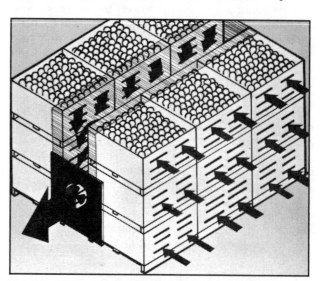

Fig. 8.2. Diagrammatic view of a forced-air cooling tunnel. Either bins or palletized containers can be placed to form a tunnel from which air is exhausted. The negative pressure then causes cold air from the room to pass through ventilation slots to directly contact the warm product.

Fig. 8.3. Forced-air cooling tunnel is in operation, cooling packaged produce on unitized pallets. Air circulating fan circulates air through fruit and over cooling coils. Canvas plenum cover is designed to fit varying cooling loads.

Fig. 8.5. Cold-wall type forced-air cooler for use with stacks of flower containers. Open end-vents allow air to be pulled through the containers for rapid cooling but allow closing during shipment. *Photo courtesy of Mel Gagnon.*

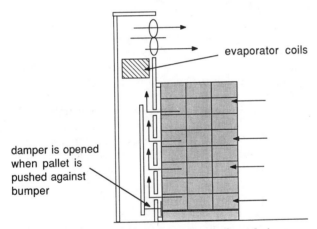

Fig. 8.4. Cross section of a cold-wall type forced-air cooler.

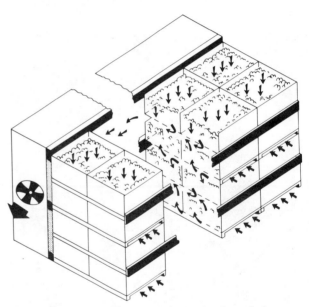

Fig. 8.6. Pattern of airflow in a serpentine forced-air cooling system. This system is specific for cooling fruit in field bins. By blocking alternate forklift openings on cold-wall and room sides, with fans operating, air is forced to pass vertically through bins to cool fruit.

loads of containers can be placed (fig. 8.5). Various damper designs ensure that air flow is blocked except when a pallet is in place. Each pallet starts cooling as soon as it is in place, thus there is no need to await deliveries to complete a tunnel. Shelves may be built so that several layers of pallets can be cooled. Different packages and even partial pallets can be accommodated by proper design of the damper system. This is a benefit in some operations where a range of commodities or varieties is handled. Each pallet must be promptly moved from the cooler as soon as it is cold in order to avoid desiccation from continued rapid air flow over the product.

## Serpentine cooling

The serpentine system is used for forced-air cooling produce in pallet bins. Bins must have bottom ventilation slots, with or without side ventilation. It requires modification of the cold-wall design to allow the forklift openings between bins to be used as air supply and return plenums. The cold air moves vertically through the product within each bin in response to the slight pressure difference between plenums (fig. 8.6). Bins may be stacked up to six rows deep against the cold wall, depending on the cooling speed desired and the available airflow. The airflow capacity of the small forklift opening plenums is the primary limitation. When side ventilation slots are used, they expand air supply capacity, and the air return plenums become the limitation. To achieve the desired airflow pattern, openings are placed in the cold wall to match alternate forklift openings, starting one bin up from the floor. On the room side of the bins, these same openings are then blocked (fig. 8.7). Thus, air flowing into an open slot between bins must pass up or down through one bin of product to reach the cold wall. This system requires no space between rows of bins and is not limited in height. Rapid cooling is achieved because air flows vertically through a relatively shallow layer of product.

## Forced-air evaporative cooling

This is simply a forced-air cooling system in which the air is cooled with an evaporative cooler instead of mechanical refrigeration. If designed and operated correctly, an evaporative cooler produces air at a few degrees above the outside wet bulb temperature and the air has a relative humidity above 90 percent. In most areas of California, product temperatures of 16°C (60°F) to 21°C (70°F) can be achieved. A typical forced-air evaporative cooler is shown in figure 8.8. This cooling method may be adequate for some chilling-sensitive products to be shipped to local or regional markets. In most cases, growers can build their own forced-air evaporative coolers. Evaporative cooling is very energy efficient compared with mechanical refrigeration.

## Container venting

Effective container venting is essential for forced-air cooling to work efficiently. Cold air must be able to pass through all parts of a container. For this to happen, container vents must remain open after stacking. If containers are palletized in-register, container side or end vents will suffice, provided they are properly located in relation to trays, pads, and so on. If cross-stacking is used, then matching side and end vents is essential. For the 400 by 300 mm (or 16 × 12 inch) container cross-stacked on the 1200 by 1000 mm (or 48 × 40 inch) pallet, vertical vent slots on 100 mm (or 4 inch) centers around the container perimeter should be considered, because they remain matched when cross-stacked.

Too little venting restricts airflow; too much venting weakens the container. A reasonable compromise appears to be about 5 to 6 percent side or end wall venting. A few large vents are more effective than many small vents for speeding the cooling rate. Locating vents midway from top to bottom is adequate unless trays or other packing materials isolate some of the product. Any type of bag, liner, or vertical divider inside the package may block vents and reduce their effectiveness. Vertical slots at least 12 mm (½ inch) wide are better than round vents.

Flower boxes are often designed with closeable vents. This allows closing after cooling so that flowers shipped on unrefrigerated transport can maintain temperatures longer than if in a box with open vents.

Fig. 8.7. Serpentine forced-air cooler in operation. Plastic straps are placed over every other forklift opening from bottom to top. These close off the openings in the room side of the cooler. Air entering the open channels then must move up and down through the product to return to the cold wall. Note that bins can be tightly stacked in rows since no center airflow plenum is needed.

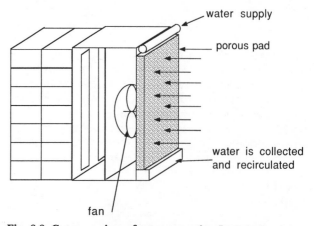

Fig. 8.8. Cutaway view of an evaporative forced-air cooler. Air is cooled by passing through the wet pad before it passes through packages and around the product.

## Hydrocooling

The use of cold water is an old and effective cooling method used for quickly cooling a wide range of fruits and vegetables in bins or in bulk before packing (figs. 8.9 and 8.10). Hydrocooling avoids water loss and may even add water to the commodity. It is less adapted to cooling packed commodities because of the difficulty of achieving sufficient water flow through the containers and because the containers must be water-tolerant. Nevertheless, several vegetables are regularly hydrocooled after packing. Hydrocoolers can use either an immersion or a shower system to bring products in contact with the cold water. Hydrocoolers can be made portable to allow them to be moved, thus extending the cooling season.

In a typical shower-type hydrocooler, cold water is pumped to an overhead perforated pan. The water showers over the commodity below, which may be in bins or boxes, or loose on a conveyer belt passing beneath. The water leaving the product may be filtered to remove debris, then passed over refrigeration coils (or ice) where it is recooled. The cooling coils may be located in a tank under or beside the conveyer or above the shower pan.

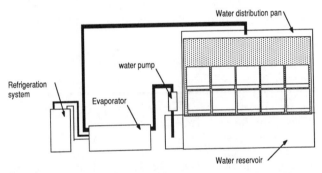

**Fig. 8.9. Side view of a batch-type hydrocooler for pallet bins.**

**Fig. 8.10. Conveyor-type bin hydrocooler in operation. Ice water is pumped into the top pan, where it runs down through the product in the bin. Dwell time in the cooler is controlled by conveyor speed.**

Efficient cooling depends upon adequate water flow over the product surface. In a shower-type hydrocooler, water flows of 15 to 25 gallons per minute per square foot (about 600 to 1000 liters/min/m$^2$) of surface area are generally desirable. Bin hydrocoolers are often designed to accommodate two-high stacking of bins on the conveyors. In such facilities it is vital that the water flow be adequate to keep the top bins partially filled with water, or excessive water channeling through the lower bin results in uneven and inadequate cooling. Both total water volume and bin vent size and location influence this.

Water is usually cooled by mechanical refrigeration, but ice may be used if it can be added fast enough to produce adequate cooling. In some areas, stream or well water is cold enough to do initial cooling or even complete cooling. Hydrocoolers should be drained and cleaned at least daily, or be equipped with special filters to clean the water.

The product and any packages and packing materials must be tolerant of wetting, not susceptible to water beating damage, and tolerant of chlorine or other chemicals that are used to sanitize the hydrocooling water. In addition to cleaning, the process maintains a low chlorine concentration in the water. The product's tolerance to chlorine must be known. Shower-pan holes must be cleaned daily to avoid plugging, which causes uneven water flow over the product. Typical cooling times are 10 minutes to 1 hour depending on the size of the product. For small products such as cherries, the cooling is fast enough to be done in-line in the packing operation.

Potential limitations of hydrocooling must also be considered. When the hydrocooler is operating at capacity, arriving warm produce must remain at ambient temperatures to await cooling. The cooled product must be moved quickly to a cold room or rapid rewarming occurs. Hydrocooling operations can require rehandling of the pallet bins before packing or storage. Energy efficiency can be good provided the hydrocooler is operated continuously at maximum capacity or is inside a cold room or an insulated enclosure.

Shower-type hydrocoolers (conveyor or batch units) are most common, but there is some use of flume-type immersion hydrocoolers. Here the product, normally in bulk, is in direct contact with the cold water as it moves down the flume. Because poor cooling would result if the product simply moved with the water, flume hydrocoolers convey the product either against (counter flow) or across (cross flow) the flow. Conveyors must be designed for positive movement of the product through and out of the water.

## Package-Icing

Some commodities are cooled by filling packed containers with ice in quantities that depend upon the initial product temperature. Initially the direct contact between product and ice causes fast cooling. However, as the ice in contact with the product melts, the cooling rate slows considerably. The ice keeps a high relative humidity around the product. Package ice may be finely crushed ice, flake-ice, or a slurry of ice and water called liquid-ice. Liquid-icing distributes the ice throughout the container, achieving better contact with the product (figs. 8.11 and 8.12). Ice can be produced during off-peak hours when electricity is cheapest and stored for daytime use.

Package-icing requires use of more expensive, water-tolerant packages. The packages should be fairly tight but with enough holes to drain meltwater. The icing process is very fast. In small operations the ice is hand raked or shoveled into containers; large operations use mechanical icers. Liquid-icers for automatically icing pallet loads of packed cartons are now used for cooling some field-packed vegetables. The iced packages must be placed into a cold environment after filling, or much of the cooling is lost.

**Fig. 8.11. Pallet liquid-icing machine in operation. The high-volume flow of the ice-water mix is pumped into chamber and flows through container vents to deposit ice throughout the package.**

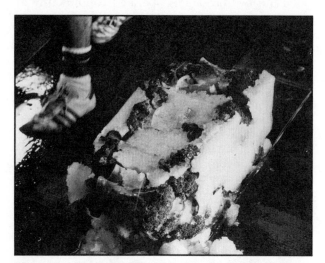

**Fig. 8.12. A package of liquid-iced broccoli is opened to show the penetration of ice throughout the package.**

The product must be tolerant of prolonged exposure to 0°C (32°F) wet conditions. Some low-density products have excess space for ice loading within the package, and ice not used for cooling can remain in the package even after transport. The excess ice can keep the product cold if the cold chain is broken. This is inefficient use of ice for cooling, and the weight of the ice can add significantly to the freight load, sometimes limiting the amount of product hauled. Also, during transport of mixed loads, water from melting ice can damage neighboring boxes that are not water-tolerant, and vehicle insulation can become wet.

## Vacuum Cooling

Vacuum cooling takes place by water evaporation from the product at very low air pressure. Products that easily release water may cool as rapidly as 20 to 30 minutes. Vegetables that have a favorable surface-to-mass ratio and that release water rapidly, such as leafy vegetables (especially iceberg lettuce), are especially suited to this method. It is also used to cool celery, some sweet corn, green beans, mushrooms, carrots, and bell peppers. Its use with carrots and peppers is primarily to dry the surface and stems, respectively, and to inhibit postharvest decay. Even boxes of film-wrapped products can cool quickly provided the film allows easy movement of water vapor.

Moisture loss is achieved by pumping air out of a large steel chamber containing the product. Reducing the pressure of the atmosphere around the product lowers the boiling temperature of its water, and as the pressure falls the water boils, quickly removing heat from the product. Vacuum cooling causes about 1 percent product weight loss (mostly water) for each 6°C (11°F) of cooling. This amount of weight loss can be objectionable for celery and some leaf lettuce. In one patented process used especially for celery, weight loss is greatly reduced by adding water in a fine spray during the vacuum cooling cycle (fig. 8.13).

A typical vacuum tube holds 800 boxes of iceberg lettuce (20 pallets). Some vacuum coolers are as small as 4 pallets. Most vacuum cooling equipment is portable and is used in two or more districts each year, thus the high cost of vacuum coolers is amortized over a longer operating season. Most coolers used today have mechanical refrigeration and rotary vacuum pumps.

## Cooling before Packing

Cooling problems with products in unitized pallets, or poly-packed products, can be avoided by cooling the produce before packing. However, this increases the basic cooling cost because products that are subsequently removed (culls or diverted products) are also cooled. With cantaloupes, this problem is avoided by removing most culls before hydrocooling. If 20

Fig. 8.13. Vacuum cooler being loaded. Batches of product are filled into the chamber, which is then closed and the vacuum drawn. This unit uses a patented process that introduces water during the cooling cycle to reduce water evaporation from the product. *Photo courtesy of Mel Gagnon.*

percent cullage occurs after cooling, the cooling cost increases 25 percent. If 50 percent is removed (diverting pears to a processor, for example), then the cooling cost per ton of packed product doubles. This cost may become prohibitive unless the cooling adds value to the diverted product.

Some rewarming occurs when produce is packed after cooling. A mild breeze can rewarm products to near ambient temperatures within 30 minutes. Some packers minimize this by only partially cooling the product before packing, followed by complete cooling after packing. One packer solves the problem another way: fruit arriving from the field is forced-air cooled in bins, and the bin dump is located in the forced-air bin cooling room. The cooled fruit moves from the cold room to the packing area, where it is sorted, sized, and volume filled into containers within 3 or 4 minutes. Packed containers are conveyed into a cold room for palletizing within 6 or 7 minutes of leaving the bin cooler. Rewarming in that system is minimal.

# III. Selecting a Cooling Method

Robert F. Kasmire and James F. Thompson

## Product Limitations

Physiological or physical characteristics of a product may limit the cooling methods that are suitable for a particular product. For example, strawberries, which cannot tolerate free moisture because of disease and injury problems, cannot be cooled by hydrocooling or package-ice. They require fast cooling after harvest, so room cooling is not suitable. Vacuum cooling is a fast cooling method for some leafy commodities, but it is not suited for berries. Thus, forced-air cooling is the only effective cooling method for strawberries. Many commodities, such as some deciduous fruits and many vegetables, are suited to several cooling methods. Table 8.1 lists the cooling methods commonly used for various types of fruits, vegetables, and flowers.

## Product Mix

If a cooling facility will be used for several types of commodities, it may or may not be possible to use the same method for all products. Table 8.1 shows that vacuum cooling, package icing, and room cooling are used for only a few products; hydrocooling is suited to a much wider variety; and forced-air cooling is adaptable to most types of products and is ideal for operations where a wide variety of products must be cooled. This is why forced-air cooling is most often recommended for small-scale operations, which typically handle many commodities and may change

the products they handle as the market changes from year to year. In some cases the product mix may require that more than one cooling system be used.

## Product Temperature Requirements

A facility that must handle products with very different optimum storage temperatures usually needs separate cooling facilities. Keeping chilling-sensitive commodities below their critical threshold temperature for too long will cause damage. Table 8.2 lists the optimum temperatures for many types of perishable products.

If product temperature requirements are not very different, then careful cooler management may allow a common cooler to be used. For example, summer squash can be forced-air cooled in a 0°C (32°F) room if it is removed from the cooler at 7°C (45°F) flesh temperature. It should then be stored at 7°C (45°F). Many chilling-sensitive commodities can be kept for short periods below their chilling threshold temperature.

## Costs of Operating Coolers

Capital costs can vary significantly between different types of coolers. Liquid-ice coolers are the most expensive to purchase, followed by vacuum coolers, forced-air coolers, and hydrocoolers. Figure 8.14 shows the capital cost, expressed in cost per daily cooling capacity, of four types of coolers based on 1988 data. The wide cost range for liquid-icing reflects the var-

**Table 8.1. Cooling methods suggested for horticultural commodities**

| Commodity | Size of operation | | Remarks |
|---|---|---|---|
| | Large | Small | |
| *Tree fruits* | | | |
| Citrus | R | R | |
| Deciduous | FA, R, HC | FA | Apricots cannot be HC |
| Subtropical | FA, R | FA | |
| Tropical | FA, R | FA | |
| Berries | FA | FA | |
| Grapes | FA | FA | Require rapid cooling facilities adaptable to SO₂ fumigation |
| *Leafy vegetables* | | | |
| Cabbage | VC, FA | FA | |
| Iceberg lettuce | VC | FA | |
| Kale, collards | VC, R, WV | FA | |
| Leaf lettuces, spinach, endive, escarole Chinese cabbage, bok choy, romaine | VC, FA, WV, HC | FA | |
| *Root vegetables* | | | |
| with tops | HC, PI, FA | HC, FA | Carrots can be VG |
| topped | HC, PI | HC, PI, FA | |
| Irish potatoes, sweet potatoes | R w/evap coolers, HC | R | With evap coolers, facilities should be adapted to curing |
| *Stem and flower vegetables* | | | |
| Artichokes | HC, PI | FA, PI | |
| Asparagus | HC | HC | |
| Broccoli, brussels sprouts | HC, FA, PI | FA PI | |
| Cauliflower | FA, VC | FA | |
| Celery, rhubarb | HC, WV, VC | HC, FA | |
| Green onions, leeks | PI, HC | PI | |
| *Mushrooms* | FA, VC | FA | |
| *Pod vegetables* | | | |
| Beans | HC, FA | FA | |
| Peas | FA, PI, VC | FA, PI | |
| *Bulb vegetables* | | | |
| Dry onions | R | R, FA | Should be adapted to curing |
| Garlic | R | | |
| *Fruit-type vegetables* | | | |
| Cucumbers, eggplant | R, FA, FA-EC | FA, FA-EC | Fruit-type vegetables are chilling sensitive but at varying temperatures |
| Melons | | | |
| cantaloupes, muskmelons, honeydew, casaba | HC, FA, PI | FA, FA-EC | |
| crenshaw | FA, R | FA, FA-EC | |
| watermelons | FA, HC | FA,R | |
| Peppers | R, FA, FA-EC, VC | FA, FA-EC | |
| Summer squashes, okra | R, FA, FA-EC | FA, FA-EC | |
| Sweet corn | HC, VC, PI | HC, FA, PI | |
| Tomatillos | R, FA, FA-EC | FA, FA-EC | |
| Tomatoes | R, FA, FA-EC | | |
| Winter squashes | R | R | |
| *Fresh herbs* | | | |
| not packaged | HC, FA | FA, R | Can be easily damaged by water beating in HC |
| packaged | FA | FA, R | |
| *Cactus* | | | |
| leaves (nopalitos) | R | FA | |
| fruit (tunas or prickly pears) | R | FA | |
| *Ornamentals* | | | |
| Cut flowers | FA, R | FA | When packaged, only use FA |
| Potted plants | R | R | |

R = Room Cooling; FA = Forced-Air Cooling; HC = Hydrocooling; VC = Vacuum Cooling; WV = Water Spray Vacuum Cooling; FA-EC = Forced-Air Evaporative Cooling; PI = Package Icing.

## Table 8.2. Optimum storage temperatures for perishable products*

| Temperature after cooling or for short-term storage | Fruits | Vegetables | Flowers |
|---|---|---|---|
| 0°C (32°F) or below (but above freezing point) | Apples†, apricots, most berries (except cranberries), cherries, dates, figs, grapes, kiwifruit, loquats, nectarines, peaches, pears, persimmons, plums, prunes, quinces | Artichokes, asparagus, beans, beets, broccoli, brussels sprouts, cabbage, carrots, cauliflower, celeriac, celery, chard, chicory, collards, corn, endive, escarole, garlic, leafy greens, horseradish, kale, kohlrabi, leeks, lettuce, mushrooms, onions, parsley, parsnips, green peas, radishes, rutabagas, salsify, spinach, turnips, watercress | Carnation, chrysanthemum, iris, lily-of-the-valley, dry rose, sweetpea, tulip |
| 0°–2°C (32°–35°F) | Apples†, oranges | Asparagus, fullslip cantaloupe, water chestnuts | Allium, aster, bouvardia, crocus, freesia, gardenia, gerbera, hyacinth, narcissus, cymbidium orchid, ranunculus, rose in preservative |
| 2°–7°C (35°–45°F) | Apples†, avocados (ripe), cranberries, guavas, oranges, pomegranates, tangerines, mandarins | Green beans, lima beans, cassava, ¾ slip cantaloupe, southern peas, summer squash, tamarillos | Acacia, alstromeria, anemone, aster, bird-of-paradise, buddleia, calendula, calla, candytuft, columbine, cornflower, dahlia, daisy, delphinium, gerbera, gladiolus, gypsophilia, heather, lily, lupine, marigolds, cymbidium orchid, poppy, phlox, primrose, protea, snapdragon, statice, stephanotis, stock, strawflower, sweet william, violet, zinnia, florists' greens |
| 7°–13°C (45°–55°F) | Avocados, carambolas, lemons, limes, papayas, passion fruit, pineapples | Cucumbers, eggplant, muskmelons (casaba, crenshaw, honeydew), watermelon, okra, sweet peppers, pumpkins, winter squash, summer squash, taro, ripe tomatoes | Bird-of-paradise, heliconia, cattleya orchid, sweet william |
| 13°C (55°F) and above | Bananas, grapefruit, mango, plantain | Ginger, jicama, watermelon, sweet potatoes, green tomatoes | Anthurium, ginger, vanda orchid, poinsettia |

*See Hardenburg et al. 1986 for complete details of proper temperatures.
†Apple cultivars vary in their susceptibility to chilling injury.

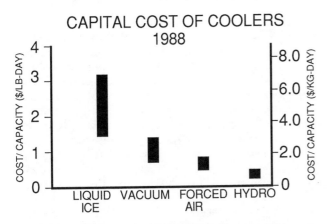

**Fig. 8.14. Capital cost (in 1988 dollars) of commonly used cooling systems.**

iation in the amount of ice that is put in the carton. If just enough ice is used, then less refrigerating capacity is needed. However, many broccoli shippers add more ice to handle refrigerating needs in transport plus an extra 4.5 kg (10 lbs) of ice in the box when it arrives at the market. The other three cooling methods cost less to buy than liquid-icing and have less variability in cost.

Capital cost per unit cooled can be minimized by using the equipment as much as possible. Vacuum-cooling equipment is very compact and is often portable. In California, vacuum coolers are moved as harvest locations change during the year. It is common for portable vacuum coolers to be used more than 10 months per year. Forced-air cooling facilities can be used for short-term storage of product during the harvest season and for long-term storage of product after the season ends.

## Energy costs

The energy cost of cooling varies greatly among coolers (fig. 8.15). Energy use is expressed in terms of an energy coefficient (EC), defined as:

$$EC = \frac{\text{cooling work done (kWh)}}{\text{electricity purchased (kWh)}}$$

High EC numbers indicate an energy efficient operation. The range of EC for each type of cooler reflects differences in design and operation procedures between coolers of the same type.

Actual energy costs for operating a cooler can be calculated assuming a value for EC with the formula below. Energy costs can be less than 5 percent of total costs in efficient cooling systems.

In customary units:

$$\text{Electricity cost} = \frac{W \times TD \times R}{3413 \times EC}$$

where: W = weight cooled (lbs)
TD = temperature reduction in product (°F)
R = electricity rate ($/kWh)
EC = energy coefficient
3413 Btu/kWh

In SI units:

$$\text{Electricity cost} = \frac{W \times TD \times R \times Cp}{3.6 \times EC}$$

where: W = weight cooled (kg)
TD = temperature reduction in product (°C)
R = electricity rate ($/kWh)
EC = energy coefficient
Cp = 4184 J/kg-°K
3.6 J/kWh

Labor and other equipment costs must be included in calculating total operating costs. Although no specific data are available for these costs, they can vary significantly. For example, a hydrocooler built into a packing line requires very little labor and no other equipment, but stand-alone coolers used in field packing operations require an operator and lift trucks for moving product in and out of the cooler.

If a cooling method requires that the product be packaged in a special carton, the extra cost of the carton should be included in a comparison of cooler types. For example, package icing, hydrocooling, and water-spray vacuum cooling need water-resistant packaging. This can increase the cost of an individual box by 25 cents to $1 (1988) depending on the design, size, and quantity of boxes purchased.

## Other considerations

Marketing tradition may dictate the choice of a cooling method. For example, some markets require that broccoli cartons arrive with ice in them. People selling in a market like this must select a package-ice cooling system.

Existing facilities may determine the type of cooler to be used. An existing cold-storage room can often be used for forced-air cooling small amounts of product by installing a small portable fan. Larger amounts of product usually require installation of more refrigeration capacity and a permanent air handling system.

A short harvesting season in a particular location may cause an operator to consider using portable cooling equipment. All types of coolers except room cooling operations can be made portable. They can be moved to a farmer's other production areas or leased to shippers in other areas and eliminate the cost of buying separate permanent coolers for each location. Some types of portable coolers can be leased or jointly owned by shippers and cooler manufacturers eliminating or reducing the need for capital expenditure.

Some growers contract with commercial companies to do their cooling. This requires no direct capital investment and no operating or management costs. But the grower loses some control over the product and loses the chance to make a profit from the cooling operation. Cooling cooperatives can give a grower some of the advantages of owning a cooler while reducing individual investment costs.

## Estimating Refrigeration Capacity

After deciding upon the cooling method(s) to use, the operator must estimate the amount of refrigeration capacity needed. This will help determine how large a cooler is needed. Coolers requiring less than 40 kW (12 tons) of refrigeration capacity (4 kW of refrigeration requires about 1 kW of compressor capacity, or 1 ton requires about 1 horsepower of compressor capacity) can often be farm built. Larger systems should be designed by an engineer.

The refrigeration capacity needed for large systems must be determined by a refrigeration engineer. The engineer will consider a number of factors such as: (1) amount of product cooled, (2) rate at which the product is received at the cooler, (3) required speed of cooling, (4) variety of products cooled and their unique cooling requirements, (5) building design and its affect on heat gain to the refrigerated

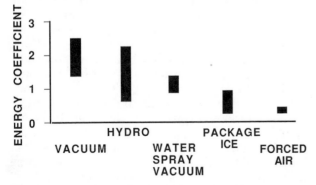

## ENERGY USE OF COOLERS

**Fig. 8.15. Energy use of commonly used cooling systems.**

volume, (6) heat input from lights, motors, forklifts, and people.

An estimate of the amount of refrigeration capacity needed for small-scale facilities does not require detailed calculations. Figure 8.16 can be used to estimate the refrigeration capacity needed for a cooler handling up to 800 kg per hour (2000 lb/hr). For example, if a product is cooled from 24°C (75°F) to 2°C (35°F), a temperature drop of about 22°C (40°F) and the cooler must handle a maximum of 500 kg per hour (1100 lb/hr), then 25 kW (7 tons) of refrigeration capacity are necessary. Estimates from this figure as based on reasonably fast cooling. Slow cooling, as is achieved by room cooling will require slightly less refrigeration capacity. The figure is also based on the assumption that heat input to the cooler from sources other than the product are less than 25 percent of the total.

Some small scale coolers use purchased ice for cooling. Figure 8.17 can be used to estimate the daily amount of ice needed to operate a small cooler. For example, if 800 kg per day (2000 lb/day) of product are cooled by about 22°C (40°F), then a little more than 500 kg (1100 lb) of ice would be melted. The figure is based on 50 percent of the ice being used for product cooling and the rest of the cooling potential lost to outside heat gain. This efficiency level is common for ice cooled hydrocoolers.

## Effective Cooler Management

Proper management of a cooler involves effective product cooling at minimum cost. Records of cooler operation are vital to enable a manager to evaluate the cooler. Good records should include (1) a sampling of incoming and outgoing product temperature for each lot and type of product cooled, (2) the temperature of the cooling medium during each cooling cycle, (3) the length of cooling cycles, (4) the quantity of product cooled in each cycle, (5) operating conditions of refrigeration system, such as suction and head pressures, and (6) monthly energy use.

Incoming product temperature is helpful in estimating the cooling time required. Outgoing product temperature is essential for determining the quality of the cooling process. It should be within acceptable tolerances, and, just as important, warmest product temperatures should be within acceptable tolerances. A good operator checks outgoing temperatures in various parts of the load to determine where the warmest product tends to be, and then always measures temperature in this area and attempts to get the product properly cooled.

Other factors are useful in determining the long-term performance of the cooler. For example, if cooling times begin to increase and the temperature of the cooling medium does not change, then there is a good chance that flow of the cooling medium through the product is being restricted (assuming that the type of product and its incoming temperature remain constant). If cooling medium temperature shows a trend of increasing during the cooling cycle, there may be problems in the refrigeration system. Changes in operating conditions of the refrigeration system are the clue to possible problems.

Regular maintenance is important for all types of coolers. In vacuum coolers, door seals need to be checked regularly and pressure gauges recalibrated about once per year. Daily cleaning is vital in proper hydrocooler operation. Trash screens, the water distribution pan, and the water reservoir must be cleaned each day and chlorine levels checked several times a day. Fluid levels and other features of the refrigeration system should be checked daily.

## Cold Rooms

Unless cooled products are loaded directly into precooled transport vehicles or placed in cold rooms, they will quickly warm up. Rewarming wastes the benefits of cooling. Cooled products left in a warm

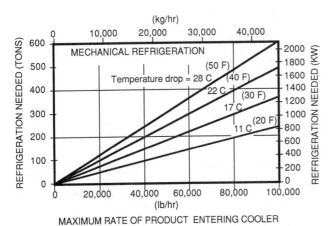

Fig. 8.16. Approximate mechanical refrigeration requirements for small scale coolers based on maximum hourly product input and product temperature drop.

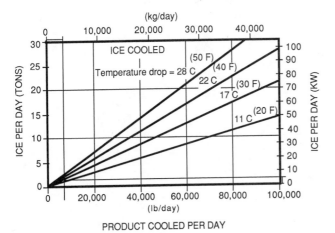

Fig. 8.17. Approximate amount of ice needed to operate small scale coolers based on amount of product cooled per day and product temperature drop.

environment are also subject to condensation, which may lead to disease problems. Therefore, a cooler probably needs a cold room with it. Sometimes the cooler can be a part of the cold room, as with forced-air coolers. Small cold rooms can be commercially constructed, purchased in prefabricated form and erected by growers, constructed by growers, or purchased as used refrigerated transport vehicles (rail cars, trailers, or marine containers). The cost of the cold room should be added to the total capital cost of a cooling facility.

## Summary

Effective cooling and temperature management in the changing horticultural crops handling system requires a complete understanding of product and market requirements, and of the cooling methods now available.

**1.** Rapid thorough cooling and good product temperature management are increasingly recognized as essential for successful marketing. This results from increased knowledge of product temperature requirements together with changing handling and marketing requirements.

**2.** Cooling is part of the total perishables handling system. Effects on cooling rate must be considered whenever a change is made in any part of the system.

**3.** Because requirements for cooling and cold storage differ, they should be considered as two separate operations.

**4.** Four cooling methods and variations are available to achieve reasonably fast cooling, even under the conditions imposed by newer handling procedures. Effective use of these options requires a complete understanding of the principles involved.

**5.** Introduction of field bins and unitized handling of packed containers has made rapid cooling more difficult to achieve. Contemporary cooling operations often involve large product masses with restricted access. Modifications to cooling systems can provide effective cooling even under such conditions.

**6.** Cooling efficiency can often be improved by attention to details of air management, package design, or pallet stacking patterns.

**7.** The increased costs involved in achieving faster cooling may be relatively small when the total cost of the cooling system is considered.

**8.** Fast cooling can often be achieved through minor modifications of existing cooling facilities. Requirements should be determined by a qualified refrigeration engineer after evaluating the complete refrigerations system.

## REFERENCES

1. Isenberg, F. M. R., R. F. Kasmire, and J. E. Parson. 1982. *Vacuum cooling vegetables*. Cornell Univ. Coop. Ext. Bull. 186. 10 pp.

2. Hardenburg, R. E., A. E. Watada, and C. Y. Wang. 1986. *The commercial storage of fruits, vegetables, and florist and nursery stocks*. U.S. Dept. Agric. Handbook 66. 130 pp.

3. Jeffrey, J. J. 1977. *Engineering principles related to the design of systems for air cooling of fruits and vegetables in shipping containers. Proc. 29th Intl. Conf. on Handling Perishables Agricultural Commodities*, 151-64. Mich. State Univ.

4. Kasmire, R. F., and F. G. Mitchell. 1974. *Perishables handling issue 36*. Univ. Calif. Coop. Ext. 14 pp.

5. Kasmire, R. F., and J. E. Parson. 1979. *Operator's guide to effective vacuum cooling*. Univ. Calif. Coop. Ext. (unnumbered).

6. Mitchell, F. G., R. Guillou, and R. A. Parsons. 1972. *Commercial cooling of fruits and vegetables*. Univ. Calif. Agric. Exp. Sta. Ext. Serv. Manual 43. 44 pp.

7. Mitchell, F. G., and R. F. Kasmire. 1978. *Perishables handling issue 39*. Univ. Calif. Coop. Ext. 12 pp.

8. Rij, R. E., J. F. Thompson, and D. S. Farnham. 1979. *Handling, precooling, and temperature management of cut flower crops for truck transportation*. Univ. Calif. Coop. Ext. Leaf. 21058.

9. Sargent, S. A., M. T. Talbot, and J. K. Brecht. 1989. Evaluating precooling methods for vegetable packinghouse operations. *Proc. Fla. State Hort. Soc.* 101:175-82.

10. Thompson, J. F., and R. F. Kasmire. 1981. An evaporative cooler for vegetable crops. *Calif. Agric.* 35(3&4):20-21.

# 9

# Storage Systems

James F. Thompson

Orderly marketing of perishable commodities often requires some storage to balance day-to-day fluctuations between product harvest and sales; for a few products, long-term storage is used to extend marketing beyond the end of harvest season. The goals of storage are:

**1.** To slow biological activity of product by maintaining the lowest temperature that will not cause freezing or chilling injury and by controlling atmospheric composition

**2.** To slow growth and spread of microorganisms by maintaining low temperatures and minimizing surface moisture on the product

**3.** To reduce product moisture loss and the resulting wilting and shrivel by reducing the difference between product and air temperatures and maintaining high humidity in the storage room

With some commodities, the storage facility may also be used to apply special treatments. For example, potatoes and sweet potatoes are held at high temperature and high relative humidity to cure wounds sustained during harvest, oranges may be degreened before shipment, and pears may be treated to ripen more quickly and uniformly. This chapter describes the equipment and techniques commonly used to control temperature, relative humidity, and atmospheric composition in a storage facility.

## Storage Considerations

### Temperature

The temperature in a storage facility normally should be kept within about 1°C (2°F) of the desired temperature for the commodities being stored. For storage very close to the freezing point, a narrower range may be needed. Temperatures below the optimum range for a given commodity cause freezing or chilling injury; temperatures above it shorten storage life. In addition, wide temperature fluctuations can result in both water condensing on the stored products and more rapid water loss. USDA Agricultural Handbook 66 (Hardenburg, Watada, and Wang 1986) lists recommended storage temperatures and humidities for horticultural products.

Maintaining proper storage temperatures within the prescribed range depends on several important design factors. The refrigeration system must be sized to handle the maximum expected heat load.

Undersized systems allow the air temperature to rise during peak heat load conditions. But an oversized system is unnecessarily expensive. The system should also be designed so that the air leaving the refrigeration coils is close to the desired temperature in the room. This prevents large temperature fluctuations as the refrigeration system cycles on and off. Large refrigeration coils allow a small temperature difference to be maintained between the air leaving them and the air in the room while still having adequate refrigeration capacity. A small temperature difference also increases relative humidity in the room and may reduce frost buildup on the coils.

Temperature variation is minimized with adequate wall and ceiling insulation and adequate air circulation. The system should be designed to provide 0.06 to 0.12 $m^3$/min of air per metric ton (20 to 40 cfm per ton) of product, based on the maximum amount of product that can be stored in the room. These low circulation rates require that the system be designed to move air uniformly past all of the stored product. Containers must be stacked to form air channels past one or two sides of each unit. The fans and ductwork must move the air past the product and not allow shortcuts through unused areas of the room or over the top of the product. Higher airflow is required if these conditions cannot be met or if the product releases large amounts of heat because it is not completely cooled or has a high respiration rate. Higher airflow rates cause more weight loss from the product than lower rates.

Thermostats are usually placed 1.5 m (5 feet) above the floor (for ease of checking) in representative locations in the room. They should not be placed near sources of heat such as doors or walls with an exterior surface. Nor should they be placed in a cold area such as near the air discharge of the refrigeration unit. A calibrated thermometer should be used to periodically check the thermostat. Remember, errors of only a few degrees can affect product quality.

### Humidity

For most perishable commodities, the relative humidity in a storage facility should be kept at 90 to 95 percent. Humidities below this range result in unacceptable moisture loss. Humidities close to 100 percent may cause excessive growth of microorganisms and surface cracking on some products, although it is unusual for a storage facility to have relative humidities that are too high.

Refrigeration equipment must be designed to maintain high relative humidity. In systems not designed for horticultural commodities, the evaporator coils (which produce the cold air) operate at a temperature about 6°C (11°F) lower than the desired air temperature in the room. This causes an excessive amount of moisture to condense on the coils, resulting in relative humidities of 70 to 80 percent in the storage room. Coils with a large surface area achieve the same refrigeration capacity as smaller coils but can operate at a higher temperature, thus reducing the moisture removed from the air. The coils should be large enough to operate 3°C (5°F) colder than the room air temperature. Mechanical humidifiers, fog spray nozzles, or steam systems are sometimes used to add moisture to the storage room and reduce the drying effect of the evaporator coils. However, this added moisture results in more frequent defrosting of coils and increases operating problems.

Some refrigeration systems use a wet coil to maintain humidity. In this system, water is cooled to 0°C (32°F) or a higher temperature if higher room temperatures are desired. The water is sprayed down through a coil, and the storage area air is cooled and humidified to nearly 100 percent as it moves upward through the coil. However, as the air moves through the storage area it picks up heat, and the rise in temperature reduces relative humidity. This system is usually limited to air temperatures above 0.2°C and does not work well for commodities that are held just below 0°C.

# Refrigeration

## Mechanical refrigeration

Most storage facilities use mechanical refrigeration to control storage temperature. This system utilizes the fact that a liquid absorbs heat as it changes to a gas. The simplest method for using this effect is to allow a controlled release of liquid nitrogen or liquid carbon dioxide in the storage area. As they boil, they cause a cooling effect in the storage area. However, this method requires a constant outside supply of refrigerant and is only used to a limited extent with highway vans and rail cars. The more common mechanical refrigeration systems use a refrigerant such as ammonia or a variety of halide fluids (sometimes referred to by the trade name "freon") whose vapor can be easily recaptured by a compressor and heat exchanger.

Figure 9.1 shows the components of a typical vapor recompression (or mechanical) refrigeration system. The refrigerant fluid passes through the expansion valve, where the pressure drops and the liquid evaporates at temperatures low enough to be effective in removing heat from the storage area. Heat from the material to be cooled is transferred to the room air, which is then forced past the evaporator (cooling coil located in the room), usually a finned

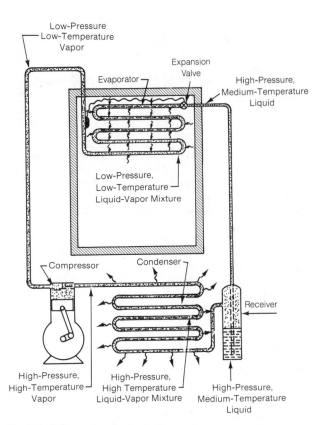

Fig. 9.1. Schematic of a typical vapor recompression or mechanical refrigeration system.

tube heat exchanger, which transfers the heat from the air to the refrigerant causing it to evaporate. After fully changing to a gas, it is repressurized by the compressor and then passes through a condenser where it is cooled to a liquid. The condenser is located outside the storage area and releases heat. Liquid is stored in the receiver and is metered out as needed for cooling.

## Expansion valves

Small mechanical refrigeration systems are controlled primarily by the expansion valve, which regulates the pressure of the refrigerant in the evaporator. Low pressures cause the liquid refrigerant to evaporate at low temperatures. The valve also controls the flow of refrigerant, which affects the amount of refrigeration capacity available. Capillary tubes and thermostatic expansion valves are the most common types of expansion valves.

The capillary tube is used with very small refrigeration equipment (less than 1 horsepower). It is a tube 0.6 to 6 m (2 to 20 feet) long with a very small inside diameter of 0.6 to 2.3 mm (0.025 to 0.090 inch). The resistance of the liquid flowing through the tube creates the needed pressure drop between the low-pressure and high-pressure sides of the system and regulates the flow of refrigerant. A capillary is inexpensive and has no moving parts to maintain, but it cannot be adjusted, is subject to clogging, and requires a relatively constant weight of refrigerant in the system.

The thermostatic expansion valve regulates the flow of refrigerant to maintain a constant temperature difference between the evaporator inlet or evaporating temperature and the coil outlet (maintains a constant super heat). It allows the low-side pressure to vary, so that when high refrigeration loads are required, the temperature of the evaporator coil increases.

Large refrigeration systems may use an evaporator coil that is designed to always have liquid refrigerant in it, called a flooded coil. It has a greater heat transfer efficiency than a nonflooded coil of equal size. Refrigerant flow is controlled primarily with a float control that ensures a constant level of refrigerant in the coil. The float control may operate in parallel with a thermostatic expansion valve.

Other controls such as suction pressure regulators may be used in conjunction with float controls. These are especially useful in maintaining the highest possible evaporator coil temperature to maintain high humidity in the storage room.

### Evaporators

Modern cold storages usually use finned tube evaporators. Air from the storage is forced past the tubes by fans, which are a part of a complete evaporator unit. Evaporators operating near 0°C (32°F) build up frost that must be removed to maintain good heat transfer efficiency. Defrosting may be done by periodically flooding the coils with water, by electric heaters, by directing hot refrigerant gas to the evaporators, or by continuously defrosting with a brine or glycol solution.

### Compressors

The most common types of refrigeration compressors are reciprocating (piston) and rotary screw (fig. 9.2). Reciprocating compressors come in a wide range of sizes and can be set up to operate efficiently at varying refrigerant flow rates. Rates are varied by shutting off pairs of cylinders in a unit which may have 6 to 12 cylinders. The main disadvantage of reciprocating compressors is their fairly high maintenance costs. Rotary screw compressors have low maintenance costs but are not available in sizes smaller than about 23 kW (30 HP) and operate efficiently only at near-maximum refrigerant flow rates. Some facilities use screw compressors for base load refrigerant needs and reciprocating compressors for the portion of the load that varies significantly during the day.

### Condensers

Condensers are categorized as air-cooled or water-cooled. Small systems usually use an air-cooled unit. Many home refrigerators, for instance, have a coiled tube in the back that allows a natural draft of air to flow past. Larger systems use a fan to provide airflow

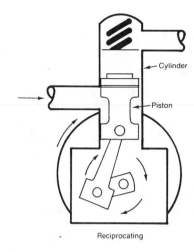

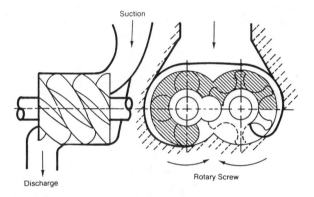

**Fig. 9.2. Common types of refrigeration compressors.**

past the condenser. Large condensers are more likely to be water cooled. Water is a better heat conductor than air, allowing water-cooled condensers to be smaller than forced-air units of equal capacity. However, water-cooled units may require large quantities of water, which can be expensive to obtain and dispose of. Evaporative condensers reduce water consumption by recycling the heated condenser water; they require close attention to water quality to maintain efficiency and to prevent damage to the heat exchanger.

### Refrigerants

The choice of which refrigerant to use in a vapor recompression system is based on the following factors:

**1. Cost of refrigerant.** Halide refrigerants are more expensive than ammonia. (Environmental concerns may cause restrictions on the availability of some halide refrigerants we now use.)

**2. Compatibility.** Ammonia cannot be used with metals that contain copper; halide refrigerants cannot be used with magnesium and may damage plastic materials.

**3. Toxicity.** Ammonia at very low concentrations can injure perishable commodities. It is toxic to humans.

### Control systems

A large refrigeration system requires a good control system and a system for displaying the system's operating condition. At a minimum, panel lights should be installed to indicate the operating status of fans and compressors and the fluid levels in surge and receiver tanks. Controls should be set up to allow manual operation of motors.

Microcomputers and programmable controllers allow even more precise control of large refrigeration systems. They are especially valuable in reducing electricity use during peak rate periods. Defrost cycles can be programmed to take place at night and unnecessary fans and compressor motors turned off during peak rate periods.

Absorption refrigeration, used in a few cold storage operations, differs from mechanical refrigeration in that the vapor is recovered primarily through use of heat rather than mechanical power. It is less energy efficient than mechanical refrigeration and usually is used only where an inexpensive source of heat is available.

The capacity of a refrigeration system is based on adding all the heat inputs to a storage area. Heat inputs include: (1) heat conducted through walls, floor, and ceiling; (2) field and respiration heat from the product; (3) heat from air infiltration, and (4) heat from personnel and equipment such as lights, fans, and forklifts.

Refrigeration equipment for storage facilities generally is not designed to remove much field heat from the product because a large capacity is required; a separate cooling facility is used for this purpose.

## Alternative Refrigeration Sources

In many developing countries, where mechanical refrigeration is prohibitively expensive to install and maintain, a number of other techniques can be used to produce refrigeration. In some cases these can provide nearly the recommended storage conditions. In others, they are a compromise between proper storage conditions and costs for equipment, capital, and operations.

### Evaporative cooling

Evaporative cooling techniques are very energy efficient and economical. A well-designed evaporative cooler produces air with a relative humidity greater than 90 percent. Its main limitation is that it cools air only to the wet-bulb temperature. During the harvest season in the U.S., wet-bulb temperatures vary from 10° to 25°C (50° to 77°F) depending on location, time of day, and weather conditions. This temperature range is acceptable for many chilling-sensitive commodities.

Minimum temperatures from an evaporative cooling system can be reduced by using a multiple-stage system. In a two-stage system, an evaporative system cools water to the wet-bulb temperature which is, in turn, used in a water-to-air heat exchanger to cool outside air. The cooled air has a reduced wet bulb temperature—a 10°C (18°F) drop in dry bulb temperature produces a 3° to 4°C (5° to 7°F) drop in the wet bulb—and can be passed through an evaporative cooler for further cooling. This system can raise the temperature reduction produced by an evaporative cooler by as much as 6°C (11°F) compared to a single-stage system. A system can be set up to use more than two stages. The theoretical minimum temperature a multi-stage system can produce is the dew point temperature of the air.

The water for cooling in the systems mentioned above comes from domestic sources. It is also practical to cool by evaporating water from the commodity. Snap beans have been cooled in transit by erecting an air scoop above the cab of the truck that forces outside air through a bulk load of beans. This system prevents heat buildup and keeps the beans at or below the outside air temperature. It is not advisable to use this system for any great length of time to avoid excessive water loss.

### Nighttime cooling

In some parts of the world, significant differences between night and day temperatures allow nighttime ventilation to be a means of refrigeration. In dry Mediterranean or desert climates the difference between daily maximum and minimum temperatures can be as great as 22°C (40°F) during the summer. Nighttime cooling is commonly used for unrefrigerated storage of potatoes, onions, sweet potatoes, hard-rind squashes, and pumpkins. As a rule, night ventilation effectively maintains a given product temperature when the outside air temperature is below the product temperature for 5 to 7 hours per day.

Low nighttime temperatures can be used to reduce field heat simply by harvesting produce during early morning hours. Some growers in California use artificial lighting to allow nighttime harvest.

It is theoretically possible to produce air temperatures below nighttime minimums by radiating heat to a clear nighttime sky. A clear night sky is very cold, and a good radiating surface such as a black metal roof can cool below the air temperature. Simulations have indicated that this method could cool air about 4°C (7°F) below night air temperatures. This concept has not yet been widely used.

### Well water

In some areas, well water can be an effective source of refrigeration. The temperature of the ground greater than about 2 m (6 feet) below the surface is equal to the average annual air temperature. Well water is often very near this temperature.

### Naturally formed ice

Before the development of mechanical refrigeration, refrigeration was provided by natural ice harvested from shallow ponds during the winter. The ice was stored in straw and hauled to cities as needed during spring and summer. Energy costs today make it unfeasible to transport ice any significant distances. However, cooling facilities in appropriate climates can store ice nearby for summer use. In some cases, it may be feasible to transport perishable commodities to the ice for storage. This would be especially practical where the ice is located between the sites of production and consumption.

### High-altitude cooling

High altitude can also be a source of cold. As a rule of thumb, air temperatures decrease by 10°C with every kilometer (5°F per 1,000 ft) increase in altitude. It is not possible to bring this air down to ground level because it naturally heats by compression as it drops in altitude. However, in some cases it may be possible to store commodities at high altitudes in mountainous areas. For example, in California most perishable commodities are grown in the valley floors near sea level. However, much of the production is shipped east across the Sierra Nevada over passes about 1,800 m (6,000 feet) high. Air temperature has the potential of being 18°C (32°F) cooler, and it may reduce energy costs to store perishables there rather than on the valley floor.

### Underground storage

Cellars, abandoned mines, and other underground spaces have been used for centuries for storage of fruits and vegetables. As mentioned earlier, underground temperature is near the average annual air temperature. Underground spaces work well for storing already cooled produce but not for removing field heat. The soil has a poor ability to transfer heat. Once the refrigeration effect is depleted from an area, it does not regenerate rapidly. This can be overcome by installing a network of buried pipes around the storage. Cooled air is pumped from the pipes to the storage area, allowing the harvest of cooling capacity from a greater soil volume.

## The Storage Building

The storage must be sized to handle peak amounts of product. The floor area can be calculated knowing the volume of the produce and maximum storage height and allowing for aisle ways, room for forklift maneuvering, and staging areas. Maximum storage height can be increased by use of shelves or racks and forklifts with suitable masts. Multistory structures are generally not used because of the difficulty and expense of moving the product between levels.

The building ideally should have a floor perimeter in the shape of a square. A rectangular configuration has more wall area per square foot of floor area, resulting in higher construction cost and higher heat loss than a square configuration. Entrances, exits, and storage areas should be arranged so that the product generally moves in one direction through the facility, especially if the storage facility is used in conjunction with a cooler to remove field heat.

### Site selection

Good utility service must be available for the facility. Extending roads and energy utilities to a facility can be very expensive. Three-phase electrical power is needed to operate refrigeration equipment motors. In some areas a backup power supply may be advisable. There should be enough water to supply the evaporative condensers, personnel needs, and the needs of a packing house, if it is a part of the project. Consider the availability of fire protection services, gas supply, and sewer utilities.

The area should have good drainage and room for future expansion. There should be enough space around the facility for smooth movement of large highway trucks.

### Building layout

Room layouts with an interior corridor offer better operating conditions for cold storages and better control of controlled atmosphere storages than designs that allow room access only through exterior doors. The interior corridor, seen in layouts 1 to 3 in figure 9.3, allows easy access to piping and controls. Doors and equipment are shielded from the elements, and product observation is easier. Layout 1 is common in small operations where storage and packing are done in one building. Layout 2 allows better product flow compared with layout 3, but layout 3 has less area devoted to corridor. Layout 4 is the least expensive of the designs because none of the cold storage building is used as a permanent corridor.

Refrigerated facilities can be constructed from a wide variety of materials. The floor and foundation is usually a concrete slab. A vapor barrier is installed to prevent moisture movement through the slab, and rigid insulation is sometimes placed above the barrier and below the concrete. Walls can be made of concrete block, tilt-up concrete, insulated metal panels, metal frame, or wood frame construction. Wood frame and concrete block are losing favor to metal and tilt-up concrete construction in the United States.

Walls are insulated with fiberglass batts, rigid urethane foam boards, or sprayed-on foam. Batt and board insulation must be protected with a vapor barrier on the warm side. Properly applied sprayed-on foam can be moistureproof. Foam insulation must be coated with a fire retardant. Some storages combine insulation types; for example, the interior can be sprayed with foam to form a vapor barrier, then covered with batt or board insulation for appearance and fire protection. Total insulation level on the walls

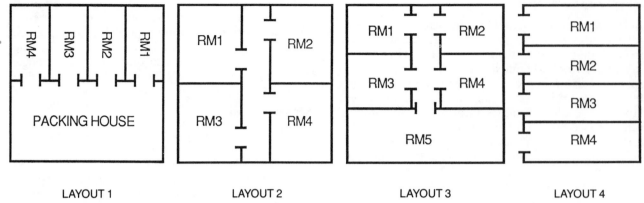

LAYOUT 1      LAYOUT 2      LAYOUT 3      LAYOUT 4

*Fig. 9.3. Typical layouts of CA facilities.

is often in the range of R20 to R40. Ceilings can be insulated with rigid board or foam materials, or built separately from the roof and insulated with loose fill or batts. Ceiling R-values of at least 60 are common in new construction.

If modified atmosphere techniques are used in the storage facility, the vapor barrier may also serve as a gas barrier, and special precautions must be taken to ensure a gastight seal.

## Small-Scale Cold Rooms

Cold rooms for small-scale operations can be purchased from commercial suppliers, self-built, or made from used refrigerated equipment such as rail cars, marine containers, or highway vans. The choice of the best system is based on cost and availability of equipment in the area and the amount of time available to invest in cold room installation.

### Rail cars

Refrigerated rail cars are very sturdy and well insulated. Refrigeration is powered by an electric motor powered in turn by a diesel generator. The generator set can be salvaged and the refrigeration connected to the farmer's electric utility. Cars in the U.S. have a 9-foot, 4-inch (2.84 m) ceiling, which limits the height that produce can be stacked. The most significant problem and greatest cost of using railcars is getting them from the railroad to the farm site.

### Highway vans

The one unique advantage of highway vans is that they are portable if the wheels are left on. The refrigeration system is powered directly with a diesel engine. This can be a benefit if utility electricity is not available at the site. In some areas, it may be less expensive to operate the refrigeration unit if it is converted to operate with an electric motor, but a considerable conversion cost must be added to the project. Highway vans are built as light as possible to maximize the load weight they can carry; this often means that used vans are in fairly poor condition.

Their insulation, which is limited to begin with, may be deteriorated and they may have poorly sealed doors that permit a lot of air leakage. Also, old vans often have fairly small fans that may not provide adequate air circulation.

### Marine containers

These are available in lengths of 20, 24, and 40 feet (6.1, 7.3, and 12.2 m). Their built-in refrigeration units are powered with 220- or 440-volt, three-phase electricity, and they can be plugged directly into utility power. They are usually well built and have deep T-beam floors and sufficient fan capacity to provide good air circulation; in fact, air circulation is good enough to allow adequate room cooling.

A disadvantage of all transport vehicles is that their refrigeration systems are usually not designed to produce high relative humidity. Drying of the product due to low humidity results in weight loss and poor quality. This is particularly a problem if the cold room is to be used for long-term storage. About the only way to reduce the drying is to keep the floor and walls of the cold room wet, but this causes increased corrosion, reduced equipment life, and increased need for defrosting.

Refrigerated transport vehicles rarely have enough refrigeration capacity to rapidly precool produce. If rapid cooling is needed, extra capacity must be added. Moreover, transportation vehicles are too narrow for the frequent product movement needed in a precooling facility. A separate self-constructed room is much more convenient for precooling operations.

### Self-constructed cold rooms

For many producers, a self-built cold room is the least expensive option. It is usually built with a concrete or wood frame floor. Walls and roof are of wood frame construction and insulated with fiberglass batts. Care must be taken to install a tight vapor barrier on the warm side of the insulation. Refrigeration is provided by a small mechanical refrigeration system. If the room is kept above 10°C (50°F), it may be possible to use a room air conditioner. These cost about half as much as a packaged refrigeration system.

## Controlled Atmosphere Storage

Controlled atmosphere (CA) storage utilizes oxygen and carbon dioxide concentrations of about 1 to 5 percent for each gas. Normal room air has an oxygen concentration of about 21 percent and carbon dioxide levels near 0.03 percent. Low oxygen and high carbon dioxide levels slow ripening processes, stop the development of some storage disorders, such as scald in apples, and slow the growth of decay organisms. All of these effects increase storage life of fresh produce compared with refrigerated air storage. More details about the potential benefits and hazards of CA storage are presented in Chapter 11.

### Simple CA system

CA storage has all of the design requirements of conventional refrigerated storage plus gastight rooms, equipment to obtain the desired gas concentrations, and equipment to measure and control atmospheric composition. The simplest system for obtaining gastight storage uses a plastic tent inside a conventional refrigerated storage room. The tent is made of 3- to 5-mil polyethylene sheeting supported by a wood framework. The sheeting is sealed to the concrete floor by pressing the plastic into a narrow trough and forcing tubing into the trough to keep the plastic in place. A better gas barrier at the floor can be obtained by laying a sheet of plastic on the floor and protecting it with wood panels. A seal is obtained by joining the tent to the floor sheet. A fan inside the tent provides air circulation. The oxygen level is reduced initially by allowing fruit respiration to consume oxygen or by using CA generators. Oxygen is kept above the minimum by allowing a controlled amount of outside air to enter the tent. The $CO_2$ level is maintained by placing bags of fresh hydrated lime in the tent to absorb excess $CO_2$.

### Permanent CA facilities

Permanent facilities require that the storage plant be designed specifically for controlled atmosphere storage. Usually a CA facility costs about 5 percent more to build than a conventional refrigerated storage facility. The extra cost is in building the storage rooms to be airtight and, in some cases, designing smaller individual rooms than are needed in common refrigerated storage.

**Room size.** Individual rooms should be sized to allow them to be filled in a short time. Many apple CA rooms are built to hold a week's fruit harvest. If an operator wishes to use rapid CA, the room should be small enough to be filled in three days. Many facilities have several room sizes to allow for variations in incoming fruit volume, fruit varieties, and marketing strategy.

Modern storages are single-story designs, with individual rooms often tall enough to allow fruit to be stacked 10 bins high, including enough height between the bins and the ceiling for air from the evaporator coils to mix with the room air and to travel easily to the far end of the room.

Many facilities are enlarged as business grows, and the building should be located to allow expansion.

**Gas-tight construction.** Three main types of interior wall and ceiling construction are used in recently built CA storages, (1) foamed-in-place urethane over concrete or steel walls, (2) plywood-covered stud walls with fiberglass insulation, and (3) insulated, metal-covered panels.

*The urethane system.* It serves as a gas seal, insulation, and vapor barrier. But it is expensive, primarily because the urethane must be covered with a fire barrier. Foamed-in-place urethane can be applied directly to any type of masonry, wood, or primed metal. It should not be applied over rigid board insulation as the foam can distort the board, and board materials that are not tightly attached cause the gas seal to fail over time. Figure 9.4 shows a cross section of a typical CA storage sealing system with urethane form insulation.

*Plywood cover over fiberglass insulation.* This is a common method of insulating and sealing CA storages. It is usually the least expensive of the three gas barrier systems. The plywood sheets are sealed with a butyl rubber compound applied between the sheet and the wood framing. Sheets are separated with a 3-mm (⅛ inch) gap to allow for expansion. The gap is filled with butyl rubber and the joint covered with fabric and an elastomeric sealer. If regular plywood is used, the whole board must be covered with the sealer. High-density plywood does not need to be treated for gas tightness.

A vapor barrier is usually installed on the outside (warm side) of the insulation. But an outside vapor barrier is not recommended with the plywood system, because there is already a gastight seal on the inside of the insulation. A vapor barrier on the outside will trap any moisture that might get into the insulation, ruining its insulating value.

*Insulated panels.* These are usually a sandwich design. Three to 6 inches of rigid foam insulation is covered on both sides with painted metal or fiberglass sheets. Panels are 1.2 to 2 m (4 to 6 ft) wide and usually extend from floor to ceiling. The panels are installed inside a building shell and are held together with mechanical fasteners and flexible sealing materials. Joints are sealed with polyvinylacetate copolymer or latex emulsion sealers. Large gaps are backed with a nonwoven fabric.

*Floors.* Floors can be made gastight with two main systems. An insulated floor is usually a sandwich design, with board insulation placed between two slabs of concrete. With this design, a gas seal of two layers of hot-mopped asphalt roofing felt is applied

76

to the subfloor (fig. 9.4). A single-slab floor is used if only perimeter insulation is used. In this case the floor is sealed by applying special materials to the top surface. Chlorinated rubber compounds are sometimes used.

In all designs, the wall-to-floor seal is the area that is most likely to fail. Prevent the floor from moving with respect to the wall by carefully backfilling and thoroughly compacting the subgrade. The floor can be tied to the walls with rebar or the floor set on a 10-cm (4 inch) ledge built into the foundation wall.

*Doors.* Many different door designs can be used in CA facilities. In all of them the door is constructed of a solid frame that can be clamped tightly against a gasketed door frame without warping. The frame can be covered with well-sealed plywood or metal. Some of the more expensive doors have aluminum sheets welded to an aluminum frame. The bottom of the door is usually sealed with caulking compound after the door is closed. Most doors are 2.4 to 3 m (8 to 10 ft) wide and tall enough to allow a lift truck with two bins to pass through.

Each room should also have a 60 × 75 cm (24 × 30 inch) access door that allows for entry for checking fruit and making repairs without opening the main door. Many storages also have a clear acrylic window near the top of a wall to allow the fruit to be inspected without entering the room. It is usually of a concave shape, allowing all areas of the room to be seen.

## Pressure relief

A pressure difference between the cold room and the outside can develop because of changes in weather or room temperature. This difference can damage the gas seal if it is not relieved. A water trap as pictured in figure 9.5 is usually used to allow pressures to equalize. The trap is often filled with glycol to avoid problems with the water evaporating. A spring-loaded or weight-loaded check valve can be used, but they are much more expensive than a water trap. Ten square centimeters of vent opening should

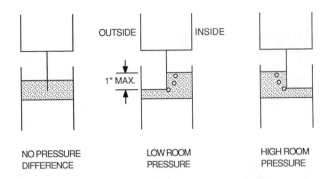

Fig. 9.5. Schematic of water trap pressure relief system.

be provided for each 40 m³ of room volume (1 in² per 1000 ft³).

Small changes in pressure can be relieved by using breather bags. These have the advantage of capturing the gas mixture in the room and allowing it to reenter the room at a later time. Bags should have 0.35 to 0.4 m³ of capacity per 100 m³ of room volume (3.5–4 ft³/1000 ft³), although some are designed with several times this capacity. If a CA room is so tightly sealed that air must be regularly bled into the room to maintain $O_2$ level, breather bags are not necessary.

## Pressure test

The overall gas-tightness of a CA room is tested by pressurizing it and measuring the rate of pressure drop. After all doors and openings are sealed, a small fan is used to increase the static pressure in the room to 25 cm (1 inch) of water column. The fan is then turned off and sealed, and the tester determines how long it takes for the pressure to drop by half. If this takes 20 minutes or longer, it is considered tight enough for a CA system that uses hydrated lime to maintain $CO_2$ levels. At least 30 minutes are required for rooms that use carbon or water scrubbers to control $CO_2$. Some authorities believe that rooms should be even more gastight than these standards and require 45 minutes to 2 hours for the pressure to drop by half. Because gas seals deteriorate, new rooms should be much tighter than prevailing standards.

If the room does not meet the desired standard, check for leaks. Leaks are most common around doors, at the wall and floor junctions, near poorly sealed penetrations, and at unsealed electrical boxes. Test for leaks by putting the room under a slight vacuum and listening for air leaking or spraying suspect areas with soapy water and watching for bubbles. Smoke sticks or bee smokers can be used to detect streams of outside air entering the room.

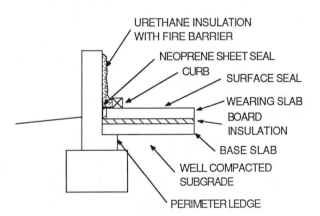

Fig. 9.4. Gastight seal using urethane foam insulation.

## Atmospheric modification

The least expensive, but slowest method of modifying the storage room atmosphere is to let the product do it through natural respiration. Fruit and vegetables use oxygen and release carbon dioxide. The product filling a sealed room eventually lowers the oxygen to the level needed for CA storage; if oxygen drops too low, outside air can be added to restore it to the desired range. However, respiration causes carbon dioxide levels to rise well above required levels. Bags of hydrated lime are used to absorb excess $CO_2$. Lime requirements are 1 to 3 kg per 100 kg of product, depending on the product being stored, storage time, surface area, and quality of the lime. Bags can be placed either in single layers in the CA room or in an adjacent lime room that is connected to the CA room with a fan and ductwork. Rooms should be sized to hold about 15 kg of lime per metric ton of product (30 lbs/ton). Lime is very effective in producing the low levels of $CO_2$ that are being used more and more in apple storage.

Carbon dioxide levels can also be controlled with activated carbon adsorption systems, molecular sieves, or brine pumped over evaporator coils. The brine system is not used in newer installations because it cannot reduce $CO_2$ levels quickly enough. Molecular sieves, which tend to use more energy than activated carbon systems, are less common than activated carbon equipment.

Relying on product respiration to remove oxygen is fairly slow, and the product storage life can be increased if the oxygen is removed faster. Some operations purge the CA room with nitrogen, purchased in liquid form or produced on site. One type of nitrogen generator uses ammonia in a combustion process to consume oxygen and produce nitrogen and water. Two other systems use a molecular sieve process (pressure swing adsorption or PSA system) or semipermeable membrane to remove oxygen. Machines that remove oxygen by combustion of natural gas or propane all produce carbon dioxide, which must be removed by another process. Incomplete combustion in these machines has caused explosions in CA rooms and carbon monoxide poisoning of workers. A shift is taking place from these machines to nitrogen purging.

## Refrigeration equipment

Refrigeration equipment for controlled atmosphere facilities is the same as for any other cold storage operation. Most storages are designed to maintain 0°C (32°F) at 95 percent relative humidity. Air circulation during storage should be in the range of 0.01–0.02 $m^3$/sec per metric ton (20–40 cfm per ton) of production. Up to 0.05 $m^3$/sec per metric ton (100 cfm per ton) is required for product cooling after initial loading.

## Monitoring equipment

Oxygen and carbon dioxide levels must be monitored daily to ensure that they are within prescribed limits. Traditionally, operators have used an Orsat gas analyzer, a wet chemistry system that is fairly time-consuming to use. Automatic equipment is now widely used. It is more accurate than the traditional system, provides a log of the data, and can be connected to a controller to automatically maintain proper gas concentrations.

Temperature should also be monitored regularly. A minimum of two calibrated dial thermometers should be installed in each room. One should be near eye level with the dial located on the outside of the room. The other should be above the fruit and readable through an observation window near the ceiling. Electronic thermometers allow easier observation of room temperature, and the data can easily be printed out for a permanent record of operating conditions. Most new storages use four probes (or more) per room. Some probes are placed into bins to monitor fruit temperatures.

## Safety considerations

The atmosphere in CA rooms will not support human life, and people have died of asphyxia while working in CA rooms without breathing apparatus. A danger sign should be posted on the door. The access hatch in the door should be large enough to accommodate a person equipped with breathing equipment. At least two people with breathing equipment should work together at all times, one inside and one outside the room watching the first person.

## REFERENCES

1. ASHRAE. 1986. *ASHRAE handbook fundamentals*. Atlanta, GA: Am. Soc. Heating, Refrig. Air-Cond. Eng. Chapter 20.

2. Bartsch, J. A., and G. D. Blanpied. 1984. *Refrigeration and controlled atmosphere storage for horticultural crops.* Cornell Univ. Coop. Ext. Bull. NRAES-22. 42 pp.

3. Davis, D. C. 1980. Moisture control and storage systems for vegetable crops. In *Drying and storage of agricultural crops*, ed. C. W. Hall, 310-59. Westport, CT: AVI Publ. Co.

4. Dewey, D. H. 1983. Controlled atmosphere storage of fruits and vegetables. In *Developments in Food Preservation*, ed. S. Thorne, 1-24. London: Applied Science Publ.

5. Fokens, F. H., and H. F. Th. Meffert. 1967. Apparatus for measuring leakage from C.A. storage rooms and other insulated construction, presented at XIIth Int. Congress of Refrigeration, Madrid.

6. Hallowell, E. R. 1980. *Cold and freezer storage manual.* Westport, CT: AVI Publ. Co. 356 pp.

7. Hardenburg, R. E., A. E. Watada, and C. Y. Wang. 1986. *The commercial storage of fruits, vegetables, and florist and nursery stocks.* U.S. Dept. Agric. Handb. 66. 136 pp.

8. Hunter, D. L. 1982. C.A. "Storage Structure." In *Controlled atmospheres for storage and transport of perishable agricultural commodities*, eds. D. G. Richardson and M. Meheriuk. Beaverton, OR: Timber Press.

9. International Institute of Refrigeration. 1976. *Guide to refrigerated storage.* Paris: Internat. Inst. Refrig., chapter 1.

10. Sainsbury, G. F. 1959. *Heat leakage through floors, walls and ceilings of apple storages*. U.S. Dept. Agric. Mktg. Res. Rept. 315. 65 pp.

11. Stoecker, W. F. 1988. *Industrial refrigeration*. Troy, MI: Business News Publ. 386 pp.

12. Thompson, J. F., and R. F. Kasmire. 1988. *Small-scale cold rooms for perishable commodities*. Univ. Calif. Div. Agric. Nat. Res. Leaflet 21449. 8 pp.

13. Waelti, H., and J. A. Bartsch. 1990. Controlled atmosphere storage facilities. In *Food preservation by modified atmospheres*, eds. M. Calderon and R. Barkai-Golan, 373-89. Boca Raton, FL: CRC Press.

# 10

# Psychrometrics and Perishable Commodities

JAMES F. THOMPSON

Psychrometrics is the measurement of the heat and water vapor properties of air. Commonly used psychrometric variables are temperature, relative humidity, dew-point temperature, and wet-bulb temperature. While these may be familiar, they are often not well understood.

## Psychrometric Chart

The psychrometric chart describes the relationships between these variables. Figures 10.1 and 10.2 are psychrometric charts in English and metric units, respectively, which will help to illustrate the meaning of various terms.

Temperature, sometimes called dry-bulb temperature after the unwetted thermometer in a psychrometer, is the horizontal axis of the chart. The vertical axis is the moisture content of the air, called humidity ratio (sometimes called mixing ratio or absolute humidity). The units of humidity ratio are mass of water vapor per mass of dry air. Under typical California conditions, the humidity ratio of outside air varies between 0.004 and 0.015 kg/kg. Even though water vapor represents only 0.4 to 1.5 percent of the weight of the air, this small amount of water vapor plays a very significant role in the postharvest life of perishable commodities.

The maximum amount of water vapor that air can hold at a specific temperature is given by the leftmost, upward-curved line in the psychrometric chart. Notice that air holds more water vapor at increasing temperatures. As a rule of thumb, the maximum amount of water that the air can hold doubles for every 11°C (20°F) increase in temperature. This line is also called the 100 percent relative humidity line. A corresponding 50 percent relative humidity line is approximated by the points that represent the humidity ratio when the air contains one-half of its maximum water content. The other relative humidity lines are formed in a similar manner.

Notice that relative humidity without some other psychrometric variable does not determine a specific

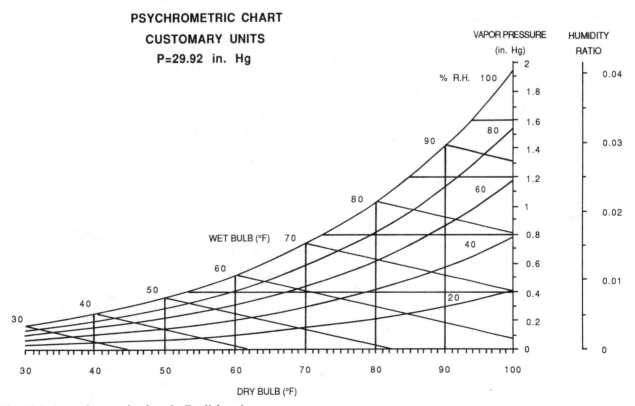

Fig. 10.1. A psychrometric chart in English units.

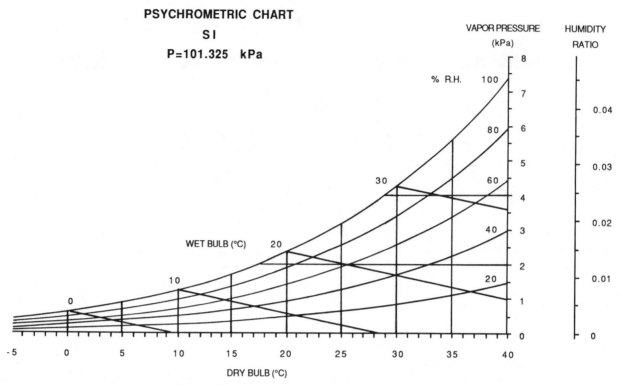

**PSYCHROMETRIC CHART**

**S I**

**P=101.325 kPa**

Fig. 10.2. A psychrometric chart in SI (metric) units.

air condition on the chart and is not very meaningful. For instance, 80 percent relative humidity at 0°C (32°F) is a much different air condition than 80 percent relative humidity at 20°C (68°F).

If a mass of air is cooled without changing its moisture content, it loses capacity to hold moisture. If cooled enough, it becomes saturated (has 100 percent relative humidity) and if cooled further, begins to lose water in the form of dew or frost. The temperature at which condensation begins to form is called the dew point temperature if it is above 0°C (32°F) or the frost point temperature if it is below 0°C (32°F).

Another commonly used psychrometric variable is wet-bulb temperature. On the chart this is represented by lines that slope diagonally upward from right to left. These lines represent the temperature and water vapor conditions of a thermometer covered with water-soaked gauze. In practice, wet-bulb lines are used to determine the exact point on the psychrometric chart that represents the air conditions in a given location as measured by a psychrometer. The intersection of the diagonal wet-bulb temperature line (equal to the temperature of a wet-bulb thermometer) and the vertical dry-bulb temperature line defines the temperature and humidity conditions of air.

Water vapor pressure is not usually shown on psychrometric charts but is an important concept in handling perishables. It is directly proportional to

humidity ratio. The following formula is used to calculate vapor pressure:

$$Vp = \frac{w \times Pa}{e}$$

where: Vp = vapor pressure (same units as Pa)
 w = humidity ratio
 Pa = atmospheric pressure
 e = a constant = 0.622

Psychrometric charts and calculators are based on a specific atmospheric pressure, usually a typical sea-level condition. Precise calculations of psychrometric variables require an adjustment for barometric pressures different from those listed on a standard chart. Consult the ASHRAE Handbook listed in the references for more information on this. Most field measurements do not require adjustment for pressure.

## Effect of Psychrometric Variables on Perishable Commodities

### Temperature

Air temperature is the most important variable because it tends to control the flesh temperature of perishable commodities. All perishables have an optimum range of storage temperatures. Above the optimum, they respire at unacceptably high rates and are more susceptible to ethylene and disease damage. In fact, horticultural commodities respire at rates that double, triple, or even quadruple for every 10°C (18°F) increase in temperature. Temperatures below the optimum result in freezing or chilling damage. Accurate control of temperature is vital in maintaining maximum shelf life.

## Vapor pressure

The rate of moisture loss from a perishable is primarily controlled by the difference in water vapor pressure between the air in the intercellular spaces of plant material and the air surrounding it. The air in fresh plant material is nearly saturated or, in other words, is close to 100 percent relative humidity. The water vapor pressure of this air is therefore determined solely by the temperature of the plant material. From the psychrometric chart it is apparent that low temperatures result in low internal water vapor pressures and high temperatures cause high internal water vapor pressure.

Consider several examples of how the drying of perishables is influenced by water vapor pressure differences. An apple precooled to 0°C (32°F) and placed in a refrigerated room with saturated air at 0°C (32°F) would not lose moisture because the water vapor pressures of the air in the apple and of the surrounding air are the same. Table 10.1 lists water vapor pressures for the example cases. However, if the apple were at 20°C (68°F) because it was not precooled before being placed in the refrigerator, the air in the apple would have a high water vapor pressure compared to the refrigerated air, causing the apple to dry. If the apple were precooled to 0°C (32°F) but the refrigerated air were at 70 percent relative humidity, drying would occur because the refrigerated air is at a lower water vapor pressure than the nearly saturated air in the apple. However, the rate of moisture loss is much greater when the apple is not precooled than when the apple is at storage temperature but the storage room air is not saturated. The difference in water vapor pressure between the air in the apple and the storage air is over nine times more when the apple is not precooled than when it is cooled and put in unsaturated storage air.

Drying is reduced by decreasing the difference in water vapor pressure between the air in the perishable commodity and the air surrounding it. Both temperature of the commodity and humidity ratio in the surrounding air must be controlled.

**Table 10.1 Water vapor pressure of various storage air conditions and product temperatures**

| Variables | | Water vapor pressure (kPa) |
|---|---|---|
| Room air at: | 0°C, 100% rh | 0.61 |
| | 0°C, 70% rh | 0.43 |
| Fresh product* at: | 0°C (32°F) | 0.61 |
| | 20°C (68°F) | 2.34 |

*Assumes air in product is saturated.

## Other factors

**Relative humidity.** Relative humidity is a commonly used term for describing the humidity of the air but is not particularly meaningful without knowing the dry-bulb temperature of the air. Together these two variables allow the determination of water vapor pressure, which is a better index of the potential for desiccation.

**Dew point temperature.** Condensation of liquid water on perishables can be a factor in causing disease problems. If a commodity is cooled to a temperature below the dew point temperature of the outside air and brought out of the cold room, condensation forms. Condensation can also occur in storage if air temperature fluctuates too greatly.

## Measuring Psychrometric Variables

All psychrometric properties of air are determined by measuring two psychrometric variables (three, if barometric pressure is considered). For example, if wet- and dry-bulb temperatures are measured, then relative humidity, vapor pressure, dew point, and so on can be determined with the aid of a psychrometric chart. Many variables can be measured to determine the psychrometric state of air, but dry-bulb temperature, wet-bulb temperature, dew point temperature, and relative humidity are most commonly measured.

### Dry-bulb temperature

Dry-bulb temperature can be simply and inexpensively measured by a mercury-in-glass thermometer. The thermometer should be marked in divisions of at most 0.2°C or 0.5°F if it is used in conjunction with a wet-bulb thermometer for determining cold storage air conditions. The thermometer should be shielded from radiant heat sources such as motors, lights, external walls, and people. This can be done by placing the thermometer where it cannot "see" the warm object or by protecting it with a radiant heat shield assembly.

Hand-held thermistor, resistance bulb, or thermocouple thermometers can also be used. They are more expensive than a mercury-in-glass thermometer but are not necessarily more accurate. These instruments can be purchased with a sharp probe allowing them to be used for measuring product pulp temperature. Inexpensive alcohol-in-glass and bimetallic dial thermometers are not recommended unless their calibrations have been checked against a calibrated thermometer. In field situations, an ice-water mixture is an easy way to check calibration at 0°C (32°F).

### Wet-bulb temperature

Use of a wet-bulb thermometer in conjunction with a dry-bulb thermometer is a common method for determining the state point on the psychrometric chart. The wet-bulb thermometer is basically an ordinary glass thermometer (although electronic temperature sensing elements can also be used) with a wetted cotton wick secured around the mercury bulb. Air is forced over the wick, causing it to cool to the wet-bulb temperature. The wet- and dry-bulb temperatures together determine the state point of the air on the psychrometric chart, allowing all other variables to be determined.

An accurate wet-bulb temperature reading depends on (1) sensitivity and accuracy of the thermometer, (2) maintenance of an adequate air speed past the wick, (3) shielding of the thermometer from radiation, (4) use of distilled or deionized water to wet the wick, and (5) use of a cotton wick.

The thermometer sensitivity required to determine an accurate humidity varies according to the temperature range of the air. More sensitivity is needed at low than at high temperatures. For example, at 65°C a 0.5°C error in wet-bulb temperature reading results in a 2.6 percent error in relative humidity determination, but at 0°C that same error results in a 10.5 percent error in relative humidity. In most cases, absolute calibration of the wet- and dry-bulb thermometer is not as important as ensuring that they read the same at a given temperature. For example, if both thermometers read 0.5°C low, this will result in less than a 1.3 percent error in relative humidity at dry-bulb temperatures between 65°C and 0°C (at a 5°C difference between dry- and wet-bulb temperatures). Before wetting the wick of the wet-bulb thermometer, operate both thermometers long enough to determine if there is any difference between their readings. If there is a difference, assume that one is correct and adjust the reading of the other accordingly when determining relative humidity.

The rate of evaporation from the wick is a function of air speed past it. A minimum air speed of about 3 meters per second (500 feet per minute) is required for accurate readings. An air speed much below this will result in an erroneously high wet-bulb reading. Wet-bulb devices that do not provide a guaranteed air flow cannot be relied on to give an accurate reading.

As with the dry-bulb thermometer, sources of radiant heat such as motors, lights, and so on can affect the wet-bulb thermometer. The reading must be taken in an area protected from these sources of radiation or thermometers must be shielded from radiant energy.

A buildup of salts from impure water or contaminants in the air affects the rate of water evaporation from the wick and results in erroneous data. Distilled or deionized water should be used to moisten the wick and the wick should be replaced if there is any sign of contamination. The wick material should not have been treated with chemicals such as sizing compounds that affect the water evaporation rate.

Special care must be taken when using a wet-bulb thermometer when the wet-bulb temperature is near freezing. Most humidity tables and calculators are based on a frozen wick at wet-bulb temperatures below 0°C (32°F). At temperatures below 0°C, touch the wick with a piece of clean ice or another cold object to induce freezing, because distilled water can be cooled below 0°C without freezing. The psychrometric chart or calculator must use frost-bulb, not wet-bulb temperatures, below 0°C to be accurate with this method.

Under most conditions wet-bulb temperature data are not reliable when the relative humidity is below 20 percent or the wet-bulb temperature is above 100°C (212°F). At low humidities, the wet-bulb temperature is much lower than the dry-bulb temperature and it is difficult for the wet-bulb thermometer to be cooled completely because of heat transferred by the glass or metal stem. Water boils above 100°C (212°F), so wet-bulb temperatures above that cannot be measured with a wet-bulb thermometer.

In general, properly designed and operated wet- and dry-bulb psychrometers can operate with an accuracy of less than 2 percent of the actual relative humidity. Improper operation greatly increases the error.

### Relative humidity

Direct relative humidity measurement usually uses an electric sensing element or a mechanical system. Electric hygrometers are based on substances whose electrical properties change as a function of their moisture content. As the humidity of the air around the sensor increases, its moisture increases, proportionally affecting the sensor's electrical properties. These devices are more expensive than wet- and dry-bulb psychrometers, but their accuracy is not as severely affected by incorrect operation. An accuracy of less than 2 percent of the actual humidity is often obtainable. Sensors lose their calibration if allowed to become contaminated, and some lose calibration if water condenses on them. Most sensors have a limited life. Mechanical hygrometers usually employ human hairs as a sensing element. Hair changes in length in proportion to the humidity of the air. The response to changes in relative humidity is slow and is not dependable at very high relative humidities. These devices are acceptable as an indicator of a general range of humidity but are not suitable for accurate measurements.

### Dew point indicators

Two types of dew point sensors are commonly used today: a saturated salt system and a condensation dew point method. The saturated salt system operates at dew points between −12° and 37°C (10° to 100°F) with an error of less than ±1°C (2°F). The system costs less than the condensation system, is not significantly affected by contaminating ions, and has a response time of about 4 minutes. The condensation type is very accurate over a wide range of dew point temperatures (less than ±0.5°C from −73° to

100°C or less than ±1°F from −100° to 212°F). A condensation dew point hygrometer can be expensive.

Nondispersive infrared gas analyzers can be set up to measure water vapor. These can provide very accurate air moisture determinations in laboratory situations.

## REFERENCES

1. ASHRAE. 1986. *ASHRAE handbook—fundamentals*. Am. Soc. Heating, Refrigeration and Air Conditioning Engineers. Atlanta, GA.

2. Gaffney, J. J. 1978. Humidity: Basic principles and measurement techniques. *HortScience* 13(5): 551-55.

3. Wexler, A., and W. G. Brombacher. 1951. *Methods of measuring humidity and testing hygrometers*. National Bureau of Standards, Circ. 512.

## Sample Psychrometric Calculations

**1.** A wet-bulb thermometer reads 18°C and a dry-bulb thermometer reads 25°C. What is the relative humidity?

*Solution:* On figure 10.3 the diagonal 18°C wet-bulb (wb) line and the vertical 25°C dry-bulb (db) line intersect at point A. Point A falls on the 50 percent relative humidity (rh) line.

**2.** What is the dew point temperature of the air in problem 1?

*Solution:* If the air represented by point A is cooled without changing its moisture content, it will follow a horizontal line until it reaches 14°C. At that tem-perature, it has 100 percent relative humidity, and any further cooling will cause water to condense out of the air (dew forms). The dewpoint (dp) temperature is 14°C.

**3.** What is the humidity ratio and water vapor pressure of the air in problem 1?

*Solution:* Find the humidity ratio and water vapor pressure of the air represented by point A by reading horizontally across to the vertical axis of the psychrometric chart. The humidity ratio (w) is 0.01 kg/kg and the vapor pressure is 1.6 kPa.

**4.** If the air in problem 1 is passed through a 100 percent efficient evaporative cooler, what will be its temperature after it leaves the cooler?

*Solution:* Evaporative cooling (and spray humidification) follows the diagonal wet-bulb lines. As air passes through the cooler, it will move from point A along the 18°C wet-bulb line until it reaches 100 percent relative humidity. At this humidity, it is saturated, will not accept any more water vapor, and will stop cooling. It will leave the cooler at a temperature of 18°C.

**5.** When air represented by point A (db = 25°, wb = 18°) enters a storage room with a temperature of 0°C and a relative humidity of 95 percent, will it add moisture to the storage room or dry it out?

*Solution:* The air has a dew point temperature of 14°C (remember problem 2). When this air is cooled to just less than 14°C it will begin to lose water and

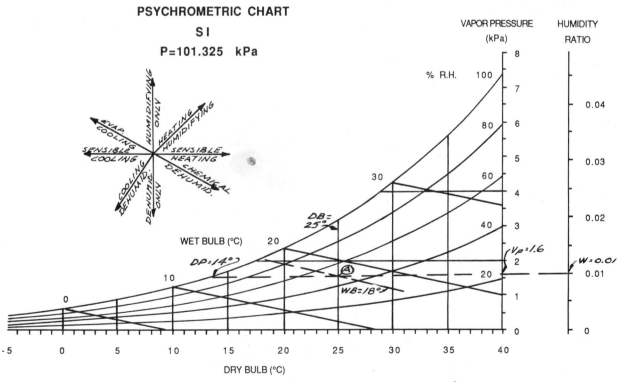

**Fig. 10.3. Example of how to use a psychrometric chart.**

will continue to lose water until it reaches the storage room temperature. If fact, each kilogram of air will lose about 0.006 kg of water as it cools. The air will add moisture to the storage room.

**6.** If air leaves a wet coil evaporator at 0°C and 100 percent relative humidity and heats 2°C before it reaches a stored product, what is the relative humidity of the air that the product is exposed to?

*Solution:* Sensible heating processes follow horizontal lines on the psychrometric chart. Air will leave the coil at point B and move horizontally to the right on the chart until it reaches 2°C. At that point, the relative humidity will be 83 percent.

# 11

# Modified Atmospheres during Transport and Storage

ADEL A. KADER

Modified atmospheres (MA) or controlled atmospheres (CA) mean removal or addition of gases resulting in an atmospheric composition around the commodity that is different from that of air (78.08 percent $N_2$, 20.95 percent $O_2$, 0.03 percent $CO_2$). Usually this involves reduction of oxygen ($O_2$) and/or elevation of carbon dioxide ($CO_2$) concentrations. MA and CA differ only in the degree of control; CA is more exact.

The use of modified or controlled atmospheres should be considered as a supplement to proper temperature and relative humidity management (fig. 11.1). The potential for benefit or hazard from using MA is dependent upon the commodity, variety, physiological age, atmospheric composition, and temperature and duration of storage. This helps explain the wide variability in results among published reports for a given commodity.

Continued efforts to develop MA or CA technology have permitted its increased use during transport, temporary storage, or long-term storage of horticultural commodities destined for fresh market or processing. A prestorage treatment with elevated $CO_2$ can also be used for some fruits. Carbon monoxide (CO) is used, to a limited extent, as an added component to MA for slowing down brown discoloration and controlling decay.

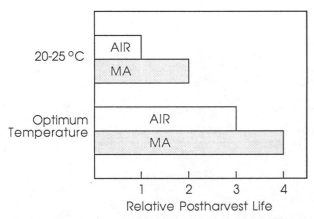

**Fig. 11.1. Relative postharvest life of a fresh commodity stored in air or in its optimum modified atmosphere (MA) at room temperature (20–25°C), or at its optimum temperature (near 0°C for nonchilling-sensitive commodities or 5–14°C for chilling-sensitive commodities).**

## Effects of Controlled Atmospheres

### Potential benefits

Used properly, MA or CA can supplement proper temperature management and can result in one or more of the following benefits, which translate into reduced quantitative and qualitative losses during postharvest handling and storage of some horticultural commodities.

**1.** Retardation of senescence (ripening) occurs, along with associated biochemical and physiological changes, i.e., slowed down respiration and ethylene production rates, softening, and compositional changes.

**2.** Reduction of fruit sensitivity to ethylene action occurs at $O_2$ levels below about 8 percent and/or $CO_2$ levels above 1 percent.

**3.** Alleviation of certain physiological disorders results such as chilling injury of various commodities, russet spotting in lettuce, and some storage disorders of apples.

**4.** Modified atmospheres can directly or indirectly affect postharvest pathogens and consequently decay incidence and severity. For example, elevated $CO_2$ levels (10 percent to 15 percent) significantly inhibit development of Botrytis rot on strawberries, cherries, and other fruits.

**5.** Modified atmospheres can be a useful tool for insect control in some commodities.

### Potential harmful effects

In most cases, the difference between beneficial and harmful MA combinations is relatively small. Also, MA combinations that are necessary to control decay or insects, for example, cannot always be tolerated by the commodity and may result in faster deterioration. Potential hazards of MA to the commodity include:

**1.** Initiation and/or aggravation of certain physiological disorders can occur, such as blackheart in potatoes, brown stain on lettuce, and brown heart in apples and pears.

**2.** Irregular ripening of fruits, such as banana, pear, and tomato, can result from $O_2$ levels below 2 percent or $CO_2$ levels above 5 percent.

**3.** Off-flavors and off-odors at very low $O_2$ concentrations may develop as a result of anaerobic respiration.

**4.** Susceptibility to decay may increase when the commodity is physiologically injured by too-low $O_2$ or too-high $CO_2$ concentrations.

**5.** Stimulation of sprouting and retardation of periderm development can occur in some root and tuber vegetables such as potatoes.

## CA/MA: Requirements and Recommendations

During the past 50 years, uses of CA and MA have increased steadily and have contributed significantly to extending the postharvest life and maintaining quality of several fruits and vegetables. This trend is expected to continue as technological advances are made in attaining and maintaining CA and MA during transport, storage, and marketing of fresh produce. Several refinements in CA storage include low $O_2$ (1.0 to 1.5 percent) storage, low ethylene CA storage, rapid CA (rapid establishment of the optimum levels of $O_2$ and $CO_2$), and programmed or sequential CA storage (e.g., storage in 1 percent $O_2$ for 2 to 6 weeks followed by storage in 2 to 3 percent $O_2$ for the remainder of the storage period). Other developments that may expand use of MA during transport and distribution include using edible coatings or polymeric films to create a desired MA within the commodity.

Fresh fruits and vegetables vary greatly in their relative tolerance to low $O_2$ concentrations (table 11.1) and elevated $CO_2$ concentrations (table 11.2). These are the levels beyond which physiological damage would be expected. These limits of tolerance can be different at temperatures above or below recommended temperatures for each commodity. Also, a given commodity may tolerate brief exposures to higher levels of $CO_2$ or lower levels of $O_2$ than those

**Table 11.1. Fruits and vegetables classified according to their tolerance to low $O_2$ concentrations**

| Minimum $O_2$ concentration tolerated (%) | Commodities |
|---|---|
| 0.5 | Tree nuts, dried fruits, and vegetables |
| 1.0 | Some cultivars of apples and pears, broccoli, mushroom, garlic, onion, most cut or sliced (minimally processed) fruits and vegetables |
| 2.0 | Most cultivars of apples and pears, kiwifruit, apricot, cherry, nectarine, peach, plum, strawberry, papaya, pineapple, olive, cantaloupe, sweet corn, green bean, celery, lettuce, cabbage, cauliflower, Brussels sprouts |
| 3.0 | Avocado, persimmon, tomato, pepper, cucumber, artichoke |
| 5.0 | Citrus fruits, green pea, asparagus, potato, sweet potato |

**Table 11.2. Fruits and vegetables classified according to their tolerance to elevated $CO_2$ concentrations**

| Maximum $CO_2$ concentration tolerated (%) | Commodities |
|---|---|
| 2 | Apple (Golden Delicious), Asian pear, European pear, apricot, grape, olive, tomato, pepper (sweet), lettuce, endive, Chinese cabbage, celery, artichoke, sweet potato |
| 5 | Apple (most cultivars), peach, nectarine, plum, orange, avocado, banana, mango, papaya, kiwifruit, cranberry, pea, pepper (chili), eggplant, cauliflower, cabbage, Brussels sprouts, radish, carrot |
| 10 | Grapefruit, lemon, lime, persimmon, pineapple, cucumber, summer squash, snap bean, okra, asparagus, broccoli, parsley, leek, green onion, dry onion, garlic, potato |
| 15 | Strawberry, raspberry, blackberry, blueberry, cherry, fig, cantaloupe, sweet corn, mushroom, spinach, kale, Swiss chard |

indicated. The limit of tolerance to low $O_2$ would be higher as storage temperature or duration increases because $O_2$ requirements for aerobic respiration of the tissue increase with higher temperatures. Depending on the commodity, damage associated with $CO_2$ may either increase or decrease with an increase in temperature. $CO_2$ production increases with temperature, but its solubility decreases; thus, $CO_2$ in the tissue can be increased or decreased by an increase in temperature. Further, the physiological effect of $CO_2$ could be temperature dependent. Tolerance limits to elevated $CO_2$ decrease with a reduction in $O_2$ level, and similarly the tolerance limits to reduced $O_2$ increase with the increase in $CO_2$ level.

Current MA/CA recommendations are summarized in table 11.3 (fruits) and table 11.4 (vegetables). Also included is an estimate of potential benefits and extent of current commercial use. There is no doubt that some of these MA combinations will change as more research is completed. The possibility of adding carbon monoxide to MA/CA for some commodities may change its potential for benefit. Hypobaric or low-pressure systems may also provide new opportunities for making CA a more useful treatment for some commodities.

Current CA use for long-term storage of fresh fruits and vegetables is summarized in table 11.5. Its use on nuts and dried commodities (for insect control and quality maintenance, including prevention of rancidity) is increasing as it provides an excellent substitute for chemical fumigants (such as methyl bromide) used for insect control. Also the use of CA on commodities listed in table 11.5 other than apples and pears is expected to increase as international market demands for year-round availability of various commodities expand.

CA/MA use for short-term storage and transport of fresh horticultural crops (table 11.6) will continue to increase supported by technological developments

**Table 11.3. Summary of recommended CA or MA conditions during transport and/or storage of selected fruits**

| Commodity | Temp. range* (C°) | CA†<br>% O₂ | CA†<br>% CO₂ | Potential for benefit‡ | Remarks§ |
|---|---|---|---|---|---|
| *Deciduous tree fruits* | | | | | |
| Apple | 0–5 | 1–3 | 1–5 | A | About 50% of production is stored under CA |
| Apricot | 0–5 | 2–3 | 2–3 | C | No commercial use |
| Cherry, sweet | 0–5 | 3–10 | 10–15 | B | Some commercial use |
| Fig | 0–5 | 5–10 | 15–20 | B | Limited commercial use |
| Grape | 0–5 | 2–5 | 1–3 | C | Incompatible with SO₂ fumigation |
| Kiwifruit | 0–5 | 1–2 | 3–5 | A | Some commercial use; C₂H₄ must be maintained below 20 ppb |
| Nectarine | 0–5 | 1–2 | 3–5 | B | Limited commercial use |
| Peach | 0–5 | 1–2 | 3–5 | B | Limited commercial use |
| Pear, Asian | 0–5 | 2–4 | 0–1 | B | Limited commercial use |
| Pear, European | 0–5 | 1–3 | 0–3 | A | Some commercial use |
| Persimmon | 0–5 | 3–5 | 5–8 | B | Limited commercial use |
| Plum and prune | 0–5 | 1–2 | 0–5 | B | Limited commercial use |
| Raspberry and other cane berries | 0–5 | 5–10 | 15–20 | A | Increasing use during transport |
| Strawberry | 0–5 | 5–10 | 15–20 | A | Increasing use during transport |
| Nuts and dried fruits | 0–25 | 0–1 | 0–100 | A | Effective insect control method |
| *Subtropical and tropical fruits* | | | | | |
| Avocado | 5–13 | 2–5 | 3–10 | B | Limited commercial use |
| Banana | 12–15 | 2–5 | 2–5 | A | Some commercial use during transport |
| Grapefruit | 10–15 | 3–10 | 5–10 | C | No commercial use |
| Lemon | 10–15 | 5–10 | 0–10 | B | No commercial use |
| Lime | 10–15 | 5–10 | 0–10 | B | No commercial use |
| Olive | 5–10 | 2–3 | 0–1 | C | No commercial use |
| Orange | 5–10 | 5–10 | 0–5 | C | No commercial use |
| Mango | 10–15 | 3–5 | 5–10 | C | Limited commercial use |
| Papaya | 10–15 | 3–5 | 5–10 | C | No commercial use |
| Pineapple | 8–13 | 2–5 | 5–10 | C | No commercial use |

*Usual and/or recommended range. A relative humidity of 90% to 95% is recommended.
†Best CA combination may vary among cultivars and according to storage temperature and duration.
‡A = excellent, B = good, C = fair.
§Comments about use refer to domestic marketing only; many of these commodities are shipped under MA for export marketing.

in transport containers, MA packaging, and edible coatings. Carbon monoxide at 5 to 10 percent, added to O₂ levels below 5 percent, is an effective fungistat that can be used for decay control on commodities that do not tolerate 15 percent to 20 percent CO₂. However, CO is very toxic to humans, and special precautions must be taken.

The major limitation to long-term storage of many cut flowers is pathological breakdown due to infection with *Botrytis cinerea* (gray mold). Atmospheres containing sufficient CO₂ to reduce fungal attack cause severe bronzing of the foliage of some cultivars. The use of CO as a fungistat is limited by its ethylene-mimicking effects. It is not yet possible to identify the best MA/CA combination for each species of ornamentals because of insufficient or inconclusive data.

CA/MA conditions, including MA packaging (MAP), can replace certain postharvest chemicals used for control of some physiological disorders, such as scald on apples. Proper use of CA can also eliminate the need for using daminozide on apples. Furthermore, some postharvest fungicides and insecticides can be reduced or eliminated where CA/MA

provides adequate control of postharvest pathogens or insects.

CA/MA may facilitate picking and marketing more mature (better flavor) fruits by slowing their postharvest deterioration to permit transport and distribution. Another potential use for CA/MA is in maintaining quality and safety of minimally processed fruits and vegetables, which are increasingly being marketed as value-added, convenience products.

The residual effects of CA/MA on fresh commodities after transfer to air (during marketing) may include reduction of respiration and ethylene production rates, maintenance of color and firmness, and delayed decay. Generally, the lower the concentration of O₂ and the higher the concentration of CO₂ (within the tolerance limits of the commodity), and the longer the exposure to CA/MA conditions, the more prominent are the residual effects.

## Carbon Monoxide as a Supplement

Beginning in 1970, carbon monoxide (CO) at 2 percent to 3 percent has been used as a supplement to MA during transit of lettuce to inhibit discoloration.

**Table 11.4. Summary of recommended CA or MA conditions during transport and/or storage of selected vegetables**

| Commodity | Temp. range* (C°) | CA† % O₂ | CA† % CO₂ | Potential for benefit‡ | Remarks§ |
|---|---|---|---|---|---|
| Artichokes | 0–5 | 2–3 | 2–3 | B | No commercial use |
| Asparagus | 0–5 | air | 5–10 | A | Limited commercial use |
| Beans, snap | 5–10 | 2–3 | 4–7 | C | Potential for use by processors |
| Beets | 0–5 | None | | D | 98–100% rh is best |
| Broccoli | 0–5 | 1–2 | 5–10 | A | Limited commercial use |
| Brussels sprouts | 0–5 | 1–2 | 5–7 | B | No commercial use |
| Cabbage | 0–5 | 2–3 | 3–6 | A | Some commercial use for long-term storage of certain cultivars |
| Cantaloupes | 3–7 | 3–5 | 10–15 | B | Limited commercial use |
| Carrots | 0–5 | None | | D | 98–100% rh is best |
| Cauliflower | 0–5 | 2–3 | 2–5 | C | No commercial use |
| Celery | 0–5 | 1–4 | 0–5 | B | Limited commercial use in mixed loads with lettuce |
| Corn, sweet | 0–5 | 2–4 | 5–10 | B | Limited commercial use |
| Cucumbers | 8–12 | 3–5 | 0 | C | No commercial use |
| Honeydews | 10–12 | 3–5 | 0 | C | No commercial use |
| Leeks | 0–5 | 1–2 | 3–5 | B | No commercial use |
| Lettuce | 0–5 | 1–3 | 0 | B | Some commercial use with 2–3% CO added |
| Mushrooms | 0–5 | air | 10–15 | C | Limited commercial use |
| Okra | 8–12 | 3–5 | 0 | C | No commercial use; 5–10% CO₂ is beneficial at 5–8°C |
| Onions, dry | 0–5 | 1–2 | 0–5 | B | No commercial use; 75% rh |
| Onions, green | 0–5 | 1–2 | 10–20 | C | Limited commercial use |
| Peppers, bell | 8–12 | 3–5 | 0 | C | Limited commercial use |
| Peppers, chili | 8–12 | 3–5 | 0 | C | No commercial use; 10–15% CO₂ is beneficial at 5–8°C |
| Potatoes | 4–12 | None | | D | No commercial use |
| Radish | 0–5 | None | | D | 98–100% rh is best |
| Spinach | 0–5 | air | 10–20 | B | No commercial use |
| Tomatoes, mature-green | 12–20 | 3–5 | 0–3 | B | Limited commercial use |
| partially ripe | 8–12 | 3–5 | 0–5 | B | Limited commercial use |

*Usual and/or recommended range. A relative humidity of 90% to 98% is recommended unless otherwise indicated under "Remarks."
†Best CA combination may vary among cultivars and according to storage temperature and duration.
‡A = excellent, B = good, C = fair, D = slight or none.
§Comments about use refer to domestic marketing only; many of these commodities are shipped under MA for export marketing.

**Table 11.5. Summary of CA use for long-term storage of fresh fruits and vegetables**

| Storage duration (months) | Commodities |
|---|---|
| More than 12 | Almond, filbert, macadamia, pecan, pistachio, walnut, dried fruits and vegetables |
| 6–12 | Some cultivars of apples and European pears |
| 3–6 | Cabbage, Chinese cabbage, kiwifruit, some cultivars of Asian pears |
| 1–3 | Avocado; olive; some peach, nectarine, and plum cultivars; persimmon; pomegranate |

**Table 11.6. Summary of CA/MA use for short-term storage and/or transport of fresh horticultural crops**

| Primary benefit of CA/MA | Commodities |
|---|---|
| Delay of ripening and avoiding chilling temperatures | Avocado, banana, mango, melons, nectarine, papaya, peach, plum, tomato (picked mature-green or partially ripe) |
| Control of decay | Blackberry, blueberry, cherry, fig, grape, raspberry, strawberry |
| Delay of senescence and undesirable compositional changes (including tissue brown discoloration) | Asparagus, broccoli, lettuce, sweet corn, fresh herbs, minimally processed fruits and vegetables |

Some additional benefits of CO are now known. These benefits as well as possible hazards of CO are summarized below.

## Beneficial effects

**1.** Carbon monoxide (1 to 5 percent) added to reduced (2 to 5 percent) O₂ atmospheres inhibits discoloration of lettuce butts and mechanically damaged tissue. Similar effects have been observed on other commodities, including lightly processed (cut, sliced, etc.) fruits and vegetables. This inhibition of discoloration is lost when the commodity is moved from MA to air during destination marketing.

**2.** Carbon monoxide (5 to 10 percent) added to MA has been found to inhibit growth of several important postharvest pathogens and to prevent decay development on several fruits and vegetables. The fungistatic effects of CO are maximized at $O_2$ levels below 5 percent.

**3.** Although CO alone was not found to be an effective fumigant for insect control in harvested lettuce, its possible use with other CA combinations merits further study.

### Possible hazards

**1.** Carbon monoxide may aggravate certain physiological disorders. For example, in a situation where $CO_2$ accumulates above 2 percent during transit of lettuce, CO increases the severity of brown stain (a carbon dioxide-induced disorder).

**2.** Carbon monoxide mimics ethylene ($C_2H_4$) effects such as enhancing ripening and inducing certain physiological disorders. However, when CO is used in combination with reduced $O_2$ or elevated $CO_2$, such effects are minimized to insignificance except for commodities that are extremely $C_2H_4$-sensitive, such as kiwifruit.

**3.** Because of its extreme toxicity to humans and flammability at concentrations between 12.5 percent and 74.2 percent in air, strict safety measures should be followed if and when CO is used.

## Prestorage Treatments with Elevated Carbon Dioxide

Tests conducted at several Experiment Stations indicated that treating apples for 2 weeks or pears for 2 to 4 weeks with 12 percent $CO_2$ at 0°C to 5°C (32°F to 41°F) before CA storage delayed fruit softening. However, this treatment resulted in varying amounts of internal and external $CO_2$ injury depending on variety, season, and production area. Its commercial application is currently limited to some Golden Delicious apples in the U.S. Northwest.

Elevated $CO_2$ treatments have also been shown to alleviate chilling injury symptoms on some subtropical and tropical fruits, but this treatment is not recommended yet for commercial application.

## Ethylene Removal in Modified Atmosphere Storage

Most researchers have assumed that removing ethylene from MA storage rooms is not necessary because its effects on fruit ripening at 0°C to 5°C (32°F to 41°F) and under MA conditions are negligible. However, the presence of ethylene at concentrations likely to occur in MA and CA rooms can enhance fruit softening during long-term storage. Thus, ethylene removal is recommended in long-term CA storage of apples and pears. It is particularly important for storage of kiwifruit, carnations, or other commodities that are extremely sensitive to ethylene.

Further research is needed to evaluate the effects of ethylene and nonethylenic volatiles on other commodities under MA conditions. Also, there is a need for effective and economical methods for removing ethylene and other volatiles from MA storage rooms.

## Atmospheric Modification

### Atmosphere generators

**Oxygen control.** The oxygen level can be controlled by recirculating air from the CA room into the generator and back into the room, or by a purge system where fresh air has its oxygen reduced in the generator, then fed into the room. Catalytic burners or converters, used in the past, have been replaced by purging with nitrogen obtained from liquid nitrogen or from separators that circulate air through molecular sieve beds or membrane systems to separate the nitrogen.

**Carbon dioxide control.** Addition of $CO_2$ is usually from pressurized gas cylinders. Dry ice is sometimes used as a source of $CO_2$ during transport. Carbon dioxide is reduced by scrubbing methods that use sodium hydroxide, water, activated charcoal, hydrated lime (Ca[OH]$_2$), or a molecular sieve. The most common scrubber uses brine, which is pumped over the evaporator coil where it absorbs $CO_2$. A lime box is also commonly used adjacent to the CA room with a circulating system to pass the room atmosphere through it. The box is usually sized to hold about 12 kg lime per ton of fruit, and the spent lime is replaced with fresh lime. The amount of lime needed to absorb $CO_2$ may be placed inside the CA room. Several types of activated carbon scrubbers are currently used.

**Carbon monoxide addition.** Carbon monoxide can be added from pressurized gas cylinders, blending it with nitrogen to avoid exceeding 10 percent CO.

**Ethylene removal.** A few methods can be used to remove ethylene from cold storage facilities. Ventilation (one air exchange per hour) to reduce $C_2H_4$ concentration cannot be used in CA storage. Also, the use of ozone to oxidize $C_2H_4$ requires $O_2$ levels above those available in CA storage. Use of ethylene absorbers, such as potassium permanganate alone or in combination with activated and brominated charcoal, can be effective in CA storage facilities provided that the air is circulated through these materials, which must be replaced when spent. Use of catalytic burners for ethylene removal appears promising.

### Hypobaric or low-pressure system (LPS)

Reducing the total pressure (under partial vacuum conditions) results in reducing the partial pressures of individual gases in air. This can be an effective method for reducing $O_2$ tension, and for accelerating the escape of $C_2H_4$ and other volatiles. LPS has the advantages over other methods of atmospheric modification of (1) more exact control of $O_2$ concentrations that permit the use of lower $O_2$ tensions than is possible with CA, and (2) removal of ethylene and other volatiles.

However, LPS has limitations when $CO_2$ or CO addition is important for a given commodity. Transit vehicles with LPS were tested to a limited extent on a commercial scale for transporting some animal and plant products. Stationary storage structures with LPS were being developed by Grumman Dormavac before it terminated its efforts related to hypobaric storage.

### Commodity-generated modified atmosphere

In some cases the commodity itself, through respiration, is used to reduce $O_2$ and increase $CO_2$ with restricted air-exchange conditions and barriers, as shown in figure 11.2. If elevated $CO_2$ is not desirable, scrubbers are used. Restricted air exchange may be achieved by use of airtight cold storage rooms, packaging in film wraps or bags, use of polyethylene liners in shipping containers, use of pallet shrouds (plastic

covers), manipulation of shipping container vents, application of waxes and other surface coatings, or use of plastic covers with diffusion windows (polymeric membranes). Atmospheric modification by these methods is usually slow, and much of the benefit of MA may be lost.

### Atmospheres during transit

**Modified atmospheres in rail cars, trucks, and seavans.** Gastight transit vehicles, essential to the maintenance of MA during transit, are a limiting factor to expanded use of MA. The Tectrol system (TransFresh Corp.), used in rail cars and marine containers, is based on (1) reduced $O_2$ achieved by $N_2$ flushing, (2) $CO_2$ and/or CO added using gas blending manifolds, (3) $CO_2$ removal by placing bags of fresh hydrated lime in the transit vehicle, and (4) breather bags to compensate for barometric pressure fluctuations.

Some systems carry a tank of liquid $N_2$ along with the container van. The van is equipped with an $O_2$ sensor for controlling $N_2$ release or introducing fresh

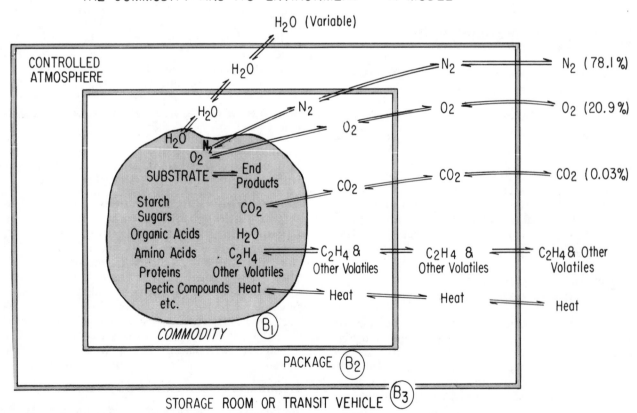

Fig. 11.2. Types of barriers that can be used to establish a modified atmosphere.
$B_1$—Natural epidermis, skin, peel, or rind; wax coating, film wrap.
$B_2$—Package—Wood, paperboard, plastic (may include additional liner in package)
$B_3$—Storage room wall or vehicle wall, may be sealed against gas exchange.
Additional barriers may include—consumer packages inside the master package and pallet covers over several packages.

air as needed to maintain the desired $O_2$ concentration. Scrubbers for $CO_2$ or $C_2H_4$ may be used in some container vans. There are efforts to use small nitrogen separators in container vans.

**Modified atmosphere in a pallet cover.**Polyethylene pallet covers (or shrouds) are used to cover all shipping containers on a pallet and are sealed by various means (tape, heat seal, and so on) onto a plastic sheet placed on the wooden pallet base (fig 11.3). A partial vacuum is established within the pallet cover and the desired gas mixture is introduced. This method is in common use on strawberries and increasing use on cane berries, cherries, figs, and other commodities. It can facilitate mixing of commodities that require different MA conditions during transit at the same temperature. Potential problems are primarily related to loss of the seal due to tearing of the pallet cover or an imperfect seal at the base.

**Modified atmospheres in individual shipping containers.**Examples of commercial use of commodity-generated MA during transit include polyethylene liners in cherry boxes, polyethylene bags for bananas destined for distant markets (Banavac system), and for cut lettuce and other vegetables. Lightly processed (shredded or chopped) lettuce may be packaged into 5-mil plastic bags, then partial vacuum is established and a gas mixture (30 to 50 percent $O_2$ + 4 to 6 percent CO) is introduced into the bag, which is then sealed. This procedure is currently in limited commercial use.

## Modified atmosphere packaging (MAP)

Modified atmospheres can be created either passively by the commodity or intentionally, as described below.

**Commodity-generated or passive MA.**If commodity and film permeability characteristics are properly matched, an appropriate atmosphere can passively evolve within a sealed package through consumption of $O_2$ and production of $CO_2$ by respiration. The gas permeability of the selected film must allow $O_2$ to enter the package at a rate offset by the consumption of $O_2$ by the commodity. Similarly, $CO_2$ must be vented from the package to offset the production of $CO_2$ by the commodity. Furthermore, this atmosphere must be established rapidly and without creating anoxic conditions or injuriously high levels of $CO_2$.

**Active modified atmosphere.**Because of the limited ability to regulate a passively established atmosphere, it is likely that atmospheres within MAP will be actively established and adjusted. This can be done by pulling a slight vacuum and replacing the package atmosphere with the desired gas mixture. This mixture can be further adjusted through the use of absorbing or adsorbing substances in the package to scavenge $O_2$, $CO_2$, or $C_2H_4$.

Although active modification implies some additional costs, its main advantage is that it ensures the rapid establishment of the desired atmosphere. In addition, ethylene adsorbers can help delay the climacteric rise in respiration for some fruits. Carbon dioxide absorbers can prevent the buildup of $CO_2$ to injurious levels, which can occur for some commodities during passive modification of the package atmosphere.

Many plastic films are available for packaging, but relatively few have been used to wrap fresh produce, and fewer have gas permeabilities that make them suitable to use for MAP. Because $O_2$ content in a MA package is typically reduced from an ambient 21 percent to 2 to 5 percent, there is a danger that $CO_2$ will increase from ambient 0.03 percent to 16 to 19 percent in the package. This is because there is normally a one-to-one correspondence between $O_2$ consumed and $CO_2$ produced. Because such high levels of $CO_2$ would be injurious to most fruits and vegetables, an ideal film must let more $CO_2$ exit than it lets $O_2$ enter. The $CO_2$ permeability should be about 3 to 5 times the oxygen permeability, depending upon the desired atmosphere. Several polymers used in film formulation meet this criterion (table 11.7). Low-density polyethylene and polyvinyl chloride are the main films used in packaging fruits and vegeta-

**Fig. 11.3. A pallet of strawberries and raspberries covered with 5-mil polyethylene pallet cover and ready for introduction of air enriched with 15 to 20 percent $CO_2$ as a fungistat.**

**Table 11.7. Permeabilities of films available for packaging fresh produce**

| Film type | Permeabilities (cc/m²/mil/day at 1 atm) | | $CO_2:O_2$ Ratio |
|---|---|---|---|
| | $CO_2$ | $O_2$ | |
| Polyethylene: low density | 7,700–77,000 | 3,900–13,000 | 2.0–5.9 |
| Polyvinyl chloride | 4,263–8,138 | 620–2,248 | 3.6–6.9 |
| Polypropylene | 7,700–21,000 | 1,300–6,400 | 3.3–5.9 |
| Polystrene | 10,000–26,000 | 2,600–7,700 | 3.4–3.8 |
| Saran | 52–150 | 8–26 | 5.8–6.5 |
| Polyester | 180–390 | 52–130 | 3.0–3.5 |

bles. Polystyrene has been used, but Saran and polyester have such low gas permeabilities that they would be suitable only for commodities with very low respiration rates.

## Monitoring atmospheric composition

Accurate monitoring of $O_2$ and $CO_2$ concentrations is essential to successful CA or MA storage. It is required for certification of CA storage in some states and by some insurance companies. The various methods of gas sampling and analysis are discussed in chapter 12.

## REFERENCES

1. Blankenship, S. M., ed. 1985. *Controlled atmospheres for storage and transport of perishable agricultural commodities. Proc. Fourth Natl. Controlled Atmos. Res. Conf.* July 23-26, 1985, Dept. of Hortic., Univ. of North Carolina, Raleigh, NC. 512 pp.

2. Brecht, P. E. 1980. Use of controlled atmospheres to retard deterioration of produce. *Food Technol.* 34(3):45-50.

3. Brody, A. L., ed. 1989. *Controlled/modified atmosphere/vacuum packaging of foods.* Trumbull, CT: Food & Nutrition Press. 179 pp.

4. Burton, W. G. 1974. Some biophysical principles underlying the controlled atmosphere storage of plant material. *Ann. Appl. Biol.* 78:149-68.

5. Calderon, M., and R. Barkai-Golan, eds. 1990. *Food preservation by modified atmospheres.* Boca Raton, FL: CRC Press. 402 pp.

6. Dalrymple, D. G. 1967. *The development of controlled atmosphere storage of fruits.* U.S. Dept. Agric. Div. Mktg. Utiliz. Sci. 56 pp.

7. Dewey, D. H., ed. 1977. *Controlled atmospheres for the storage and transport of perishable agricultural commodities. Proc. 2nd Natl. CA Res. Conf.* April 1977. Mich. State Univ. Dept. Hortic. Rept. 28. 301 pp.

8. Dewey, D. H., R. C. Herner, and D. R. Dilley, eds. 1969. *Controlled atmospheres for the storage and transport of horticultural crops. Proc. 2nd Natl. CA Res. Conf.* January 1969. Mich. State Univ. Dept. Hortic. Rept. 9. 155 pp.

9. El-Goorani, M. A., and N. F. Sommer. 1981. Effects of modified atmospheres on postharvest pathogens of fruits and vegetables. *Hortic. Rev.* 3:412-61.

10. Fellman, J.K., ed. 1989. *Proceedings of the fifth international controlled atmosphere research conference. June 14-16, 1989, Wash-ington State Univ. Research and Extension Center, Wenatchee, WA).* Vol. 1, 515 pp.; vol. 2, 374 pp.

11. Isenberg, F. M. R. 1979. Controlled atmosphere storage of vegetables. *Hortic. Rev.* 1:337-94.

12. Jamison, W. 1980. Use of hypobaric conditions for refrigerated storage of meats, fruits, and vegetables. *Food Technol.* 34(3):64-71.

13. Kader, A. A. 1980. Prevention of ripening in fruits by use of controlled atmospheres. *Food Technol.* 34(3):51-54.

14. ———. 1985. *Modified atmospheres: an indexed reference list with emphasis on horticultural commodities, supplement 4* (Jan. 1981-Mar. 1985). Univ. Calif., Davis Postharvest Hortic. Ser. 3. 31 pp. (391 titles).

15. ———. 1986. Biochemical and physiological basis for effects of controlled and modified atmospheres on fruits and vegetables. *Food Technol.* 40(5):99-100, 102-104.

16. Kader, A. A., and L. L. Morris. 1977a. *Modified atmospheres: an indexed reference list with emphasis on horticultural commodities, supplement 2* (May 1974 - Feb. 1977). Univ. Calif. Davis, Veg. Crops Ser. 187. 28 pp. (386 titles).

17. ———. 1977b. Relative tolerance of fruits and vegetables to elevated $CO_2$ and reduced $O_2$ levels. Mich. State Univ. *Hortic. Rept.* 28:260-65.

18. ———. 1981. *Modified atmospheres: an indexed reference list with emphasis on horticultural commodities, supplement 3* (Mar. 1977 - Dec. 1980). Univ. Calif. Davis, Veg. Crops Ser. 213. 36 pp. (467 titles).

19. Kader, A. A., D. Zagory, and E. L. Kerbel. 1989. Modified atmosphere packaging of fruits and vegetables. *CRC Crit. Rev. Food Sci. Nutr.* 28:1-30.

20. Lipton, W. J. 1975. Controlled atmospheres for fresh vegetables and fruits: why and when. In *Postharvest biology and handling of fruits and vegetables,* ed. N. F. Haard and D. K. Salunkhe, 130-43. Westport, CT: AVI Publ. Co.

21. Lougheed, E. C. 1987. Interactions of oxygen, carbon dioxide, temperature, and ethylene that may induce injuries in vegetables. *HortScience* 22:791-94.

22. Lougheed, E. C., D. P. Murr, and L. Berard. 1978. Low pressure storage for horticultural crops. *HortScience* 13:21-27.

23. Morris, L. L., L. L. Claypool, and D. P. Murr. 1971. *Modified atmospheres: an indexed reference list through 1959, with emphasis on horticultural commodities.* Univ. Calif. Div. Agric. Sci. 115 pp. (2326 titles).

24. Murr, D. P., A. A. Kader, and L. L. Morris. 1974. *Modified atmospheres: an indexed reference list with emphasis on horticultural commodities, supplement 1* (Jan. 1970–Apr. 1974). Univ. Calif. Davis, Veg. Crops Ser. 168. 39 pp. (395 titles).

25. Richardson, D. G., and M. Meheriuk, eds. 1982. *Controlled atmospheres for storage and transport of perishable agricultural commodities. Proc. 3rd Natl. CA Res. Conf.* July 1981. Beaverton, OR: Timber Press. 390 pp.

26. Smith, S., J. Geeson, and J. Stow. 1987. Production of modified atmospheres in deciduous fruits by the use of films and coatings. *HortScience* 22:772-76.

27. Smock, R. M. 1979. Controlled atmosphere storage of fruits. *Hortic. Rev.* 1:301-36.

28. Solomos, T. 1987. Principles of gas exchange in bulky plant tissues. *HortScience* 22:766-71.

29. Weichmann, J. 1986. The effect of controlled atmosphere storage on the sensory and nutritional quality of fruits and vegetables. *Hortic. Rev.* 8:101-27.

30. Zagory, D., and A. A. Kader. 1989. *Modified atmospheres: an indexed reference list with emphasis on horticultural commodities, supplement 5* (Apr. 1985 - May 1989). Univ. Calif. Davis, Postharvest Hort. Ser. 6. 49 pp. (561 titles).

# 12

# Methods of Gas Mixing, Sampling, and Analysis

Adel A. Kader

## Principles of Mixing Gas

In postharvest research and technology we are usually concerned with monitoring atmospheric composition and with mixing two or more of the following gases: air, nitrogen, oxygen, carbon dioxide, ethylene, and carbon monoxide. Procedures for gas mixing are based on mass, volume, or pressure relationships. Remember the following laws and definitions for gases:

1. **Avogadro's Law:** One mole of any compound contains $6.0228 \times 10^{23}$ molecules. This quantity of a gas will occupy 22.414 liters at standard temperature ($0°C = 273°K$) and pressure (760 mm Hg).

2. **Boyle's Law:** $V = K(1/P)$; $P_1 V_1 = P_2 V_2$
   where: V = volume (in liters)
   K = proportionality constant
   P = pressure (in atmospheres)

3. **Charles' Law:** $PV = KT$; thus: $\dfrac{P_1 V_1}{T_1} = \dfrac{P_2 V_2}{T_2}$
   where: P = pressure (in atmospheres)
   V = volume (in liters)
   T = temperature (°K)
   K = constant

4. **Ideal Gas Law:** If K is proportional to the number of moles of gas (n), then: $PV = nRT$
   where: R = molar gas constant
   P, V, and T = same as above

5. Density = $\dfrac{mass}{volume} = \dfrac{P \times M}{R \times T}$
   where: P = pressure
   M = molecular weight
   R = molar gas constant
   T = temperature

6. The rate of diffusion of a gas is inversely proportional to the square root of its density (**Graham's Law of Diffusion**).

7. **Dalton's Law of Partial Pressures:** the total pressure of a mixture of gases is the sum of the partial pressures of the component gases.

8. Concentration ($C_a$), in percent (ml/100 ml) or ppm ($\mu$l/l):
   $$C_a\,(\%) = \frac{100 \times V_a}{V_a + V_b + \ldots + V_n} = \frac{100 \times P_a}{P_a + P_b + \ldots + P_n}$$

   where:
   $V_a$ to $V_n$ = volume of components
   $P_a$ to $P_n$ = partial pressures of components
   $$C_a\,(ppm) = \frac{10^6 \times V_a}{V_D + V_a} = \frac{10^6 \times P_a}{P_D + P_a}$$
   where: $V_D$ = volume of diluent gas
   $P_D$ = partial pressure of diluent gas

9. Gas diffusion (or rate of transfer from a region of high concentration to a region of lower concentration) is represented by **Fick's First Law of Diffusion:**

   $$Flux = -D \times A \times 1/T \times (C_i - C_o)$$

   where:
   Flux = the rate of transfer
   D = the diffusion coefficient (the negative sign indicates the substance is moving in the direction of decreasing concentration)
   A = the area of the barrier to diffusion
   T = the thickness of the barrier to diffusion
   $C_i$ = the initial, or inside concentration
   $C_o$ = the later, or outside concentration
   The diffusivity D of most gases is related inversely to the square root of their molecular weight, the pressure, and the absolute temperature.

10. **Henry's Law:** the mass of any gas that will dissolve in a given volume of liquid is directly proportional to the pressure of the gas. The various components of a gas mixture behave independently of each other.

## Gas Mixing Techniques

### Static system

1. Gravimetric procedure (mixing by weight): This method is independent of temperature, pressure, and compressibility. It involves weighing components into a gas cylinder.

2. Mixing by volume: Evacuate cylinder to 0.1 mm Hg, flush with diluent gas, evacuate again, inject component gas using a gastight syringe, allow

diluent gas to pressurize cylinder to the desired pressure.

3. Mixing by pressure: Because the partial pressure of each component equals its mole fraction (MF) times total pressure ($P_t$) of the mixture, a mix of 10 percent A and 90 percent B at a total cylinder pressure of 2,000 psia can be prepared as follows:

$P_A = MF_A \times P_t = 0.10 \times 2,000 = 200$ psia
Add 200 psia of A, then 1,800 psia of B.

4. Homogenizing the gas mixture: Homogeneity depends upon the densities and the relative amounts of the components. Homogenize gas mixtures by rolling cylinders or by thermal convection; temperatures above 50°C (122°F) should be avoided. Once the mixture is homogeneous, it remains so and does not separate except in the case of liquefied gases. Liquefied components may partially condense in the cylinder if subjected to low temperatures.

5. Gas mixtures should then be calibrated (analyzed) using chemical and gravimetric techniques (for some primary standards) or other gas analysis methods mentioned later in this chapter.

6. Accuracy, purity, and tolerances: Commercially available gas mixtures vary in their accuracy (see table below). Even the purest gases and gas mixtures may contain impurities. This is of more concern in research work than in technological practice.

**Commercial gas mixtures: Accuracy**

| Designation | Accuracy limits |
|---|---|
| Primary standards | Within 0.02% absolute or 1% of the component, whichever is smaller |
| Certified mixtures | Within 2 to 5% of component |
| Unanalyzed (commercial grade) | Same as certified but without certificate of analysis |

7. Storage and handling of compressed gas cylinders:

□ Gas cylinders should be tested by hydrostatic pressure for their suitability for use with compressed gases

□ For cylinder filling, the pressure limit is 2,000 psi at 21°C (70°F)

□ Cylinder contents and whether the cylinder is full or empty should always be identified clearly

□ Cylinders must be well secured and are best stored at 21°C (70°F)

□ Proper transportation procedures should be followed

□ Proper valves and regulators relative to the standardized outlets for various families of gases should be used to prevent interchange of regu-lator equipment between gases that are not compatible

□ Safety procedures required for toxic and flammable gases (e.g., CO at 12.5 to 75 percent and $C_2H_4$ at 3 to 30 percent) must be adhered to in handling gas cylinders

### Dynamic system

Gases are mixed (continuous flow mixing) as needed by volume at constant pressure and temperature using flow control devices such as capillary tubing and needle valves.

## Gas Sampling

### Sampling and sample containers

1. Syringes of various volumes—but most commonly used ones are between 1 and 10 ml

2. Plastic film gas-impermeable bags with sealable gas inlet and septum for withdrawing subsamples for analysis

3. Glass containers of various capacities with gas inlet and sampling port

4. Vacuum containers: Evacuated 150- to 250-ml cans with septum or vacuotainers—evacuated 20-ml test tubes (commonly used for blood sampling).

Important points to consider:

1. Make sure sample containers are gastight, to ensure no leaks, and are clean before use to minimize errors.

2. When vacuum containers are used, vacuum should be determined on each container before use, and appropriate correction factors should be applied to the analysis data.

3. Samples should be representative of the atmosphere to be analyzed.

## Gas Analysis Methods

### Methods for immediate and on-the-spot analysis

1. Volumetric gas analyzers for $O_2$ and $CO_2$ (Orsat, Fyrite, and so on)

2. Kitagawa gas sampler and detector tubes for $C_2H_4$, CO, $SO_2$, and other gases

3. Portable gas analyzers ($O_2$, $CO_2$, CO, $C_2H_4$, $SO_2$, $NH_3$, and other gases)

### Laboratory gas analysis instruments

The following methods are much more accurate than the methods mentioned for on-the-spot analysis. They can be used to monitor atmospheric composition in controlled atmosphere storage facilities, ripening rooms, and $SO_2$ fumigation chambers.

| Gas | Instruments |
|---|---|
| $O_2$ | Oxygen analyzers (paramagnetic; polarographic; electrochemical) |
| | Gas chromatography (thermal conductivity detector) |
| $CO_2$ | Infrared $CO_2$ analyzer |
| | Gas chromatography (thermal conductivity detector) |
| CO | Gas chromatography (thermal conductivity detector) |
| $C_2H_4$ | Gas chromatography (flame ionization detector) |
| $SO_2$ | Infrared $SO_2$ analyzer |

## Methods of measuring respiration rates

1. To determine $O_2$ consumed: For tissue slices or organelles—Warburg method or $O_2$ electrode; for intact plant organs—laboratory gas analysis methods for $O_2$ mentioned above.

2. To determine $CO_2$ produced: Colorimetric method (Claypool and Keefer, 1942), and methods mentioned above for laboratory gas analysis of $CO_2$.

3. Results are usually expressed as ml $O_2$ (or $CO_2$) per kg-hr and calculated as follows:

$$\frac{\Delta O_2\% \text{ or } CO_2\%}{100} \times \frac{\text{flow rate (ml/hr)}}{\text{sample weight (kg)}}$$

4. To convert ml $CO_2$ to mg $CO_2$, multiply by appropriate factor for temperature used:

| °C | (°F) | mg/ml $CO_2$ |
|---|---|---|
| 0 | (32) | 1.98 |
| 10 | (50) | 1.90 |
| 20 | (68) | 1.84 |
| 30 | (86) | 1.78 |

5. Conversion factors for calculating heat production:
mg $CO_2$/kg-hr 61.2 = kcal/metric ton - day
mg $CO_2$/kg-hr 220 = Btu/ton - day

## REFERENCES

1. Barmore, C. R., and T. A. Wheaton. 1978. Diluting and dispensing unit for maintaining trace amount of ethylene in a continuous flow system. *HortScience* 13:169-71.

2. Claypool, L. L., and R. M. Keefer. 1942. A colorimetric method for $CO_2$ determination in respiration studies. *Proc. Am. Soc. Hortic. Sci.* 40:177-86.

3. Jeffery, P. G., and P. F. Kipping. 1972. *Gas analysis by gas chromatography*. 2nd ed. New York: Pergamon Press. 196 pp.

4. Leshuk, J. A., and M. E. Saltveit, Jr. 1990. A simple system for the rapid determination of the anaerobic compensation point of plant tissue. *HortScience* 25:480-82.

5. McNair, H. M., and E. J. Bonelli. 1967. *Basic gas chromatography*. Walnut Creek, CA: Varian Aerograph. 306 pp.

6. Nelson, G. O. 1972. *Controlled test atmospheres—principles and techniques*. Ann Arbor, MI: Ann Arbor Sci. Publ. 247 pp.

7. Peterson, S. J., W. J. Lipton, and M. Uota. 1989. Methods for premixing gases in pressurized cylinders for use in controlled atmosphere experiments. *HortScience* 24:328-31.

8. Pratt, H. K., and D. B. Mendoza, Jr. 1979. Colorimetric determination of carbon dioxide for respiration studies. *HortScience* 14:175-76.

9. Pratt, H. K., M. Workman, F. W. Martin, and J. M. Lyons. 1960. Simple method for continuous treatment of plant material with metered traces of ethylene or other gases. *Plant Physiol.* 35:609-11.

10. Saltveit, M. E. 1978. Simple apparatus for diluting and dispensing trace concentrations of ethylene in air. *HortScience* 13:249-51.

11. ———. 1982. Procedures for extracting and analyzing internal gas samples from plant tissues by gas chromatography. *HortScience* 17:878-81.

12. Saltveit, M. E., Jr., and T. Strike. 1989. A rapid method for accurately measuring oxygen concentrations in milliliter gas samples. *HortScience* 24:145-47.

13. Watada, A. E., and D. R. Massie. 1981. A compact automatic system for measuring $CO_2$ and $C_2H_4$ evolution by harvested horticultural crops. *HortScience* 16:39-41.

14. Young, R. E., and J. B. Biale. 1962. Carbon dioxide effects on fruit respiration; 1. Measurement of oxygen uptake in continuous gas flow. *Plant Physiol.* 37:409-15.

# 13

# Ethylene in Postharvest Technology

MICHAEL S. REID

Ethylene's role as a potent plant growth regulator, affecting many phases of plant growth and development, was established only in the last 50 years, but its effects have been known for centuries. The use of ethylene to hasten the ripening of fruits dates to antiquity. Examples include the ripening of sorb apples in southern Italy, using emanations from ripe quinces, and the ripening of mangos in India in an atmosphere created by burning straw. In biblical times, farmers scarified the skin of young sycamore figs (Amos 7:14) to induce rapid growth and ripening of the fruit, a response now known to be due to increased ethylene production by the wounded fruit.

Ethylene plays a role in the postharvest life of many horticultural crops—often deleterious, speeding senescence and reducing shelf-life, and sometimes beneficial, improving the quality of the product by promoting faster, more uniform ripening before retail distribution. This chapter is concerned with the properties of this gas and with ways to harness its beneficial effects and avoid its deleterious effects during postharvest handling of perishable commodities.

## PROPERTIES OF ETHYLENE

The remarkable effects of ethylene on plants were first noted when flammable gas, used for lighting and heating, was piped through the streets of Europe. This gas contained added ethylene to ensure that lamps burned with a yellow flame, increasing illumination. It was soon noticed that plants growing in the vicinity of leaky pipes showed various abnormalities in growth and development, including premature leaf fall and death of flowers. A Russian graduate student, Neljubow, showed that the cause of these bizarre effects was ethylene. With this finding began the widespread research into the effects of ethylene on plant growth and development that continues today.

### Ethylene in plant growth and development

Over the years since Neljubow's study, researchers have shown that many phases of plant growth and development are affected by ethylene: it stimulates germination of some dormant seeds, changes the direction of seedling growth to bypass obstacles in the soil, stimulates growth of special aerating roots in waterlogged soil, causes abscission of leaves in plants under drought stress, may stimulate flowering, and is often the trigger for fruit ripening and abscission. In early studies, the effects of ethylene were thought to be an interesting example of growth regulation by a synthetic chemical. In the 1930's, however, it was discovered that ethylene is produced by plants, and that therefore the responses to ethylene are part of normal growth and development. Ethylene is now considered a plant hormone, an important part of the mechanisms controlling plant growth and development.

### Ethylene biosynthesis

The unravelling of the biochemical pathway of ethylene biosynthesis in plants (fig. 13.1) has been one of the most interesting biochemical stories of recent years. Researchers in Europe and North America competed to find each step in the pathway. Lieberman, at the Beltsville laboratories of the USDA, showed that application of the amino acid methionine greatly stimulated ethylene production in apples, and this compound was then considered to be the starting point for ethylene biosynthesis. Researchers at Davis identified SAM (S-adenosyl-methionine) as another key compound in the pathway and then, almost simultaneously, Amrhein in West Germany and Adams and Yang at Davis discovered that SAM was converted to an unusual cyclic amino acid, ACC (1-aminocyclopropane-1-carboxylic acid) which is now thought to be the immediate precursor for ethylene. The interesting biochemistry in this pathway had some practical implications. The enzyme which controls the rate at which the pathway operates, ACC synthase, is activated by a common enzyme co-factor, pyridoxal phosphate. Inhibitors of enzymes that require pyridoxal phosphate, such as AVG (aminoethoxyvinyl glycine) and AOA (aminooxyacetic acid) can be used to inhibit ethylene production. Cobalt ion and low $O_2$, which inhibit the final step in the pathway, the ethylene forming enzyme (EFE), can also reduce ethylene production.

Current research is investigating the way in which ethylene induces such a range of effects. The favored model is that ethylene binds to a protein, called a binding site (fig. 13.2), thus stimulating release of a so-called second message instructing the DNA (the molecules carrying all the information in the plant

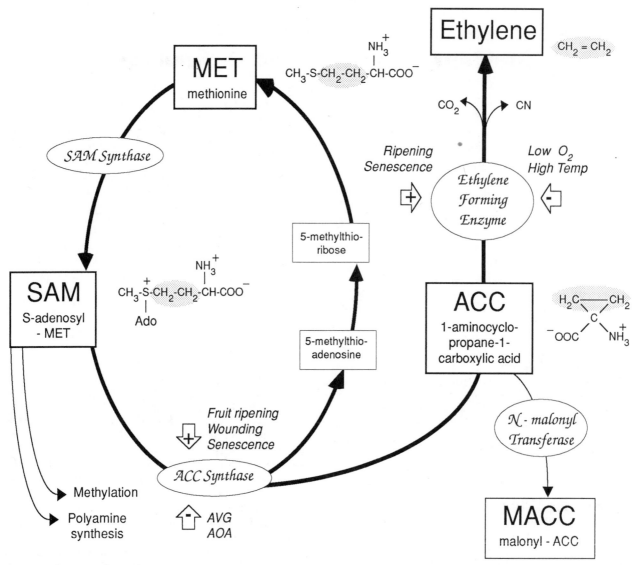

**Fig. 13.1. Pathway of ethylene biosynthesis.** *Redrawn from Yang, 1987.*

cell) to form mRNA (messenger RNA) molecules specific for the effects of ethylene. These molecules are "translated" into proteins by polyribosomes, and the proteins so formed are the enzymes that cause the actual ethylene response.

## Other properties

**Physical.** Ethylene is the first member of the unsaturated or olefin series of hydrocarbons, and its properties are summarized in table 13.1.

**Toxicological.** Ethylene is a gas with a characteristic suffocating, sweetish odor. It is both an anaesthetic and asphyxiant. High vapor concentrations can cause rapid loss of consciousness and perhaps death by asphyxiation. Removal to fresh air usually results in prompt recovery if the person is still breathing. When the gas is handled in liquefied form, skin and eye burns can result from contact with the liquid. Cases in which liquid ethylene contacts the eye must be seen to by a physician.

**FDA status.** Use of ethylene gas to promote ripening of fruits and vegetables is sanctioned under FDA Regulation 120,1016. Ethylene is exempted from the requirement of a residue tolerance when used as a plant regulator either before or after harvest.

**Explosive.** Mixtures of ethylene gas and air are potentially explosive when the concentration of ethylene rises above 3.1 percent by volume. This concentration is at least 30,000 times the concentration required to initiate ripening of most fruits and vegetables. Above 32 percent by volume, ethylene-air mixtures are not explosive.

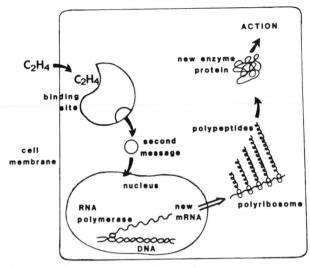

**Fig. 13.2. Mechanism of ethylene action.**

**Table 13.1. Physical properties of ethylene gas**

| Appearance | Colorless, hydrocarbon gas with a faint, sweetish odor that is easily detected in parts per million concentration |
|---|---|
| Molecular weight | 28.05 |
| Boiling point | |
| at 760 mm Hg | $-103.7°C$ |
| at 300 mm Hg | $-118°C$ |
| at 10 mm Hg | $-153°C$ |
| $\Delta$b.p./$\Delta$p at 750 to 770 mm Hg | 0.022°C per mm Hg |
| Freezing point at saturation pressure (triple point) | $-169.2°C$ |
| Surface tension at $-103.7°C$ | 16.4 dynes/cm |
| Flammable limits in air* | |
| lower | 3.1% by vol |
| upper | 32% by vol |

*All compositions between the upper and lower limits are flammable and can be explosive.

## MEASUREMENT OF ETHYLENE

A remarkable feature of the effects of ethylene on plants is the minute concentration required. Fruit ripening, for example, typically occurs at the maximum rate at levels of 1 part of ethylene in 1 million parts of air (1 ppm). Effects on opening of roses can be seen at ethylene concentrations as low as 10 parts per billion parts of air (10 ppb). The development of the gas chromatograph and the flame ionization detector, which made it possible to rapidly measure ethylene at such low concentrations, was crucial to our present understanding of the role of ethylene. New techniques, such as laser-acoustic devices, offer even more sensitivity. Unfortunately, sensitive measurement of ethylene is still expensive; satisfactory chromatographs cost from $7,000 to $10,000. For the postharvest technologist, gas-sampling tubes giving a colorimetric reaction can be read reasonably easily down to 1 ppm. These are satisfactory for occasional use by ripening-room operators but can-

not monitor the low levels of ethylene that may be of concern in storage rooms and marketing outlets.

## POSTHARVEST USES

The wide range of approved uses for ethephon, an ethylene-releasing chemical, in agriculture (table 13.2) indicates the utility of this growth regulator. Fruit ripening is by far the largest application of ethylene gas in postharvest technology, but other responses are also in use for some crops and will be described briefly first.

### Flower and Sprout Induction

The stimulation of flowering of pineapples by ethylene treatment is critical to that industry. Less well known is the spectacular response of some flowering bulbs to ethylene. Japanese bulb growers discovered that iris bulbs from fields that had been burned at the end of the season to control leaf diseases flowered earlier and more prolifically than controls. It was found that smoke did the same for bulbs that had been harvested, and smoking of bulbs is still practiced in Japan. The active ingredient in the smoke is ethylene, and it has now been shown that ethylene treatment of the propagules of a number of flowering crops stimulates flowering. Perhaps the most remarkable example is narcissus. Medium-sized (6 cm circumference) narcissus bulbs normally do not flower; treatment with ethylene for a few hours just after lifting induces almost 100 percent flowering (fig. 13.3).

**Table 13.2. Approved uses for ethephon in U.S. agriculture***

| Use | Approved crops and states (no parentheses = all States) |
|---|---|
| Postharvest fruit ripening | Bananas, tomatoes (FL) |
| Preharvest fruit ripening | Peppers, tomatoes |
| Fruit removal | Apples, carob, crabapples, olive |
| Defoliation | Apple, buckthorn, cotton, roses |
| Fruit loosening | Apples, blackberries (WA, OR), cantaloupes, cherries (CA, AZ, TX), tangerines |
| Maturity and/or color development | Apples, cranberries (MA, NJ, WI), figs (CA), filberts (OR), grapes, peppers, pineapple, tomatoes |
| Degreening (preharvest) | Tangerines |
| Degreening (postharvest) | Lemons |
| Dehiscence | Walnuts |
| Leaf curing | Tobacco |
| Flower induction | Pineapple and other bromeliads |
| Sex expression | Cucumber, squash |
| Flower bud development | Apple |
| Plant height control | Barley, daffodils, hyacinth, wheat |
| Stimulate lateral branching | Azaleas, geraniums |

*Adapted from Kays and Beaudry 1987

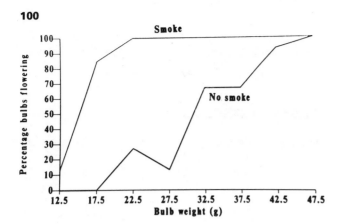

**Fig. 13.3. Effect of ethylene treatment of bulbs on flowering of narcissus.**

Now being used commercially, this treatment is applied either using ethephon or using ethylene gas in rooms similar to those used for ripening fruits (see below). Ethylene is also used as a short treatment to enhance the sprouting of seed potatoes—the gas breaks the dormancy of the buds, but prolonged treatment inhibits their extension growth.

## Shuck Loosening and Fruit Release

Although physiologists knew for years that ethylene induced abscission of leaves, flowers, and fruit from many plants, it was not until ethylene-releasing compounds such as ethephon and silaid (fig. 13.4) became available that these responses could be tested in horticulture. Ethylene-releasing chemicals are now approved for a variety of uses of interest to postharvest technologists. Preharvest application to walnut and pecan trees induces shuck loosening and improves harvest efficiency. Similarly, these chemicals are used to loosen the abscission zone on the stalk of fruits that are mechanically harvested, notably sour cherries, thereby improving harvest yields.

## Chlorophyll Destruction

In many plant tissues, ethylene treatment results in rapid loss of chlorophyll, the green color in leaves and unripe fruit. This response was used to blanch celery in the days before green celery was generally accepted in the U.S. and is still used to accelerate the curing of tobacco. An important example is the degreening of citrus, where the orange color is revealed as the chlorophyll is destroyed during ethylene treatment.

## Fruit Ripening

### Physiology

The concentrations of ethylene required for the ripening of various commodities vary (table 13.3) but in most cases are in the range of 0.1 to 1 ppm. The time

$$Cl-CH_2-CH_2-\overset{\overset{\textstyle O}{\|}}{\underset{\underset{\textstyle OH}{|}}{P}}-OH$$

(2-chloroethyl)phosphonic acid

Ethrel

(2-chloroethyl)methylbis(phenylmethoxy)silane

Silaid

(2-chloroethyl)tris(2-methoxyethoxy)silane

Alsol

**Fig. 13.4. Molecules of ethephon, silaid, and alsol, three ethylene-releasing chemicals.**

**Table 13.3. Threshold for ethylene action in various fruits**

| Fruit | Threshold concentration (*ppm*) |
|---|---|
| Avocado (var. Choquette) | 0.1 |
| Banana (var. Gros Michel) | 0.1 - 1.0 |
| (var. Lacatan) | 0.5 |
| (var. Silk fig) | 0.2 - 0.25 |
| Cantaloupe (var. P.M.R. No. 45) | 0.1 - 1.0 |
| Honeydew melon | 0.3 - 1.0 |
| Lemon (var. Fort Meyers) | 0.1 |
| Mango (var. Kent) | 0.04 - 0.4 |
| Orange (var. Valencia) | 0.1 |
| Tomato (var. VC-243-20) | 0.5 |

of exposure to initiate full ripening may vary, but for climacteric fruits exposures of 12 hours or more are usually sufficient. Full ripening may take several days after the ethylene treatment.

### Technical considerations

The effectiveness of ethylene in achieving faster and more uniform ripening depends on the type of fruit being treated, its maturity, the temperature and relative humidity of the ripening room, ethylene concentration and duration of exposure to ethylene. In general, optimum ripening conditions for fruits are:

Temperature: 18° to 25°C (65° to 77°F)
Relative humidity: 90 to 95 percent
Ethylene concentration: 10 to 100 ppm
Duration of treatment: 24 to 72 hours depending on fruit kind and maturity stage
Air circulation: Sufficient to ensure distribution of ethylene within the ripening room
Ventilation: Require adequate air exchanges to prevent accumulation of $CO_2$, which reduces the effectiveness of $C_2H_4$

**Amount of gas needed.** The recommended treatment concentration is 10 to 100 ppm (1 cubic foot of $C_2H_4$ in 10,000 cubic feet of room space). Lower concentrations are used in well-sealed rooms that will maintain the ethylene concentration, or in rooms where the trickle system (see below) is being used. Higher concentrations are used in leaky rooms to compensate for the fall in concentration during treatment. Concentrations higher than 100 ppm do not speed up the ripening process. Adding too much ethylene may create an explosive air-gas mixture.

**Temperature.** Control of temperature is critical to good ripening with ethylene. Optimum ripening temperatures are 18° to 25°C (65° to 77°F). At lower temperatures ripening is slowed; at temperatures over 25°C (77°F), bacterial growth and rotting may be accelerated, and above 30°C (86°F) ripening may be inhibited. Fruit that have been cool-stored must be warmed to 20°C (68°F) to ensure that ripening proceeds rapidly. As ripening starts, the burst of respiration that accompanies it (the climacteric) generates a burst of heat. Adequate refrigeration equipment controlled with proper thermostats (fig. 13.5) is essential to ensure that this heat does not increase pulp temperatures to the point where ripening is inhibited (above 30° to 35°C = 86° to 95°F, depending on the commodity). Many ripening room operators "pigeon-hole" stack banana boxes to ensure that heat generated during ripening is carried away efficiently. This is especially important if boxes of bananas are packed with a polyethylene liner, which restricts airflow and heat removal from the carton. The principles of forced air cooling can also be applied to maintaining temperature control in the ripening room and thus eliminate the hand labor required to unload pallets, construct pigeon-hole stacks, and reload pallets when ripening is complete.

**Safety precautions.** Because of the explosion hazard of ethylene mixed with air at concentrations between 3 percent and 30 percent, the rules listed below must be followed stringently to prevent buildup of these concentrations, and to prevent ignition if they should form.

*1.* Do not permit open flames, spark-producing devices, fire, or smoking in or near a room containing ethylene gas, or near the cylinder.

*2.* Use an approved meter for accurately measuring the gas when discharging ethylene from the cylinder.

*3.* Ground all piping to eliminate the danger of electrostatic discharge.

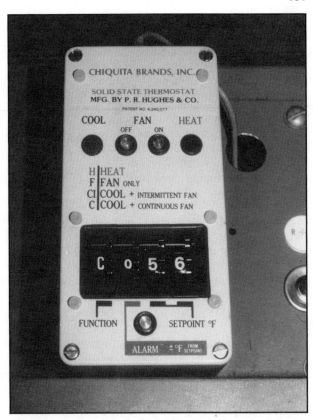

**Fig. 13.5. Thermostat for temperature control in a banana ripening room.**

*4.* Store ethylene cylinders in accordance with all instructions and standards of the National Board of Fire Underwriters.

*5.* All electrical equipment, including lights, fan motors, and switches should comply with the National Electric codes for Class 1, Group D equipment and installation.

*6.* Instruments that detect the concentration of ethylene in air can be set to sound an alarm if the concentration approaches explosive levels.

## Treatment Systems

Handlers can equip existing rooms for use as ripening rooms, or they can install specially built chambers with automatic control of temperature, humidity, and ventilation. It is not essential that the rooms be hermetically sealed, but they should be as tight as practicable to prevent leakage. If pure ethylene is being used to provide the ethylene treatment, it is essential that the rooms adhere to the safety standards above.

Rooms should be heated with hot water or steam pipe systems or with indirect gas or electric heaters that have been examined and Underwriters Laboratories (U.L.) listed, never with an open flame. Because of the rapid increase in respiratory heat production following ethylene treatment, ripening rooms should be equipped with refrigeration systems ade-

102

quate to hold the temperature in the desired range. Room temperature should be continuously monitored using a distant-reading thermometer.

Several methods, varying in sophistication, are used to provide the proper ethylene concentration in the ripening room.

**1. The "shot" system.** In the shot system, measured quantities of ethylene are introduced into the room at regular intervals. The shots may be applied by weight (rarely used in the U.S. today), or by flow, using a gauge that registers the discharge of ethylene in cubic feet per minute. The required ethylene application is made by adjusting the regulator to give an appropriate flow rate, then timing the delivery of the gas. Any piping leading into the ripening room should be grounded to prevent possible electrostatic ignition of the explosive concentrations of ethylene that are always present near the orifice when ethylene is being introduced.

The gas needed for a room is calculated, using the following formula, where:

C is the ppm of ethylene required (usually between 10 and 100),

V is the room volume (in thousands of cubic feet),

F is the flow rate of gas (measured from a flowmeter) in cubic feet per minute (CFM), and

T is the time (in minutes) for which the gas is allowed to flow.

If the ethylene is administered by weight, the number of pounds of ethylene needed is given by the formula:

$$C \times V/13,000,000$$

*Example.* How many pounds of ethylene are required to provide 100 ppm in a room 20 feet high, 100 feet long, and 50 feet wide?

C = 100
Room volume (V) = 20 × 100 × 50 = 100,000 ft³

From the equation, the number of pounds of ethylene required is then:

$(100 \times 100,000)/13,000,000 = $ ¾ lb of ethylene

It is hard to measure a ¾-pound change in an 80-pound gas cylinder, so one can see why the weighing method is now considered old-fashioned.

Using the more usual system of metering the flow of ethylene into the room, the required time (in minutes) for which the gas should flow is given by the formula:

$$(C \times V)/(F \times 1,000,000)$$

For the same room (V = 100,000), a desired ethylene concentration of 100 ppm, and an ethylene flow rate of 20 cfm, the gas should flow for:

$(100 \times 100,000)/(20 \times 1,000,000) = $ ½ minute

This is easy to measure with a stopwatch. If the flowmeter is calibrated in milliliters per minute (ml/

min), then the formula is a little different. The time, in this case, is given by:

$$(C \times V)/(36 \times F)$$

For a flowmeter with a flow rate of 5000 ml/min, the time to get the same concentration (100 ppm) in the same room (V = 100,000) is:

$(100 \times 100,000)/(36 \times 5000) = $ 55 minutes

The long time taken is why many ripening rooms have several flowmeters in parallel. In this situation, add the flows to get the total flow, and use that as F in the equations above.

Because the room containing the product being ripened is sealed in the shot system, $CO_2$ accumulates in the room and may inhibit the ripening process. It is customary to apply a shot of ethylene twice each day. The room should be well ventilated before each application, particularly if it is well sealed, by opening the doors for about half an hour. In large ripening rooms, a ventilating fan should be provided. Where the ripening rooms are near rooms used for storage or handling of ethylene-sensitive commodities (for example, in a wholesale distribution center), the rooms should be ventilated to the exterior to prevent contamination.

**2. The trickle or flow-through system.** The ethylene is introduced into the room continuously, rather than intermittently. As the flow of ethylene is very small, it has to be regulated carefully. This is usually done by reducing the pressure using a two-stage regulator and passing the gas into the room through a metering valve and flowmeter (fig. 13.6).

To prevent a buildup in either $CO_2$ or $C_2H_4$, fresh air is drawn into the ripening room at a sufficient rate to ensure a change of air every 6 hours. The air is vented through a small exhaust port to the rear of the room. The fan size in CFM (cubic feet per minute) is calculated by:

Volume of room (cubic feet)/360.

The ethylene flow rate (in CFM) needed to maintain 100 ppm in the room is calculated by: Ventilation fan delivery (CFM) × 0.0001.

In ml/min, the flow rate is: ventilation fan delivery (CFM) × 2.8.

A convenient way of monitoring gas being supplied in a trickle system is a simple "sight glass" in which the ethylene bubbles through a water trap on its way to the ripening room (fig. 13.7). As in the shot system, correct temperature maintenance and adequate air circulation are essential for good ripening.

## SOURCES OF ETHYLENE

Ethylene gas is a relatively inexpensive industrial chemical, but it is often more convenient or safer to provide ethylene by means other than the gas bottles assumed in the above discussion of treatment systems. Regardless of the source of ethylene, the treat-

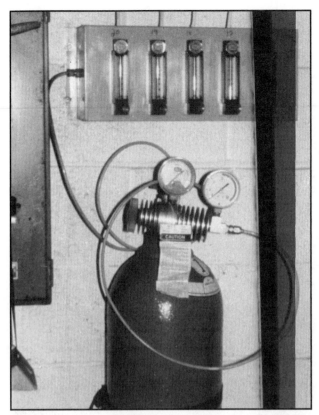

**Fig. 13.6. Pressure regulators and flowmeters used for controlling and monitoring flow of ethylene into ripening rooms.**

**Fig. 13.7. A simple "sight-glass" can be used to monitor flow of ethylene into the ripening room.**

ment conditions outlined above are still important to a good out-turn in the ripening process.

**Explosion-proof ethylene mixtures.** The danger of explosions from oversupply of ethylene to a ripening room can be eliminated by using mixtures of ethylene with inert gases. The proportion of the inert gas should be such that at high concentrations of ethylene not enough oxygen remains in the ripening space to provide an explosive mixture. For example, one commercial formulation, Ripegas, contains 6 percent $C_2H_4$ in $CO_2$ by weight. When using these mixtures, the calculation of volumes of gas required must be modified to reflect the composition of the mixture. In the case of Ripegas, for example, the weights, volumes, or flow rates calculated using the formulas above would be increased by a factor of 100/6, or 17, to give the required concentration.

**Ethylene generators.** Ethylene generators, in which a liquid produces ethylene when heated in the presence of a catalyst, are now widely used for supplying ethylene in ripening rooms. The liquid, a proprietary product, comprises ethanol and agents that catalyze its dehydration:

$$C_2H_5OH - H_2O \rightarrow C_2H_4$$

The generator combines a simple heater with a system for attaching a bottle of the generator liquid. The liquid comes in 1-pint and 1-quart bottles and is used up by the generator at the rate of 1 pint every 8 hours. The generator delivers about 14 liters (0.5 cubic foot) of ethylene gas per hour, adjustable on newer models. These figures can be used to determine the number of generators to be used in a given size of room, knowing the air leakage or ventilation rate (gas exchanges per hour). For example, in a 5,000-cubic-foot ripening room where there is one air exchange per hour, a 1-quart bottle will generate 100 ppm for 16 hours.

**Ethephon.** Ethephon (2-chloroethane phosphonic acid) is strongly acidic in water solution. When in solutions above a pH of about 5, the ethephon molecule spontaneously hydrolyses, liberating ethylene. Ethephon is commercially available (Ethrel, Florel, Cepa) and is registered for preharvest use on a variety of crops for controlling developmental processes, or inducing ripening. As a material for enhancing postharvest ripening, it has the disadvantage that it has to be applied to the fruit in a water solution as a spray or as a dip, an extra step in handling with attendant dangers of microbial infection. In contrast to ethylene treatment, however, no special facilities are required to ripen fruit with ethephon, provided the ambient temperatures are within the range required to ripen the commodity. Ethephon is approved for postharvest use on only a few commodities (table 13.2).

On a small scale, commodities can be treated using the "shot" method with ethylene liberated from ethephon. Place the calculated amount of ethephon (approximately 7 fluid ounces of active ingredient to release one cubic foot of ethylene gas) in a stainless steel bowl, then, just before closing the room, add enough caustic soda pellets (approximately 3 ounces for each 7 fluid ounces a.i. of ethephon) to completely neutralize the ethephon. CARE! CAUSTIC SODA AND ETHEPHON ARE CORROSIVE. WEAR SAFETY GLASSES AND RUBBER GLOVES.

**Calcium carbide.** Calcium carbide, a grayish solid, is readily produced by heating calcium oxide with charcoal under reducing conditions. When hydrolyzed, calcium carbide produces acetylene, containing trace amounts of ethylene that are sufficient to be used in fruit ripening. Simple generators that are used to provide acetylene for lamps can be used in partially vented spaces to ripen or degreen fruits under conditions where ethylene is not available. In some instances $Ca_2C$ wrapped in newspaper can be used as the generator. Water vapor from the fruit releases sufficient ethylene from $Ca_2C$ to cause ripening.

**Use of fruits.** Traditionally, ripening has often been stimulated by enclosing unripe fruit with other fruits that are already ripe. This technique forms the basis for a cheap and simple method of fruit ripening. Table 3.3 (in chapter 3) shows the range of ethylene production known for fruits and vegetables. Ripe fruits with high ethylene production can be used in very small scale commercial operations or at home to ripen or degreen other fruits in much the same way as any other ethylene generating system.

## UNDESIRABLE EFFECTS OF ETHYLENE

Given the wide range of physiological effects of ethylene in plants and the common occurrence of ethylene and other gases with ethylenelike effects (table 13.4) as air pollutants, it is not surprising that ethylene-mediated growth responses result in quality reduction in a range of commodities.

**Accelerated senescence.** In green tissues, ethylene commonly stimulates senescence, as indicated by loss of chlorophyll, loss of protein, and susceptibility to desiccation and decay. Ethylene pollution can result in yellowing of leafy vegetables (spinach), fresh herbs (parsley; see fig. 13.8), and other green vegetables (broccoli). The senescence of some flowers is stimulated by ethylene at very low concentrations (fig. 13.9). These effects occur in flowers where increased ethylene production is part of natural senescence

Table 13.4. Comparative effectiveness of ethylene and related analogues in peas stem-section assay*

| Compound | Relative activity (moles/unit) |
|---|---|
| Ethylene | 1 |
| Propylene | 130 |
| Vinyl chloride | 2,370 |
| Carbon monoxide | 2,900 |
| Acetylene | 12,500 |
| 1-Butene | 140,000 |

*From Burg and Burg 1966.

Fig. 13.8. Parsley yellows rapidly at room temperature when exposed to low concentrations of ethylene.

Fig. 13.9. Carnations senesce rapidly when exposed to low concentrations of ethylene.

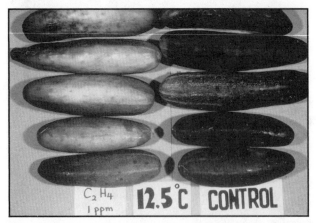

Fig. 13.10. Cucumbers yellow prematurely when exposed to ethylene.

(e.g., carnations, sweet peas) and in some where it is not (roses, brodiaeas).

**Accelerated ripening.** Although acceleration of ripening is a beneficial use of ethylene, it can also be undesirable, as in the premature yellowing of cucumbers in the presence of ethylene (fig. 13.10). Most climacteric fruits respond to ethylene in the atmosphere, even at storage temperatures, so ethylene in the store reduces storage life. The firmness of kiwifruit in storage is dramatically reduced if the ethylene concentration in the cool store is more than 20 ppb.

**Induction of leaf disorders.** In many plants, exposure to ethylene results in darkening or death of portions of their leaves. This response is commonly seen in foliage plants and is of major economic consequence in lettuce, where ethylene causes the disorder known as russet spotting (fig. 13.11). In lettuce, the browning results from collapse and death of areas of cells following increased synthesis of phenolic compounds in response to ethylene.

**Isocoumarin formation.** In carrots, ethylene exposure causes the biosynthesis of bitter isocoumarins, which make the carrots bitter. Recently, it has been shown that ethylene concentrations as low as 0.5 ppm cause significantly increased bitterness of carrots within 2 weeks, even when they are stored at 2.5°C (fig. 13.12).

**Sprouting.** The ethylene-stimulated sprouting which is useful in propagules is, of course, undesirable in commodities intended for consumption. Sprouting of potatoes, for example, is unsightly and increases water loss, leading to early shriveling.

**Abscission of leaves, flowers, and fruits.** Ethylene-induced abscission is most often a problem in ornamental plants, where low concentrations can cause complete loss of flowers or leaves. As an example, *Schlumbergera*, the Christmas cactus, is sold when the first flowers are open, but it often arrives at the market with all the flowers in the bottom of the box due to exposure to ethylene during transportation.

**Toughening of asparagus.** Ethylene stimulates the lignification of xylem and fiber elements in the growing asparagus spear, leading to undesirable toughness and reducing the portion of the spear that is edible.

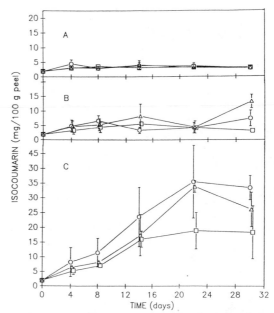

Time course of isocoumarin accumulation in carrots stored at 0 C (□), 2.5 C (△), or 5 C (○) in air (A), 0.1 ppm (B), or 0.5 ppm (C) ethylene.

**Fig. 13.12. Exposing carrots to ethylene turns them bitter due to the synthesis of isocoumarin, an intensely bitter compound (Lafuente et al, 1989).**

**Induction of physiological disorders.** Ethylene sometimes induces or hastens the appearance of physiological disorders of stored commodities. Rapid ripening of apples with low calcium contents induces high levels of the bitter pit storage disorder. Similarly, high ethylene levels in the storage chamber reduce the effectiveness of controlled atmospheres in maintaining quality of apples. While useful in inducing flowering in bulbs and other propagules, ethylene damages these propagules after the flowers have started to develop. Ethylene pollution during marketing of tulip bulbs, for example, results in failure of the flowers to develop, a condition called "blasting."

## SOURCES OF ETHYLENE IN THE ENVIRONMENT

Ethylene is produced whenever organic materials are stressed, oxidized, or combusted. There are many sources of ethylene pollution during postharvest handling of perishables, but the most important are internal combustion engines, ripening rooms, and ripening fruits. Other sources are aircraft exhaust, fluorescent ballasts, decomposing produce and sometimes fungi growing on it, cigarette smoke, rubber materials exposed to heat or UV light, and virus-infected plants. The sources are not always obvious: supermarkets in Texas traced the problems they were having with their flowers to ethylene contamination from propane-powered floor polishers that were only used at night.

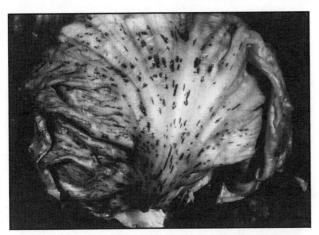

**Fig. 13.11. Dark brown spotting of the midribs, "russet spotting," caused by exposure of lettuce to ethylene.**

# OVERCOMING UNDESIRABLE ETHYLENE EFFECTS

A number of techniques have been developed to protect sensitive commodities from the effects of ethylene. Selection of the appropriate method obviously depends on the commodity and the handling techniques used in its marketing.

Removing ethylene from the atmosphere around the commodity is the preferable method of preventing deterioration of ethylene-sensitive produce. Of the available methods, the simplest and cheapest are in many cases the most effective.

## Eliminating Sources of Ethylene

In a great majority of cases, high levels of ethylene in storage and handling areas can be avoided by removing sources of ethylene. In particular, commodities sensitive to ethylene should be handled using electric forklifts. Internal combustion engine vehicles should be isolated from handling and storage areas, and engines should never be left idling in an enclosed space during loading and unloading operations. Where these techniques are not feasible, it is possible to fit combustion engine exhausts with catalytic converters, which will reduce $C_2H_4$ emissions by 90 percent. Rigorous attention to sanitation will remove overripe and rotting produce which can be a source of ethylene.

## Ventilation

Where the air outside storage and handling areas is not polluted, simple ventilation of these areas can reduce ethylene concentrations. An exchange rate of one air change per hour can readily be provided by installing an intake fan and a passive exhaust. The cost of using such a system (ignoring the small initial capital investment and assuming a power cost of 7 cents per kilowatt-hour) can be determined using the equation:

Cost/year (dollars) = 0.001 × cooler volume (ft$^3$) × [outside temperature − cooler temperature (°F)]

## Chemical removal

Ethylene can be removed by a number of chemical processes; the most important are described below:

**1. Potassium permanganate.** Commercial materials, such as Purafil, utilize the ability of potassium permanganate ($KMnO_4$) to oxidize ethylene to $CO_2$ and $H_2O$. The requirements for such materials are a high surface area coated with the permanganate, and ready permeability to gases. Many porous materials have been used to manufacture permanganate absorbers, including vermiculite, pumice, and brick. The type of material may depend on the purpose for which the absorber is required. For removing ethylene from room air, the absorber should be spread out in shallow trays, or air should be drawn through the absorber system. Attempts to develop liquid scrubbers using $KMnO_4$ have been unsuccessful.

**2. Ultraviolet lamps.** Australian researchers have developed an effective method for ethylene removal using ultraviolet lamps. In commercial equipment now available using this method, air from the storage room is drawn past the lamps. Ultraviolet lamps produce ozone, which was thought to be the active ethylene removing agent. It now appears that the ethylene is oxidized by a much more reactive intermediate in the formation of ozone. Whatever the mechanism, the ozone produced by the lamps is very toxic to fresh produce, and must be removed. Fortunately this is easily achieved by a filter in the unit, of rusty steel wool.

**3. Activated or brominated charcoal.** Charcoal air purifiers, especially if brominated, can absorb ethylene from air. These systems are largely confined to use in the laboratory, as potassium permanganate absorbers are cheaper and more widely available.

**4. Catalytic oxidizers.** If ethylene and oxygen are combined at high temperature in the presence of a catalyst (such as platinized asbestos) the ethylene will be oxidized. Ethylene scrubbers utilizing this effect are now available commercially (fig. 13.13), and overcome the difficulty of heating the incoming air, by clever use of a bed of ceramic used as a heat sink, and reversible flow of gas through the bed. These scrubbers are very efficient, reducing the ethylene concentration in the air to 1 percent of the input concentration. Because of the small air volume they process, they are most suited to small spaces or long-term CA storage systems.

**5. Bacterial systems.** Approximately 30,000 metric tons of ethylene is liberated into the atmosphere each day from internal combustion engines, but the concentration of ethylene in air remains very low (nearly undetectable in fresh rural air). This implies that

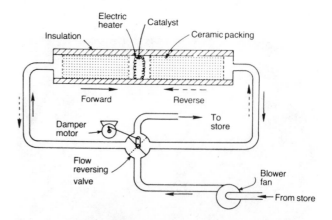

Fig. 13.13. This "Swingtherm" unit removes ethylene from air using a catalytic oxidizer and is equipped with heat exchangers to reduce energy consumption. *Reproduced, with permission, from Blanpied, Bartson, and Turk 1985.*

something removes ethylene from the atmosphere. Bacteria that use ethylene as a biochemical substrate have been isolated from soils. It seems possible that a scrubber may be developed in which the bacteria grow at the expense of ethylene in the store atmosphere.

## Hypobaric storage

Removal of endogenous ethylene was the first benefit ascribed to hypobaric (low-pressure) storage: levels of ethylene inside fruit were greatly reduced, and the longer storage life obtained could be reduced by adding ethylene to the atmosphere. There appear to be few commodities where the benefits of reducing the tissue ethylene content warrant the use of this cumbersome and expensive apparatus. Many of the benefits of hypobaric storage are due more to reduction in the partial pressure of oxygen, which automatically accompanies reduction in the atmospheric pressure.

## Inhibiting the Effects of Ethylene

Sometimes it is not possible to ensure low concentrations of ethylene in the air, as, for example, in supermarkets. For these settings there are several techniques available to inhibit the effects of ethylene.

**Controlled atmospheres.** Low concentrations of $O_2$ and high concentrations of $CO_2$ in the storage atmosphere reduce the rates of respiration, ethylene production, and other metabolic processes. $CO_2$-enriched atmospheres also may inhibit the action of ethylene on tissues sensitive to it. For example, bananas packed in polyethylene-lined boxes can be transported at 15° to 25°C (59° to 77°F) in the presence of a potassium permanganate absorber. Even without the absorber, bananas transported in this way arrive in much better condition than control fruit (fig. 13.14) because of the effect of the accumulated $CO_2$ produced by the fruit on preventing the action of ethylene.

**Specific anti-ethylene compounds.** Compounds that inhibit the action of ethylene include silver ion and a gaseous olefin, 2,5-norborneadiene (NBD). While the possibility of these compounds being registered for use on comestibles is remote, ornamental commodities are already commercially treated with a complex between silver and thiosulfate (STS), which has a very low stability constant and therefore moves readily from the vase solution to the head of cut flowers. Flowers pulsed with this material last two to three times as long as control flowers (fig. 13.15). Potted flowering plants do not lose their flowers during transportation if they are first sprayed with STS (fig. 13.16).

**Inhibition of ethylene biosynthesis.** Ethylene may reduce quality even when it is not present as a pollutant if the tissue itself produces ethylene. This may occur, for example, when carnations senesce early or when fruit ripen too fast. Inhibitors of ethylene biosynthesis, such as AVG and AOA, have been used in laboratory experiments to extend flower vase life and fruit storage life. These inhibitors do not prevent the

Fig. 13.15. Treatment of carnation flowers with silver thiosulfate protects them against ethylene-mediated senescence.

Fig 13.14. Bananas shipped in sealed polyethylene bags with $KMnO_4$-type ethylene absorbers remain green during transportation.

Fig. 13.16. Pretreatment of potted Christmas cactus plants with STS prevents loss of florets during shipping.

action of ethylene that is present as an environmental pollutant.

## REFERENCES

1. Abeles, F. B. 1973. *Ethylene in plant biology.* New York: Academic Press. 302 pp.

2. Blanpied, G. D., ed. 1985. [Symposium on] Ethylene in postharvest biology and technology of horticultural crops. *HortScience* 20:39-60.

3. Blanpied, G. D., J. A. Bartson, and J. R. Turk. 1985. A commercial development programme for low ethylene controlled-atmosphere storage of apples. In *Ethylene and plant development*, (eds.) J. A. Roberts and G. A. Tucker, 393-404. London: Butterworths.

4. Burg, S. P., and E. A. Burg. 1966. Fruit storage at sub-atmospheric pressure. *Science* 153:314-15.

5. deWilde, R. C. 1971. Practical applications of (2-chloroethyl) phosphonic acid in agricultural production. *HortScience* 6:359-64.

6. Kays, S. J. and R. M. Beaudry. 1987. Techniques for inducing ethylene effects. *Acta Hortic.* 201: 77-116.

7. Lafuente, M. T., M. Cantwell, S. F. Yang and V. Rubatzky. 1989. Isocoumarin content of carrots as influenced by ethylene concentration, storage temperature and stress conditions. *Acta Hortic.* 258:523-534.

8. Liu, F. 1970. Storage of bananas in polyethylene bags with an ethylene absorbent. *HortScience* 5:25-27.

9. Pratt, H. K., and J. D. Goeschl. 1969. Physiological roles of ethylene in plants. *Annu. Rev. Plant Physiol.* 20:541-85.

10. Reid, M. S. 1987. Ethylene in plant growth, development, and senescence. In *Plant hormones and their role in plant growth and development*, ed. P. J. Davies, 257-79. The Hague: Martinus Nijhoff.

11. Scott, K. J., and R. B. H. Wills. 1973. Atmospheric pollutants destroyed in an ultra violet scrubber. *Lab. Pract.* 22:103-6.

12. Sherman, M., and D. D. Gull. 1981. *A flow-through system for introducing ethylene in tomato ripening rooms.* Univ. Florida, Veg. Crops. Fact Sheet VC-30. 4 pp.

13. Staby, G. L., and J. F. Thompson. 1978. An alternative method to reduce ethylene levels in coolers. *Flor. Rev.* 163:31, 71.

14. Union Carbide Corp. 1970. *Ethylene for coloring matured fruits, melons, and vegetables.* New York: Product Information, Union Carbide Corp. 11 pp.

15. Watada, A. E. 1986. Effects of ethylene on the quality of fruits and vegetables. *Food Technol.* 40(5):82-85.

16. Yang, S. F. 1987. Regulation of biosynthesis and action of ethylene. *Acta Hortic.* 201:53-59.

# 14

# Principles of Disease Suppression by Handling Practices

NOEL F. SOMMER

Fresh horticultural crops are increasingly marketed at great distances from the point of production. Long-distance marketing sometimes results in large economic benefits which, however, can often be achieved only by stretching the postharvest life of the commodity to its limit. Diseases and disorders ordinarily manageable during handling and transcontinental transit and marketing may be excessive when transoceanic marine transport of longer duration is involved. Similarly, storage of fruits and vegetables for weeks or months before marketing may create added disease problems.

The type of postharvest handling horticultural commodities receive determines in large measure the loss due to rot. The physiological and physical condition of these commodities is of great importance. For example, fruits in high vitality show considerable resistance to fungal attack whereas stressed or senescent fruits are disease prone. The central features of good postharvest handling procedures are to start with commodities in good condition and to emphasize methods that maintain their vitality. Debilitating environments are to be avoided and life-shortening injuries prevented or their effects minimized. At the same time, procedures must limit, directly or indirectly, the activity of pathogens.

## Physiology in Relation to Disease

Sugars translocated from photosynthesizing leaves provide the chemical energy and carbon building blocks for growth and development of plants and plant parts. Within the living cells of fruits, metabolic processes convert sugars (along with minerals and water from the soil) into the myriad compounds (carbohydrates, proteins, and fats) that compose the living cell and its storage reserves. These metabolic reactions require chemical energy obtained from compounds in the cell by respiration, a process which itself requires energy to function.

Respiration is thus a process by which the captured energy of light stored in organic compounds by photosynthesis is released by oxidation (low-temperature "burning"). The respiratory process is not entirely efficient because not all the energy produced is usable chemical energy. Some is wasted and given off as heat, the so-called vital heat or heat of respiration.

While a fruit (or other edible plant part) is still attached to the plant, the substances oxidized to $CO_2$ and $H_2O$ are easily replaced by photosynthates from green leaves or from stored reserves in stems. Once separated from the tree, the fruit is on its own. The respiratory process must continue to produce energy for cellular functions or the fruit tissues die. The only available fuel is that which is stored within the fruit itself.

Since $O_2$ is consumed and $CO_2$ is lost during respiratory oxidation, measurements of either can be used as an index of respiration. Analyses of these gases indicate the "rate of living." Perhaps it more properly should be called "the rate of dying," as the fruit or other harvested plant part will eventually consume itself through respiration if it is not earlier destroyed by pathogens. The carbon dioxide wasted to the air in the oxidation of sugars represents an irretrievable loss of carbon from the fruit.

In large part, postharvest environments are designed to reduce the rate of respiration to the minimum required to maintain vital processes. The stored reserves are thereby conserved and the postharvest life of the fruit is extended to a maximum if it is not attacked by disease organisms.

Fruits harvested and placed in respirometers at about 20°C (68°F) exhibit one of two very different respiratory patterns, as shown by $CO_2$ evolution (fig. 14.1). Sweet cherries show a gradual decrease in the respiration rate, as shown by the dotted line, as they ripen and senesce. When harvested before fully ripe,

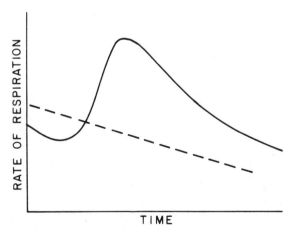

Fig. 14.1. Fruit respiration patterns. Climacteric (solid line) respiration of peaches and nonclimacteric (broken line) respiration of sweet cherries.

110

the cherries darken and become soft during the ripening process. Acids may decrease, resulting in a sweeter taste, but large increases in sugar do not occur because of the absence of large reserves of starch at harvest. In addition to sweet cherries, such fruits as grapes, citrus, strawberries, and pineapples exhibit this nonclimacteric respiratory pattern.

A climacteric respiratory pattern, shown in the solid line in figure 14.1, is exhibited by most deciduous tree fruit species (apples, pears, apricots, peaches, nectarines, plums), many tropical and subtropical fruits (banana, guava, avocado, mango), and some fruit vegetables such as the tomato. During the climacteric rise, fruits soften, and yellow colors intensify through loss of chlorophyll and increase of carotenoid pigments. Anthocyanins (red, blue, and purple colors) may be produced at this time. Evolution of ethylene increases as do other volatiles, including those associated with fruit aromas. The peak of the respiratory curve approximates the point at which fruits are considered eating ripe. Beyond that point respiration gradually decreases as the fruit senesces during its remaining life. The climacteric rise in respiration roughly coincides with a striking reduction in the fruit's resistance to certain pathogens.

Maximum postharvest life in climacteric fruits can be attained only by harvesting before the start of their respiratory climacteric rise. Studies have shown that the climacteric rise can be initiated prematurely by exposing fruit to ethylene. Thus, if some of the fruit have started to ripen, the ethylene they evolve may trigger the respiratory rise of the remainder. Similarly, rotting or badly bruised fruit may evolve sufficient ethylene to trigger ripening.

It is important to maintain fruit in a vigorous condition through lowering the respiration rate to the minimum that still permits normal cellular function. With climacteric fruits it is essential not only to reduce the respiratory rate but, if possible, to minimize and delay the climacteric rise and associated ripening processes. Low temperature is the most effective means to do this (fig. 14.2), with the possi-

bility of an important assist by use of modified atmospheres.

## Resistance to Fungal Attack

Fruit and other detached plant organs resist fungal attack in several ways. The fruit skin (cuticle and epidermis) provides protection against infection just as does the skin of animals. Most fungi are unable to penetrate sound fruit skin and must depend upon wounds. Even fungi that can penetrate the sound skin often do so with considerable difficulty and are highly dependent upon favorable environmental conditions.

Through most of the life of the developing fruit in the orchard a strong healing capability is maintained. Fruit injured by rubbing against a limb or by some other means quickly heal the wound by the formation of resistant cork cells in a periderm. The fruit thus effectively regains protection against entrance of microorganisms.

### Maturity and biochemical defense

Immature fruit may be penetrated by fungi that are unable to colonize the tissue. Similarly, no disease lesion results if the same immature fruit are punctured with needles contaminated with fungal spores. Commonly, a high degree of resistance is maintained until the fruit approaches maturity. Resistance is reduced noticeably as the fruit begins to ripen. Not only does the fruit become susceptible to its most common disease organisms when it ripens, but it succumbs to attack by fungi against which it was formerly resistant. Fruit resistance is further lost during senescence. Thus, as fruits approach maturity they are increasingly susceptible to attack by *Monilinia fructicola*, the brown rot fungus, or *Botrytis cinerea*, which causes gray mold in refrigerated storage. Other fungi, such as *Rhizopus* spp. or *Penicillium expansum*, causing blue mold rot in storage, are most likely to attack after the fruit is completely ripe or has become senescent.

### Nature of biochemical defense

It is reasonable to believe that all living organisms normally have effective mechanisms to resist disease. Susceptibility, not resistance, appears to be the exceptional condition. If this were not the case, plant and animal species would have disappeared under the onslaught of countless microorganisms. The resistance resulting from the biochemical action of tissues is not the result of a single compound or a single mechanism; instead, multiple systems are evidently involved in complex and sometimes confusing interactions.

Immature fruit appear to be resistant due in part to fungi-toxic compounds already present at infection. Most often implicated are compounds belonging to a chemical group called polyphenols, sometimes described as tannins. Members of this group

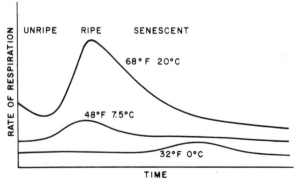

**Fig. 14.2. Effect of temperature on suppression and delay of the climacteric rise in rate of respiration.** *Redrawn from data of Fidler and North 1967.*

are responsible for the browning reaction when fruit are cut or bruised. Analyses of total polyphenols in fruit show a decrease as fruit approach maturity and ripeness, one that coincides with decreased resistance to disease. However, the total polyphenol content is not as important as the presence of specific highly fungi-toxic compounds. Furthermore, some fungi-toxic compounds not previously present are formed as a consequence of fungal attack. The plant tissue's defense is thus not merely passive but includes the ability to mount a chemical offense against the invading fungus.

Research is beginning to unravel the mysteries of the biochemical defenses of tissues. Almost none of the research has been done on fruits, probably because potato tubers, sweet potato roots, and growing plants are more amenable to study. Nevertheless, the principles being elucidated appear to have widespread applicability throughout the plant kingdom. Resistance mechanisms in fruits are likely to be similar in general to those in other plant tissues.

The evidence shows that a major portion of the resistance of tissues results when a penetrating fungus elicits a reaction from nearby cells to produce a compound toxic to the fungus. These fungi-toxic compounds, termed phytoalexins, are often polyphenolic in nature. As an example, apples have been demonstrated to produce benzoic acid in response to infection by *Cylindrocarpon mali* (*Nectria galligena*), an important storage disease in Europe. As different host tissues are studied while under attack by various pathogens, a growing list of compounds is found to be phytoalexinic.

Generally the effect of phytoalexins has been limited to areas near the site of fungal penetration. However, systemic protection conferred upon tissues by phytoalexins at considerable distances from the infection is also reported.

Little is known about the effects of postharvest environments on production of phytoalexins. What appears likely, however, is that the ability to produce phytoalexins is positively correlated with the tissue's vigor.

Although many phytoalexins are specific chemicals elicited by one or a group of pathogens, at least some can form in response to plant wounds. These "wound metabolites" appear to provide a type of wound healing that tends to prevent subsequent fungal colonization of the wound.

### Wound healing

Cuts and punctures are common avenues for infection by fungal spores. Conidia of *Monilinia fructicola*, for example, find needed moisture and nutrients for spore germination and growth in fresh wounds of a stone fruit. Pathogenic growth and rot development follow under favorable temperature conditions. Fresh wounds that escape colonization by fungi often become less subject to subsequent colonization. In particular, wounds in fruit removed from refrigerated storage may no longer be highly prone to fungal invasion. Some of this resistance might be explained by a drying of the wound area. However, studies of a wide assortment of tissue (roots, stems, tubers, leaves, fruits) suggest that a biochemical wound-healing process occurs widely throughout the plant kingdom.

When commodities are cut or otherwise damaged mechanically, they start respiring more rapidly, and ethylene production is stimulated. In plant tissues in which the effects of cuts have been studied in detail, the following events occur. Cells ruptured by the cut are killed, and cellular contents are mixed and exposed in the wound area. Enzymes, such as polyphenol oxidases, that are compartmentalized when the cell is alive are mixed with the polyphenols in the cell sap. Browning in the wound results from enzymatic oxidation of phenolic compounds. Living cells near the injury become very active metabolically even though they do not show signs of major injury. Repair is set in motion by these stressed cells. Polyphenol synthesis may lead to the accumulation of greater quantities of those already present. New compounds, often similar to those that appear after infection, may appear in the wound area. These may also be polyphenols. Compounds produced as a result of wounding include some that are highly toxic to fungi. Germinating spores deposited in such "protected" wounds are evidently killed or suppressed.

## Temperature Effects on Fungal Rots

Temperature management is so critical to postharvest disease control that all other control methods have been described as supplements to refrigeration. Without minimizing the importance of other control measures, it can be said that temperature management is central to all modern postharvest handling systems, because not only do low temperatures slow fungus development but the lowest temperature tolerated by the commodity maximizes its postharvest life potential.

### Temperature requirements of postharvest pathogens

The generalized effects of temperatures on growth of postharvest pathogens are shown in figure 14.3. Postharvest pathogens generally grow best at 20° to 25°C (68° to 77°F), depending upon the fungus species, a few responding optimally at slightly higher temperatures. The maximum temperatures for growth are typically about 32° to 38°C (90° to 100°F), but some species can grow at higher temperatures.

Fungi can be conveniently divided into those that have a minimum temperature for growth of about 0°C (32°F) or above and those that grow at lower temperatures. Non-chilling-sensitive horticultural crops can generally best be stored at the lowest temperature safe from freezing. At −1° to 0°C (30° to 32°F), only a few fungi can be expected to pose

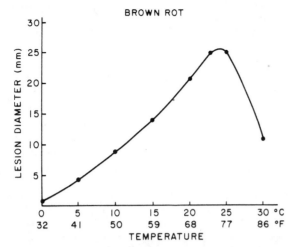

Fig. 14.3. Effects of temperature on growth of *Monilinia fructicola* in peach fruits. *Redrawn from Brooks and Cooley 1928.*

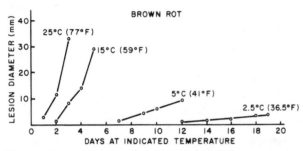

Fig. 14.4 Growth of *Monilinia fructicola* and brown rot developed in peach fruits at selected constant temperatures. *Redrawn from Brooks and Cooley 1928.*

difficulties. By far the most notable of these is *Botrytis cinerea*, particularly if the storage period extends for 3 to 4 weeks. *Penicillium expansum*, the cause of the blue mold disease of deciduous fruits, may also be of concern. Other fungi causing significant rot at 0°C (32°F) include *Alternaria alternata* and *Cladosporium herbarum*. *Monilinia fructicola* grows so slowly at 0°C (32°F) that visible brown rot of stone fruits can be seen only after excessive storage periods.

Fungi with a minimum temperature for growth of −5° to −2°C (23° to 28°F) cannot be stopped by refrigeration without freezing the fruit. Nevertheless, low temperatures are crucial to the suppression of these fungi. Although the fungi are active, their growth rate is only a minute fraction of that found at higher temperatures. Figure 14.4 shows the extent of rot development in peaches after inoculation with spores of *Monilinia fructicola* and holding at temperatures indicated.

## Significance of sigmoid growth curve

When a fungus spore lands on a medium suitable for growth (such as nutrient agar in a petri dish), the spores of many species swell, and after a few hours germ tubes protrude and growth starts. The time

taken to germinate and develop a tiny colony is called the lag phase (fig. 14.5). Growth soon achieves a rapid steady state called the log phase, which continues until growth is slowed for some reason, at which point the stationary phase of the curve is entered.

On fruit the lag phase is usually longer than on culture medium because the spore must not only germinate, but also initiate growth in the living tissues of the fruit despite considerable resistance of the tissues. Again, because of host fruit resistance, the rate of growth in the log phase may be slower than in culture, to result in a curve much less steep. The rate of growth slows when much of the fruit has become invaded.

As the temperature is lowered from near optimum to less favorable, growth curves change considerably but retain the general sigmoid (S) shape. A striking feature of the curve is that the lag phase becomes greatly extended as a consequence of very slow germination and establishment of the infection. Depending upon the fungus species, the lag phase may lengthen from a few hours to several days at optimum temperatures to weeks or months at temperatures near minimum for fungus growth. Even when a steady state has been reached, the very slow growth rate results in a much reduced slope of the log phase. This effect is illustrated by the development of brown rot in peaches (fig. 14.4). The length of the lag phase can be seen by extending the lower point of each curve to zero days.

The importance of the lag phase in fruit handling is further illustrated in figure 14.6. Data show the amount of brown rot that developed in peaches following delays in fruit cooling. After inoculation with spores of the fungus, one group of fruit was placed immediately at 0°C (32°F) while others were placed

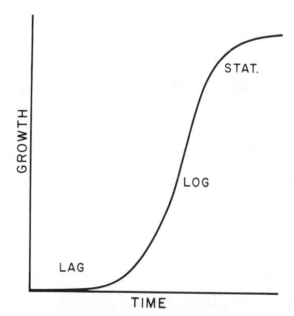

Fig. 14.5. Sigmoid curve of rot development.

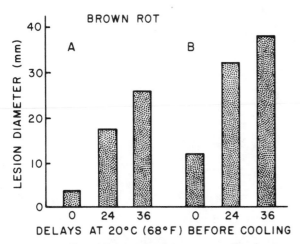

**Fig. 14.6. Cooling delays and subsequent rot development. Peaches were stored at 0°C (32°F) immediately or after delays of 24 or 36 hours. Data indicate rot development (A) 3 days and (B) 6 days after removal from cold.**

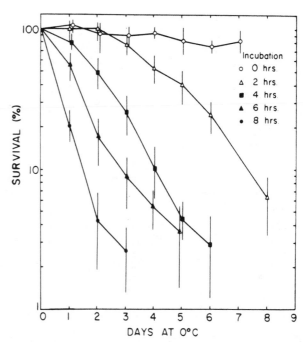

**Fig. 14.7. Killing of *Rhizopus stolonifer* spores by exposure to 0°C (32°F) after various periods of incubation at 25°C (77°F). *From Matsumoto and Sommer 1967.***

at that temperature after delays of 24 and 36 hours at 20°C (68°F). After 3 days at low temperature, obvious disease lesions had not developed. Data taken after the fourth and sixth days at 15°C (59°F) show that the effects of delayed cooling extend into the normal marketing period.

A high percentage of spore-contaminated wounds may not develop into lesions if cooling is sufficiently prompt. Spore germination is extremely slow and may fail near the minimum temperature for growth of the fungus. Processes involved in establishing the infection are also likely to be marginally functional near growth-stopping temperatures. Low temperatures while the fungus is still early in the lag phase may, consequently, result in fewer fungal lesions and delay their development.

## Fungal sensitivity to cold

Some postharvest pathogens having minimum temperatures for growth of about 5°C (41°F) or higher have developmental stages that are sensitive to low temperatures. *Rhizopus stolonifer* and *Aspergillus niger* are examples. Although ungerminated spores may not be adversely affected by low temperatures, most spores that have started to germinate are killed by several days at 0°C (32°F) (fig. 14.7). Generally, very small lesions in the fruit are sensitive to cold also, but in larger lesions the fungus is no longer killed by the cold temperature. Because only germinating species are affected, the killing is not complete. Frequently 2 or 3 percent of the spores are still viable after several days in the cold. Nevertheless, if the level of inoculum is not too high, cold can reduce disease incidence dramatically. This cold sensitivity, along with wound healing, is believed responsible for the general absence of Rhizopus rot after peaches are removed from refrigeration after a good management program. They tend to remain free of Rhizopus rot even after they are removed from refrigeration,

although similar fruit not refrigerated are heavily attacked.

The ideal objectives in the use of refrigeration for disease control are to lower the temperature (1) below the minimum temperature for growth of the fungus, (2) to a point at which the lag phase has not been completed before the fruit is consumed, or (3) in the case of cold-sensitive fungi, to kill the spores while they are germinating. These ideals are often unattainable due to better adaptability of the pathogen than the fruit host to low temperatures. The pathogen is often merely suppressed by the low temperatures that are best for maintaining the fruit in good physiological condition.

To obtain the full advantage of refrigeration, it is essential to handle fruit without delays. Field heat should be removed as soon as possible and the fruit temperature lowered to near 0°C (32°F) or the lowest temperature tolerated by the commodity. Even if subsequent transport is at about 5°C (41°F), there are advantages to cooling to 0°C because suppression of fungi is likely to be more nearly complete. The lesions of cold-sensitive fungi may be permanently halted. Further, to cool to the lowest safe temperature is advantageous from the standpoint of the physiology of the fruit.

## Modified or Controlled Atmospheres

Atmospheres around fruit are sometimes provided in which the oxygen ($O_2$) is low, carbon dioxide ($CO_2$) high, or both. If a close control of these gases is maintained, the synthetic atmosphere is commonly

**114**

called a controlled atmosphere (CA). Modified atmosphere (MA) is a term that may designate any synthetic atmosphere but often is used if there is little or no possibility of making adjustments in gas composition during storage or transportation. The purpose of these atmospheres is usually to extend the fruit's postharvest life by suppressing the rate of respiration. Another objective is to suppress diseases.

The effects of modified atmospheres on postharvest diseases can be either direct or indirect. The maintenance of the fruit in good physiological condition may result in a fruit with considerable disease resistance. A direct effect is also possible. Because the fungus pathogen respires as do the fruits, lowering the oxygen or raising the carbon dioxide can suppress growth of the fungus. In this discussion, primary attention is given to the direct effects of atmosphere modification on the fruit pathogen.

### Low oxygen

Oxygen is required for normal respiration of both the fruit and its fungus pathogen. The beneficial effects of low $O_2$ on fruit become evident as oxygen in the atmosphere is decreased to 5 percent or below; benefits increase at lower $O_2$ levels. In controlled atmosphere storages the level of $O_2$ is commonly maintained at about 2 to 3 percent. These levels are considered the lowest that can prudently be maintained with the control methods usually available in storage.

Anaerobic or fermentative respiration is the consequence of an excessively low $O_2$ level. The fruit first develops off-flavors as substances, particularly alcohols and acetaldehyde, accumulate in the tissues. Eventually tissues are irreparably damaged and fruit death results.

Suppression of fungi by a 2 percent $O_2$ atmosphere is modest, often no more than about 15 percent below the rate of growth in air (21% $O_2$) as shown for *Botrytis cinerea* (figs. 14.8 and 14.9). Significant growth reductions result if the $O_2$ level is lowered to 1 percent, but that is generally considered too low for commodity safety.

### Hypobaric atmospheres

Storage and transport under low pressure or hypobaric conditions has stirred considerable interest. Test vans have been constructed that maintain low pressure by use of vacuum pumps and regulated flow of air through the van. The refrigeration is conventional. When a 0.1 atmosphere pressure is maintained, the available $O_2$ is reduced from about 21 percent of air to about 2 percent. A 0.05 atmosphere is equivalent to 1 percent $O_2$. An added benefit to the low $O_2$ is a very effective removal of ethylene produced by commodities.

From the standpoint of disease, very few critical data are available but a comparison of results with *Botrytis cinerea* at 0.1 and 0.05 atmosphere suggests

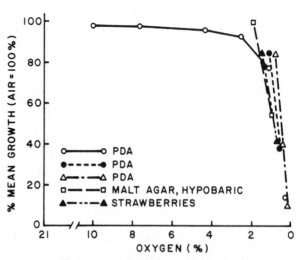

**Fig. 14.8. Effect of low oxygen on growth of *Botrytis cinerea* in culture (potato-dextrose agar [PDA], malt agar) or strawberry fruits. All tests were conducted at normal barometric pressure in an atmosphere of low oxygen composition except hypobaric was in air at 0.1 and 0.05 atmospheres. *Data from Borecka and Olak 1978, Couey et al. 1966, Follstad 1966, and Sommer et al. 1981.***

that the suppressive effect is similar to controlled atmospheres at 2 or 1 percent $O_2$.

### Carbon dioxide elevation

Air commonly contains about 0.03 percent $CO_2$. Elevation of $CO_2$ above about 5 percent noticeably suppresses fruit respiration. If the concentration of $CO_2$ is excessive, however, off-flavors develop and fruit injury results. The relationship of $CO_2$ concentration to fruit injury is time and temperature related. Fruits tolerate very high levels of $CO_2$ (more than 20 percent) for several days at transit temperatures, i.e., 3° to 5°C (38° to 41°F), but few tolerate those elevated concentrations if storage or transit in the modified atmosphere is extended for several weeks. However, species and varietal differences in $CO_2$ tolerance are important.

The addition of 10 percent to 15 percent $CO_2$ at a transit temperature of 5°C (41°F) commonly affects both host and pathogens in a manner roughly comparable to a temperature of 0°C (32°F) in air. Carbon dioxide added to air is widely used in transport of Bing cherries, primarily to suppress *Botrytis cinerea* (gray mold) and *Monilinia fructicola* (brown rot), and strawberries to suppress *B. cinerea*.

Although fungi are suppressed by elevated (10 to 20 percent) $CO_2$ levels, many fungi grow poorly in its complete absence. A number of enzymes have been implicated in $CO_2$ fixation within fungal cells.

### Combined low oxygen, high carbon dioxide

The effects of low $O_2$ and high $CO_2$ atmospheres are believed to be additive. Commonly used atmospheres of about 2 to 4 percent $O_2$ and 5 to 7 percent $CO_2$ suppress respiration and delay ripening of fruit, which could not safely be achieved with modification of the

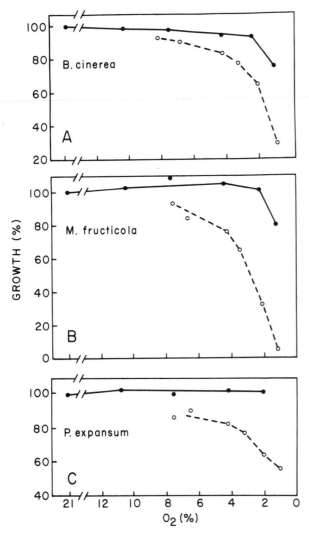

**Fig. 14.9. Suppression of stone fruit postharvest pathogens by oxygen alone (solid lines) or combined with 10 percent carbon monoxide (dotted lines).** *Sommer et al. 1981.*

atmosphere by single gases. Modification of $O_2$ alone would likely require 1 percent $O_2$ or less to achieve similar effects. $CO_2$ in air might require 15 to 20 percent, or more, to equal the combined effect.

### Controlled atmosphere with carbon monoxide

The modest suppression of fungi by low $O_2$ and elevated $CO_2$ concentrations commonly used suggests the desirability of including a fungistatic gas. Carbon monoxide (CO) may be suitable in that role. Although tests have been encouraging with some fruits, the commercial use of CO with fruits is limited.

Carbon monoxide functions physiologically as an enzyme inhibitor and a competitor of $O_2$. Carbon monoxide (10 percent) added to air results in modest reduction in growth of postharvest pathogens. When added to an atmosphere of low oxygen (3 percent or less), however, much greater suppression has been

noted (fig. 14.9). If added to a controlled atmosphere of 2.2 percent $O_2$ + 5 percent $CO_2$, even greater suppression is observed because of the additive effects of CO and $CO_2$.

In general, the suppressive effects of CO increase with lowered $O_2$ (figs. 14.9, 14.10). Further, the suppression is greater at 5.5°C (42°F) than at higher temperatures such as 12.5°C (55°F).

### Humidity

Water vapor is a gas that constitutes an important part of the atmospheric environment of harvested perishable commodities. Its composition in the atmosphere as a percentage of saturation (relative humidity or RH) varies widely with temperature changes. Although the RH of storage is ordinarily never at saturation, free water on the surface of commodities can occur. Liquid water forms if, at any time in the normal cycling of the temperature within a refrigerated storage, the temperature of the commodity surface falls below the dew point temperature of the atmosphere.

Problems of accurately measuring RH in the microclimate of the surface of fruits and large changes in RH due to minor temperature fluctuations have made studies of humidity effects difficult under postharvest conditions. With pathogens such as *Monilinia fructicola*, germination and direct penetration of commodities are aided by saturated atmospheres or water on the fruit surface. With jacketed storage or packages with moisture barriers of plastic film, the high humidity might be a factor in promoting disease if temperatures were favorable. The disease-enhancing tendency of these high-humidity environments would likely increase if the commodity were wet when packaged.

Cold commodities removed from refrigeration to ambient temperatures condense moisture on their

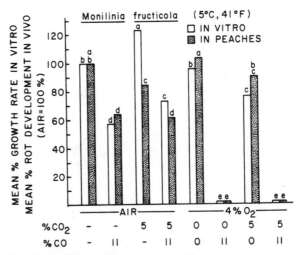

**Fig. 14.10. Effects of atmospheres of varying oxygen, carbon dioxide, and carbon monoxide content on suppression of *Monilinia fructicola* in culture and in peach fruit.** *Redrawn from Kader et al. 1982.*

surface from surrounding warm humid air. This sweating continues until the fruit has warmed to a temperature above the dew point temperature. The duration of the sweat depends on such factors as the fruit-air temperature difference, the exposure of the commodity to the air, air movement, and size or bulk of the commodity. Although moisture from sweating at the fruit surface has often been a concern among handlers, it appears likely that the warm-up period of many commodities may be too short to be an important factor.

The horticultural necessity for maintaining high humidity in commodity environments is primarily to minimize loss of moisture, which results in shrivel and loss of turgidity in tissues. With peaches, for example, a loss of 3 to 4 percent of the original weight usually results in noticeable shrivel. The effect, if any, of weight loss on disease resistance requires added studies. In some vegetables, however, weight loss has been associated with increased susceptibility to disease. In one study, carrots in storage were initially resistant to *Botrytis cinerea* but became susceptible after a moisture loss of 8 percent.

## REFERENCES

1. Borecka, H., and J. Olak. 1978. The effect of hypobaric storage conditions on the growth and sporulation of some pathogenic fungi. *Fruit Sci. Rep.* (Poland) 5:39-41.

2. Brooks, C., and J. S. Cooley. 1928. Time-temperature relations in different types of peach rot infection. *J. Agric. Res.* 37:507-43.

3. Couey, H. M., M. N. Follstad, and M. Uota. 1966. Low oxygen atmospheres for control of postharvest decay of fresh strawberries. *Phytopathology* 56:1339-41.

4. Fidler, J. C., and C. J. North. 1967. The effect of conditions of storage on the respiration of apples; I, the effects of temperature and concentrations of carbon dioxide and oxygen on the production of carbon dioxide and uptake of oxygen. *J. Hortic. Sci.* 42:189-206.

5. Follstad, M. N. 1966. Mycelial growth rate and sporulation of *Alternaria tenuis*, *Botrytis cinerea*, *Cladosporium herbarum*, and *Rhizopus stolonifer* in low oxygen atmospheres. *Phytopathology* 56:1098-99.

6. Harding, V. K., and J. S. Heale. 1980. Isolation and identification of the antifungal compounds accumulating in the induced resistance response of carrot root slices to *Botrytis cinerea*. *Physiol. Plant Pathol.* 17:277-89.

7. Kader, A. A., M. A. El-Goorani, and N. F. Sommer. 1982. Effects of carbon monoxide added to controlled atmospheres on postharvest decay, physiological responses, and quality of peaches. *J. Am. Soc. Hortic. Sci.* 107:856-59.

8. Matsumoto, T. T., and N. F. Sommer. 1967. Sensitivity of *Rhizopus stolonifer* to chilling. *Phytopathology* 57:881-84.

9. Noodén, L. D., and A. C. Leopold, eds. 1988. *Senescence and aging in plants*. Academic Press Inc., New York. 526 pp.

10. Shewfelt, P. L. 1986. Postharvest treatment for extending the shelf life of fruits and vegetables. *Food Technol.* 40(5):70-80, 89.

11. Sommer, N. F., R. J. Fortlage, J. R. Buchanan, and A. A. Kader. 1981. Effect of oxygen on carbon monoxide suppression of postharvest diseases of fruit. *Plant Dis.* 66:357-64.

12. Van den Berg, L. and C. P. Lentz. 1978. High humidity storage of vegetables and fruits. *HortScience* 13:565-69.

# 15

# Postharvest Diseases of Selected Commodities

Noel F. Sommer, Robert J. Fortlage, and Donald C. Edwards

The threat of postharvest disease influences the way most fresh horticultural crops are handled. Therefore, it is necessary to understand the nature of disease organisms, the physiology of the host commodity, and how handling methods affect them both, as well as the various environmental and handling stresses that fruits and vegetables suffer.

During handling, cuts, bruises, and punctures may facilitate entrance of a pathogen into a commodity. High or low temperatures can alter physiology, increasing its susceptibility to certain pathogens.

Relative humidity (vapor pressure deficit) and atmosphere composition are other considerations in combatting postharvest diseases. The presence of pathogens completes the disease triad (host-environment-pathogen). The level of disease the pathogen can cause and the number of fungus spores present determine disease incidence and severity.

## The Pathogen

Fungi are overwhelmingly present in postharvest diseases of fruits and vegetables. Bacteria frequently cause disease in certain vegetables, but they are generally rare in tree fruits and berries. Virus diseases may develop or intensify postharvest disease expression in certain root or tuber crops, but they do not affect fruit after harvest.

Fungal pathogens are commonly members of the class Ascomycetes and the associated Fungi Imperfecti. Phycomycetes are represented by the genus *Rhizopus* and near relatives, and by the genera *Phytophthora* and *Pythium*. Basidiomycetes, with few exceptions, are not postharvest pathogens.

Among the Ascomycetes, pathogens are usually encountered in the asexual (vegetative conidial) state in postharvest diseases. The sexual state is rarely seen in culture or in diseased commodities; in some species it is rare in nature. To use the conidial state for purposes of identification, an asexual binomial name is commonly given. Exceptions apply in certain well-known fungi, such as the stone fruit brown rot organism *Monilinia fructicola* or *Sclerotinia sclerotiorum*, which lacks a known asexual spore state.

## The Infection Process

The propagule functioning to disperse fungi is generally a spore, but other propagules exist. In fact, most living parts of the fungus are capable of growth and developing disease under favorable conditions.

Spores of postharvest fungi exist in many sizes and shapes and may be either vegetative or sexual. Sexual spores may be part of the life cycle of the fungus; sometimes, they may serve to permit survival of the fungus during drought or winter cold. It is through the vegetative spore that diseases spread.

### Spore germination

Inactive vegetative spores are usually not dormant. If a spore is deposited in a moist, fresh wound, for example, it germinates immediately if temperature and atmospheric conditions are favorable. Besides water, certain substances are required. (1) Oxygen is required for spore germination; however, low oxygen tensions suffice. (2) Commonly, spores do not germinate well in the complete absence of carbon dioxide, which may be fixed during germination. (3) The presence of metabolizable organic compounds in the liquid enhances germination and may sometimes be required.

Absorption of moisture, an initial stage of germination, is usually accompanied by spore swelling. Some spores swell manyfold; others swell relatively little. As swelling becomes noticeable, oxygen consumption increases sharply and carbon dioxide evolution marks an enhanced metabolic rate.

Before germination is triggered, spores exhibit minimal metabolic activity. Germination is associated with rapid increase in the synthesis of RNA, DNA, and protein. The amount of endoplasmic reticulum and the number of mitochondria also increase.

Spores are characteristically covered by a thick spore coat. After swelling, a germ tube protrudes through the spore coat. The wall of the germ tube is continuous with and a part of the innermost layer of the spore coat. Germ tube protrusion and possibly much of the previous spore swelling depend on protein synthesis. As the germ tube lengthens through polar growth, side branches are initiated (fig. 15.1).

Spore germination is a hazardous period in the life of the fungus. During swelling, spores are susceptible to lethal effects of gamma and ultraviolet irradiation, low and high temperatures, absence of oxygen, and exposure to toxic chemicals. Once un-

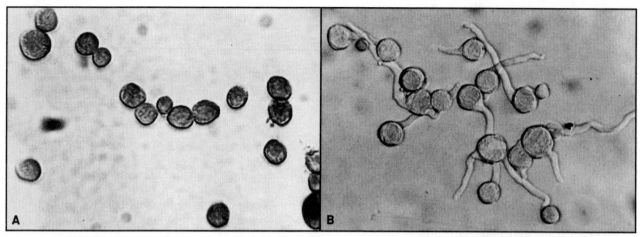

**Fig. 15.1. Sporangiospores of *Rhizopus stolonifer*: (a) ungerminated spores; (b) germinated spores. *Magnification 400×*.**

derway, germination evidently cannot stop for long without loss of the ability to resume normal growth.

### Spore dissemination

Many spores produced in exposed structures are powdery and ideally suited to wind transport. Other spores, produced in more or less enclosed structures, may be exuded to the surface in a gelatinous or mucilaginous substance. Such spores are dispersed by rain and by windborne mist often for long distances.

Some postharvest pathogens, primarily in the genus *Phytophthora*, produce sporangia structures that may germinate by germ tubes protruding like spores. Under favorable conditions, however, many motile spores may be formed instead within the sporangium. Upon emergence, the motile spores swim in soil water before resting. With favorable conditions, germination is by formation of a germ tube. These sporangia or motile spores are usually deposited on fruit in the orchard when soil is splashed on lower fruit by driving rain. Sprinkler irrigation systems, insects, small animals, birds, and humans assist in spore dissemination.

### Initial fruit penetration

Two kinds of postharvest pathogens can enter fruit through its skin. One group of decay fungi bypasses the skin through wounds: bruises, stem punctures, cuts, limb-rubbed areas, abrasions, and insect punctures. The germinating spore grows and colonizes the exposed fruit tissue. The other group forms appressoria, special structures that permit the fungus to penetrate the fruit cuticle and epidermis. Commonly, these infections are initially quiescent; rot lesions do not develop until the fruit is nearly ripe.

Spores of all postharvest pathogens require high humidity or free water for several hours to germinate. Frequently, fruit surfaces are too dry to encourage germination, but spores in wounds germinate because fruit juice is present.

Stem-end rots result from infection caused when the stem is severed at harvest. *Lasiodiplodia theobromae*

causes stem-end rots of citrus fruits, mango, papaya, and watermelon. Similarly, *Thielaviopsis paradoxa* invades stem tissues of banana fingers and pineapple. Cut peduncles of apple and pear may be colonized by *Botrytis cinerea* or *Penicillium expansum*. During storage the disease progresses from the peduncle to the body of the fruit. In some fruit species, stems may be the first part of the fruit to become senescent. Senescence is likely a factor in Alternaria stem-end rots of citrus fruits.

Fungi may colonize senescent floral parts at blossoming and only much later grow into and rot the fruit proper. For example, floral parts of California Bartlett pear are infected by *Botrytis cinerea* near the end of the blossoming period. Dead styles and stamens colonized by the fungus are retained within the floral cavity near the core. Only as the fruit become senescent at the end of storage life does the fungus successfully invade the fruit flesh.

A fungus directly penetrates fruit skin according to the following scenario. Spores landing upon a fruit, when temperature and humidity are satisfactory, germinate within a few hours by sending out a germ tube. After the germ tube is well formed, a thick-walled structure, the appressorium forms (fig. 15.2). The appressorium and germ tube adhere tightly to the fruit surface by mucilaginous material produced by the germ tube. The thickening of appressoria walls is complete except for a pore on the underside against the fruit surface, covered only by the thin germ tube wall. It is believed that enzymes are excreted through the pore to the fruit surface, including a cutinase capable of hydrolyzing the cutin that overlies the epidermis. Through the pore of the appressorium, a very fine germ-tube-like protrusion called an infection peg penetrates the cuticle where it is weakened by enzymatic action (fig. 15.3). Penetration is aided by considerable pressure from the appressorium. After penetration, the infection peg regains the normal size for mycelia of the fungus. It proceeds to branch and rebranch, thoroughly invading the fruit flesh.

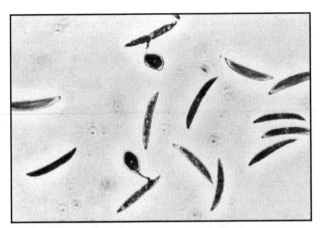

**Fig. 15.2. Appressoria produced on short germ tubes of sickle-shaped conidia of *Colletotrichum* sp. *Magnification approximately 700×.***

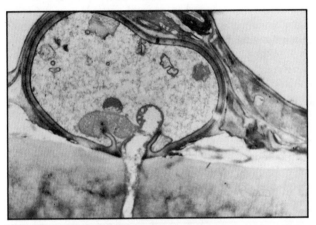

**Fig. 15.3. Electron micrograph of *Colletotrichum gloeosporioides* penetrating host. Magnification approximately 7,000×. *Courtesy of Dr. G. Eldon Brown, University of Florida, Lake Alfred Research and Extension Center, Lake Alfred, Fla.***

Latent infections result from an interruption in infection following direct penetration. If the penetrating infection peg is unable to overcome host resistance, the infection may remain quiescent until the fruit's resistance is reduced. An example is found in *Colletotrichum gloeosporioides*, a fungus causing anthracnose of many fruits, such as apple, avocado, mango, and papaya. Fruit are commonly penetrated while developing in the orchard. Before a fruit ripens, it is highly resistant, but upon ripening it becomes susceptible. Typically, anthracnose is a rot of ripe fruit.

### Tissue invasion and rotting

Once penetration succeeds, the mycelium grows and branches thoroughly invading the fruit flesh. The advancing mycelium excretes into the fruit tissue toxins that kill the cells. Extracellular enzymes are produced that break down constituents, degrading complex substances into low molecular-weight compounds that can enter the fungus cell. These com-

pounds provide the building blocks for synthesizing substances required for the fungus to grow, plus the energy for its life processes.

### Sporulation

The production of spores, the final step in the fruit rotting process, completes the vegetative spore cycle.

## Resistance to Infection

Fruits and vegetables effectively heal wounds before harvest by walling off infections with wound periderm that appears to be made mainly of ligninlike substances (fig. 15.4). In some cases, these barriers form around wounds of fruits (apple, pear, citrus) even long after harvest.

Most fruits, while still green, are highly resistant to most pathogens and insects and do not attract animals or humans. The stomachaches that afflict young children who eat green apples are lessons well learned. Fruits become attractive to animals about the time seeds mature. Then it is advantageous for the fruit to attract animals to spread the seeds. Much of the fruit's resistance to disease is lost during ripening.

Before ripening, fruit typically contain substances, often of a phenolic nature, that are toxic to fungi. These are present in the fruit at time of infection and are generally described as preformed inhibitors. Other inhibitors, formed in response to infection, are called postinfection inhibitors or phytoalexins.

## Postharvest Diseases of Temperate Fruits

### Apple and pear (pome fruits)

#### Blue mold rot

Sexual state: Unknown
Asexual state: *Penicillium expansum* Link
= *Coremium glaucum* Link
= *Penicillium glaucum* Link

*Penicillium expansum* is usually considered the causal organism of blue mold rot. Studies in Poland, South Africa, and the United States show that other species of blue or blue-green colored Penicillia may cause a similar disease, but *P. expansum* is the greatest cause of loss in stored apples.

At first, blue mold lesions are light colored and soft. As the lesion enlarges, the decayed portion can be easily separated from the surrounding sound tissue. The fungus growth on the lesion surface, at first white, becomes pale blue as sporulation occurs (fig. 15.5). Although blue mold lesions start from wound infections, the fungus in a rotting fruit can cause "nesting" by growing into sound neighboring fruit. *Penicillium expansum* produces abundant conidia that are readily airborne.

*Penicillium expansum* appears to be a strict wound parasite, or nearly so. Blue mold readily colonizes cuts or punctures. Less frequently, the fungus colonizes the peduncle, particularly peduncles that are thick and fleshy (figs. 15.6 and 15.7). Lenticels are also infected, usually following injury. Lenticels on some apple cultivars, such as 'Golden Delicious,' are thin, weakly sealed, and easily ruptured by variations in pressure. Apples handled in water are especially liable to infection of the wounded lenticels, and sprays or dips in benzimidazole fungicides may be necessary to avoid excess rot.

### Gray mold rot

Sexual state:   *Botryotinia fuckeliana* (de Bary) Whetzl
       = *Sclerotinia fuckeliana* (de Bary) Fuckel
Asexual state: *Botrytis cinerea* Pers.:Fr.
       = *Haplaria grisea* Link
       = *Botrytis vulgaris* Link:Fr.

Some isolates of *Botrytis cinerea* sporulate abundantly and sclerotia are absent; others develop abundant sclerotia with sparse spore production. Growth rates are highly variable.

The gray mold lesion, light brown to brown in color and sometimes dark brown, is characteristically less soft than blue mold lesion, and has a well-defined margin, but rotted tissue cannot be neatly separated from the surrounding healthy tissue as with blue mold. Given adequate humidity, lesions become covered with gray, sporulating mycelium.

*Botrytis cinerea* frequently rots apple and pear in storage and may colonize stems, especially in cultivars with thick, fleshy stems such as 'Yellow Newtown' apple or 'Beurre d'Anjou' pear. Stem infections may grow into the fruit proper and completely colonize it (fig. 15.8). Stem punctures and other wounds in the fruit body are readily colonized.

Calyx-end rot is common in California-grown Bartlett pear (a.k.a. Williams or bon Christien) and in 'Packham's Triumph' and 'Buerre Bosc' pears in South Africa. Infections occur during blossoming, and colonized pistils and stamens, particularly the former, are retained within the floral tube of the fruit (fig. 15.9). Fruit rot in the orchard is rare or nonexistent. Similarly, no rot occurs in storage until the fruit starts to turn from green to yellowish green. At that time changes in fruit resistance apparently permit the mycelium to grow into the fruit proper (fig. 15.10).

Germinating spores of *B. cinerea* are unlikely to penetrate unwounded apples and pears but enter the fruit mostly through wounds or injured lenticels. This fungus can directly penetrate blossoms, however. Furthermore, mycelium may grow and contact nearby fruit in containers as it can penetrate the unwounded fruit surface.

### Anthracnose rots

Sexual state:   *Pezicula malicorticis* (Jacks.) Nannf.
       = *Neofabraea malicorticis* Jacks.
       = *Neofabraea perennans* Kienholz
Asexual state: *Cryptosporiopsis curvispora* (Peck) Gremmen.
       = *Cryptosporiopsis malicorticis* (Cordl.) Nannf.
       = *Gloeopsorium perennans* Zeller & Childs.

*Pezicula malicorticis* causes serious losses in the U.S. Northwest (Oregon, Washington, Idaho) and British Columbia, as well as in Britain, France, and other northern European countries. The disease is present in California and most other apple-producing states without being a source of major concern. The conidia of *P. malicorticis* directly penetrate apple and pear fruit, often at the lenticel, or invade wounds. Infections that occur during fruit development are usually latent, becoming active as the fruit ripens. Visible lesion development occurs only after a long time in storage, when fruit change from resistant to susceptible.

Bull's-eye rot of fruit and perennial canker on trees, once thought to be caused by two organisms, are now considered the asexual and sexual state of the same organism. On fruit, lesions are usually nearly round and have a light-colored center (fig. 15.11). As lesions develop, concentric circular bands of wet, cream-colored masses of conidia may be evident, forming a target pattern (fig. 15.12). Cankers form on smaller branches usually less than 5 cm (2 inches) in diameter. The first indication of a canker is discolored bark extending inward to the cambium in late fall (fig. 15.13). At maturity the following spring, cankers are usually elliptical, 2.5 to 25 cm (1 to 10 inches) long and 5 to 7.5 cm (2 to 3 inches) across. The fungus in cankers produces spores of the asexual state. Later the sexual state develops in the canker and produces sexual spores. Fruit infections may largely result from conidia produced in the perennial cankers in the tree.

### Target or lenticel spot

Sexual state:   *Pezicula alba* Guthrie
Asexual state: *Phyctema vagabunda* Desmaz.
       = *Gloeosporium album* Oster
       = *Gloeosporium allentoidem* Peck
       = *Gloeosporium allantosporum* Fautrey
       = *Gloeosporium diervillae* Grove
       = *Gloeosporium frigidum* Sacc.
       = *Gloeosporium tineum* Sacc.
       = *Trichoseptoria fructigena* Maublanc

A disease similar to bull's-eye rot is caused by *Pezicula alba*, which has a comparable infection process. Infections characteristically occur at lenticels, although wounds serve as an alternative infection site. *Pezicula alba* is generally considered less damaging than *P. malicorticis*, and it occurs primarily in European apple-growing areas. Its presence in North America has been reported sporadically, but it has not been damaging and sometimes the organism may have been confused with *P. malicorticis*.

### Bitter rot

Sexual state:   *Glomerella cingulata* (Stonem.) Spauld. & Schrenk.
       Note: There are at least 14 synonyms.
Asexual state: *Colletotrichum gloeosporioides* (Penz.) Penz. & Sacc.
       = *Gloeosporium fructigenum* Berk.
       Note: There are believed to be several hundred synonyms.

*—Continued on p. 137*

# Postharvest Disease Symptoms

Duplicate 35mm color slides of these photographs can be purchased, either as a complete set (figs. 15.4 through 15.136) or as subsets for temperate fruits (figs. 15.4 through 15.53), subtropical fruits (figs. 15.54 through 15.67), tropical fruits (figs. 15.68 through 15.90), or vegetables (figs. 15.91 through 15.136).

For prices and ordering information, write to Cooperative Extension Visual Media, University of California, Davis, California 95616, or telephone (916) 757–8980.

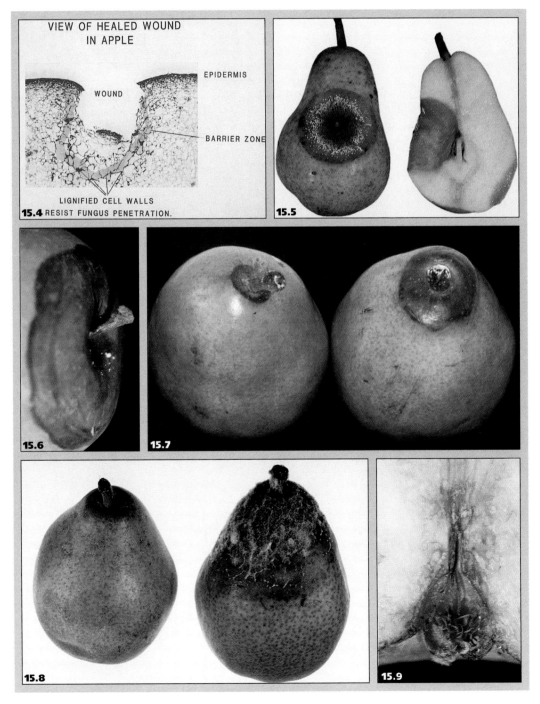

**Temperate Fruits**

**Fig. 15.4.** Wound healing in 'Yellow Newtown' apple fruit after 6 months' controlled atmosphere storage at 3.5°C (38°F). Note cell wall thickening near wound.

**Fig. 15.5.** Blue mold rot (*Penicillium expansum*) in ripe 'Bartlett' pear.

**Fig. 15.6.** Stem-end blue mold rot (*Penicillium expansum*) of 'Granny Smith' apple.

**Fig. 15.7.** Stem-end blue mold rot (*Penicillium expansum*) of 'Beurre d'Anjou' pear.

**Fig. 15.8.** Gray mold rot (*Botrytis cinerea*) of 'Beurre d'Anjou' pear.

**Fig. 15.9.** Senescent style and stamen fragments in the sub-calyx area of 'Bartlett' pear.

VIEW OF HEALED WOUND IN APPLE

EPIDERMIS

WOUND

BARRIER ZONE

LIGNIFIED CELL WALLS
**15.4** RESIST FUNGUS PENETRATION.

15.5

15.6

15.7

15.8

15.9

# Postharvest Disease Symptoms:
## Temperate Fruits

Fig. 15.10. Calyx-end rot (*Botrytis cinerea*) in 'Bartlett' pear.

Fig. 15.11. Bull's-eye rot (*Pezicula malicorticis*) in 'Yellow Newtown' apple.

Fig. 15.12. Bull's-eye rot (*Pezicula malicorticis*) showing sporulating lesion.

Fig. 15.13. Limb canker (*Pezicula malicorticis*) of apple tree. *Courtesy of Dr. A. Helton, University of Idaho, Moscow.*

Fig. 15.14. Bitter rot (anthracnose) (*Colletotrichum gloeosporioides*) in 'Golden Delicious' apple.

Fig. 15.15. Black rot (*Botryosphaeria obtusa*) in 'Yellow Newtown' apple.

Fig. 15.16. White rot (*Botryosphaeria dothidea*) in 'Yellow Newtown' apple.

Fig. 15.17. Alternaria rot (*Alternaria alternata*) in 'Yellow Newtown' apple with delayed sunscald.

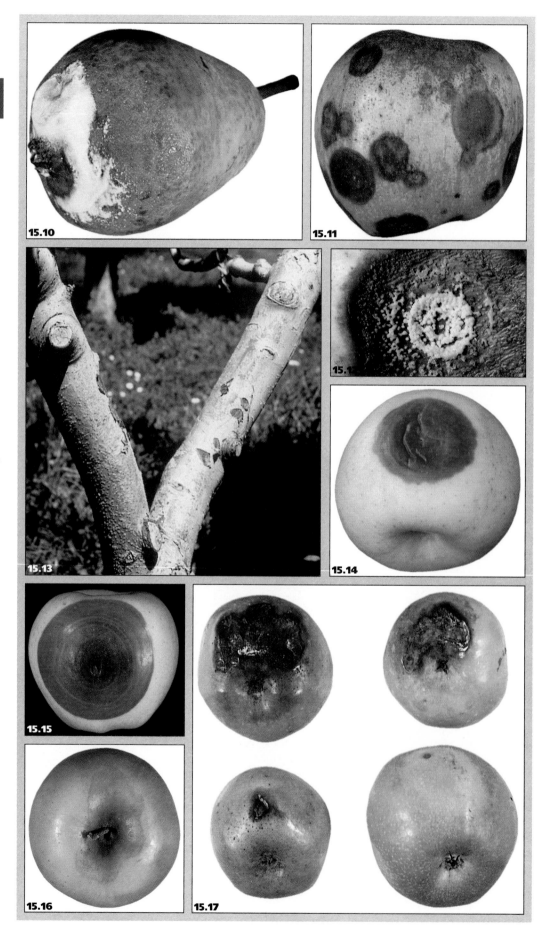

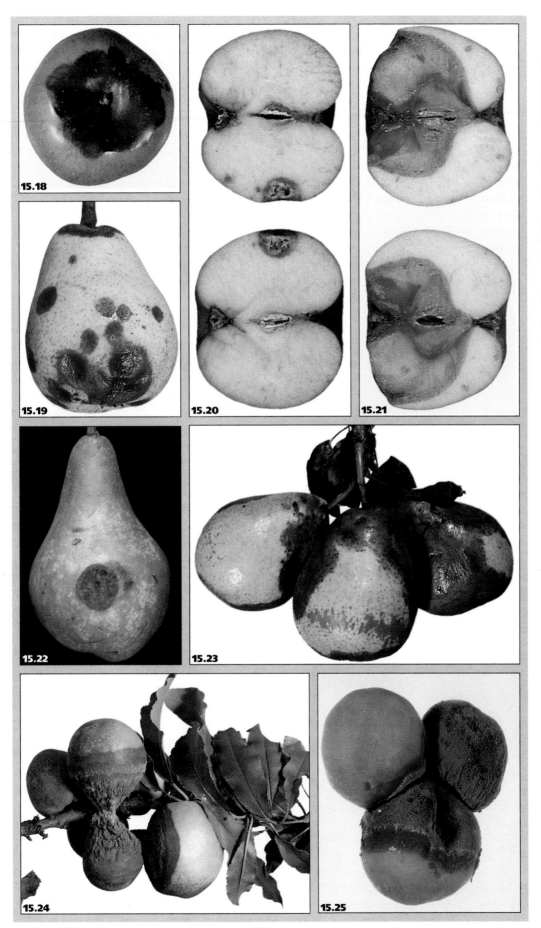

# Postharvest Disease Symptoms:
## Temperate Fruits

Fig. 15.18. Alternaria stem-end rot (*Alternaria alternata*) in chilled 'Yellow Newtown' apple.

Fig. 15.19. Cladosporium rot (*Cladosporium herbarum*) in ripe 'Bartlett' pear.

Fig. 15.20. Pleospora rot (*Stemphyllium botryosum*) in 'Yellow Newtown' apple.

Fig. 15.21. Phomopsis rot (*Phomopsis mali*) of 'Yellow Newtown' apple.

Fig. 15.22. Side rot (*Phialophora malorum*) in 'Buerre Bosc' pear.

Fig. 15.23. Phytophthora or sprinkler rot (*Phytophthora cactorum*) in 'Bartlett' pear in the orchard.

Fig. 15.24. Brown rot (*Monilinia fructicola*) in nectarines before harvest.

Fig. 15.25. Brown rot (*Monilinia fructicola*) nest of peaches.

# Postharvest Disease Symptoms:

## Temperate Fruits

Fig. 15.26. Brown rot (*Monilinia fructicola*) in fingernail wound in nectarine.

Fig. 15.27. Stem-end rot (*Monilinia fructicola*) of peach.

Fig. 15.28. Nest of *Rhizopus*-infected peaches.

Fig. 15.29. Rhizopus rot (*Rhizopus stolonifer*) in packed peaches.

Fig. 15.30. Apricots 18 months after canning. A: fruit halves disintegrated because of contamination with *Rhizopus*-infected fruit juice with pectolytic enzymes not completely inactivated by heat of canning; B: treatment with 0.5 N sodium hydroxide for 2 minutes before canning inactivated pectolytic enzymes.

Fig. 15.31. Mucor rot (*Mucor* spp.) in Chilean nectarines arriving in the United States.

Fig. 15.32. Blue mold rot (*Penicillium expansum*) of nectarines stored at 0°C (32°F).

Fig. 15.33. Alternaria rot (*Alternaria alternata*) of plum.

Fig. 15.34. Alternaria rot (*Alternaria alternata*) of Nubiana plum.

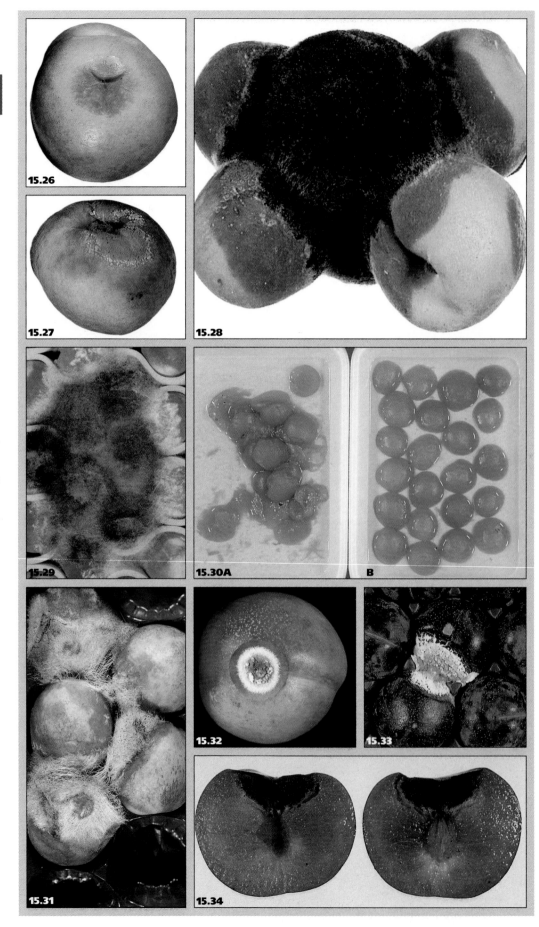

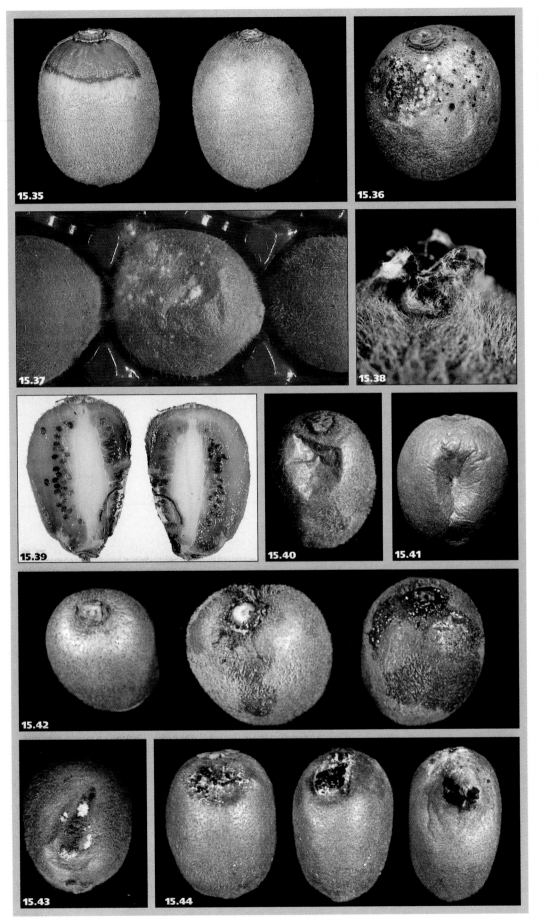

# Postharvest Disease Symptoms:
## Temperate Fruits

Fig. 15.35. Botrytis stem-end rot (*Botrytis cinerea*) of kiwifruit.

Fig. 15.36. Botrytis stem-end rot (*Botrytis cinerea*) with sclerotia on the kiwifruit.

Fig. 15.37. Nest of kiwifruit rotted by *Botrytis cinerea*.

Fig. 15.38. *Alternaria alternata* growing over kiwifruit sepals.

Fig. 15.39. Alternaria rot (*Alternaria alternata*) in sunburned kiwifruit.

Fig. 15.40. Dothiorella soft rot (*Dothiorella gregaria*) of kiwifruit.

Fig. 15.41. Phoma rot (*Phoma* spp.) of kiwifruit.

Fig. 15.42. Phomopsis stem-end rot (*Diaporthe actinidiae*) of kiwifruit with juice exudation.

Fig. 15.43. Penicillium stem-end rot (*Penicillium expansum*) of kiwifruit.

Fig. 15.44. Buckshot rot (*Typhula* sp.) of stored kiwifruit.

# Postharvest Disease Symptoms:
## Temperate Fruits

Fig. 15.45. Various aspects of Botrytis rot in strawberry.

Fig. 15.46. Nest of strawberries rotted by *Botrytis cinerea*.

Fig. 15.47. Rhizopus rot (*Rhizopus stolonifer*) of ripe or near-ripe strawberries in the field.

Fig.15.48. Mucor rot (*Mucor piriformis*) nesting in strawberry basket. Black sporangia have not yet formed on the sporangiophores of the fungus.

Fig. 15.49. Anthracnose (*Colletotrichum acutatum*) on strawberries in the field.

Fig. 15.50. Leather rot (*Phytophthora cactorum*) in green strawberries.

Fig. 15.51. Leather rot (*Phytophthora cactorum*) in ripening strawberries.

Fig. 15.52. Gray mold rot (*Botrytis cinerea*) in grapes stored at 0°C (32°F).

Fig. 15.53. Penicillium rot (*Penicillium expansum*) in grapes stored at 0°C (32°F).

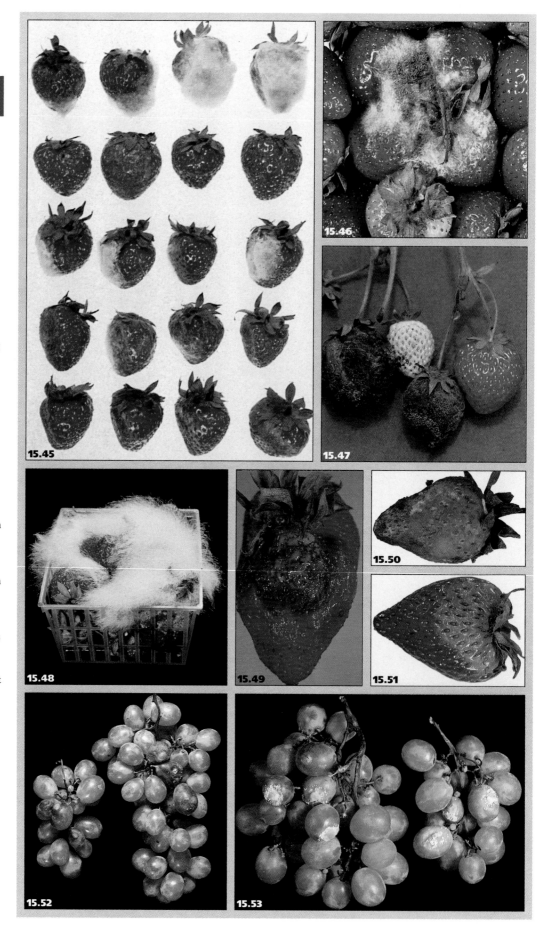

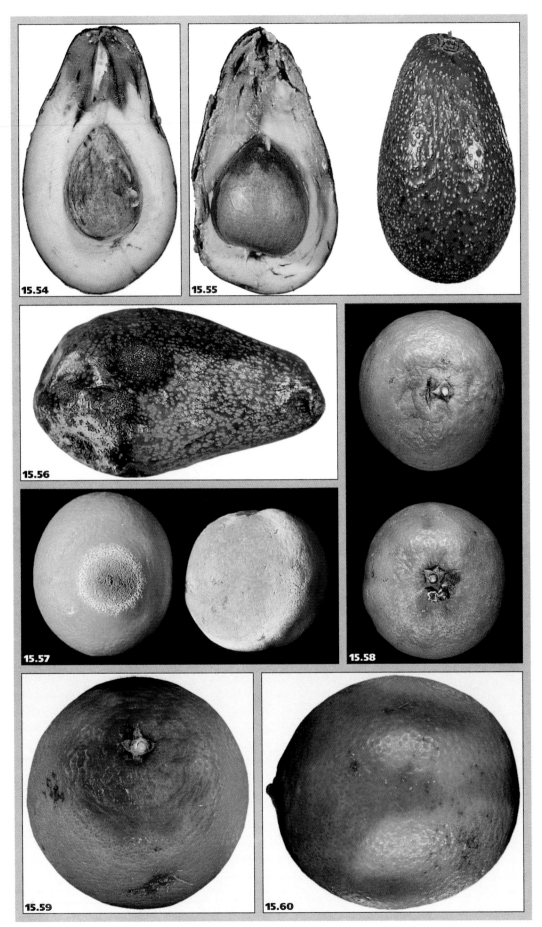

Fig. 15.54. Dothiorella stem-end rot (*Dothiorella gregaria*) of avocado.

Fig. 15.55. Lasiodiplodia stem-end rot (*Lasiodiplodia theobromae*) of avocado.

Fig. 15.56. Anthracnose (*Colletotrichum gloeosporioides*) of avocado.

Fig. 15.57. Blue and green rots (*Penicillium italicum* and *P. digitatum*) in orange.

Fig. 15.58. Citrus fruit brown rot (*Phytophthora citrophthora*).

Fig. 15.59. Stem-end rot (*Phomopsis citri*) of orange.

Fig. 15.60. Stem-end rot (*Lasiodiplodia theobromae*) of orange.

# Postharvest Disease Symptoms:
## Subtropical Fruits

Fig. 15.61. Alternaria stem-end rot (*Alternaria citri*) of grapefruit.

Fig. 15.62. Alternaria rot (*Alternaria citri*) of orange.

Fig. 15.63. Anthracnose (*Colletotrichum gloeosporioides*) of Meyer lemon.

Fig. 15.64. Sour rot (*Geotrichum candidum*) of navel orange.

Fig. 15.65. Sclerotinia rot (*Sclerotinia sclerotiorum*) of stored lemons.

Fig. 15.66. Trichoderma rot (*Trichoderma viride*) of orange.

Fig. 15.67. Botrytis rot (*Botrytis cinerea*) of orange.

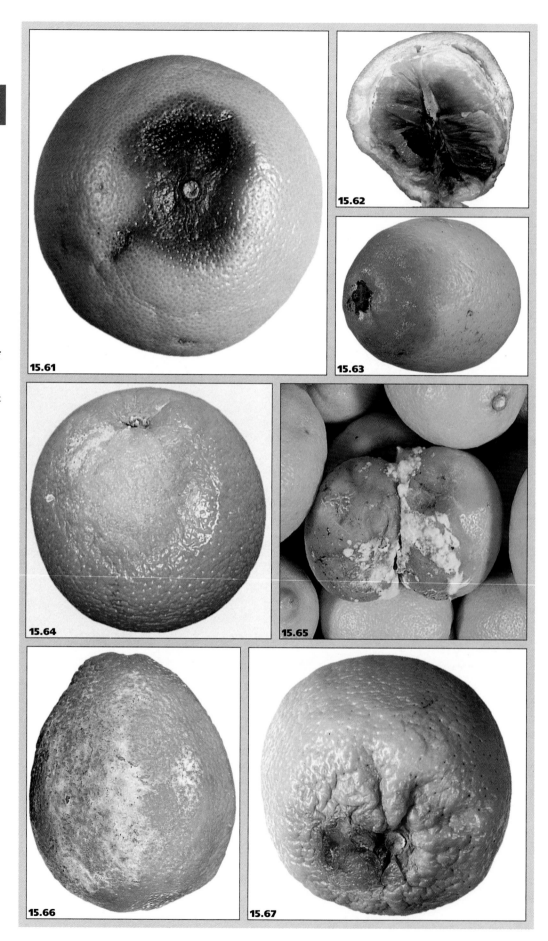

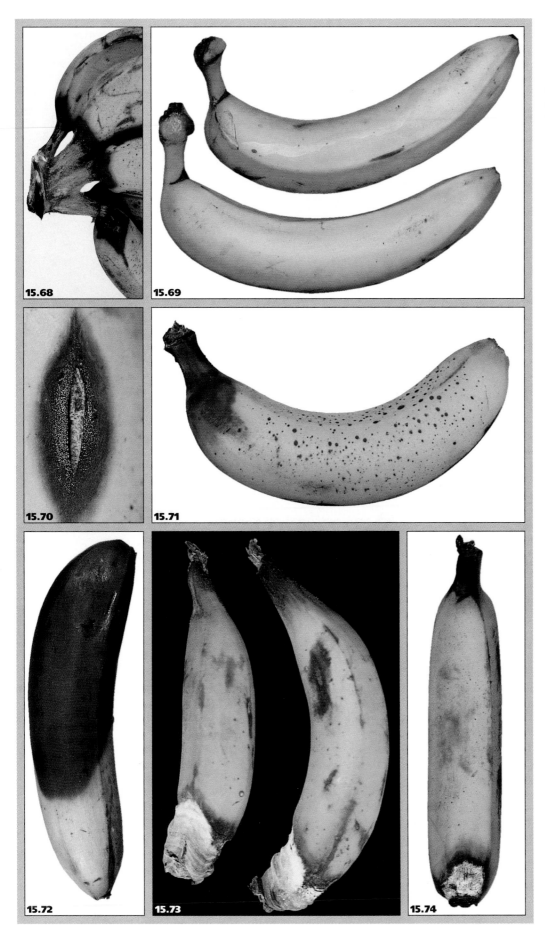

# Postharvest Disease Symptoms:

## Tropical Fruits

Fig. 15.68. Crown rot of banana caused by several fungi, often in concert.

Fig. 15.69. Creased stems of banana fingers and cut surfaces of crowns are loci of infection.

Fig. 15.70. Anthracnose (*Colletotrichum musae*) of banana.

Fig. 15.71. Lasiodiplodia stem-end rot (*Lasiodiplodia theobromae*) of banana finger.

Fig. 15.72. Thielaviopsis rot (*Thielaviopsis paradoxa*) of banana finger.

Fig. 15.73. "Cigar-end" stylar rot disease of banana.

Fig. 15.74. Fusarium stylar-end disease (*Fusarium roseum*) of banana.

# Postharvest Disease Symptoms:

## Tropical Fruits

Fig. 15.75. Anthracnose
stem-end rot
(*Colletotrichum
gloeosporioides*) of mango.

Fig. 15.76. Anthracnose
(*Colletotrichum
gloeosporioides*) of mango.

Fig. 15.77. Lasiodiplodia
rot (*Lasiodiplodia
theobromae*) of mango.

Fig. 15.78. Anthracnose
(*Colletotrichum
gloeosporioides*) of papaya.

Fig. 15.79. Black stem-
end rot (*Phoma caricae-
papayae*) of papaya.

Fig. 15.80. Black stem-
end rot (*Phoma caricae-
papayae*) of papaya
(cut fruit).

Fig. 15.81. Phomopsis rot
(*Phomopsis caricae-
papayae*) of papaya.

Fig. 15.82. Phomopsis rot
(*Phomopsis caricae-
papayae*) of papaya
(cut fruit).

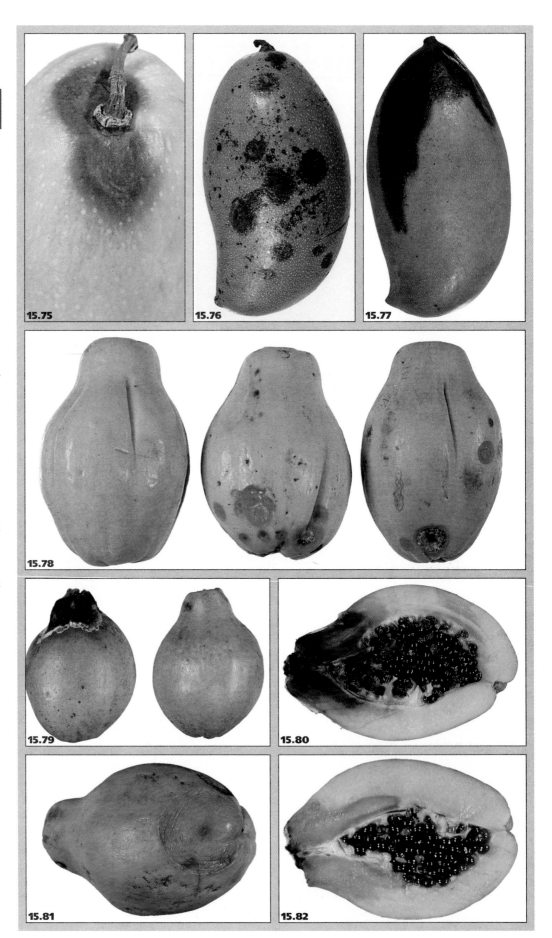

Fig. 15.83. Phytophthora stem-end rot (*Phytophthora nicotianae* var. *parasitica*) of papaya.

Fig. 15.84. Alternaria rot (*Alternaria alternata*) of papaya following chilling.

Fig. 15.85. Water blister (*Thielaviopsis paradoxa*) of pineapple, initial lesion.

Fig. 15.86. Water blister (*Thielaviopsis paradoxa*) of pineapple, developing rot.

Fig. 15.87. Water blister (*Thielaviopsis paradoxa*) of pineapple, collapsed fruit.

Fig. 15.88. Floret rot of pineapple fruit caused by *Penicillium funiculosum*, *Fusarium moniliforme*, or both.

Fig. 15.89. Yeast fermentation of pineapple.

Fig. 15.90. Yeast fermentation of pineapple, advanced stage.

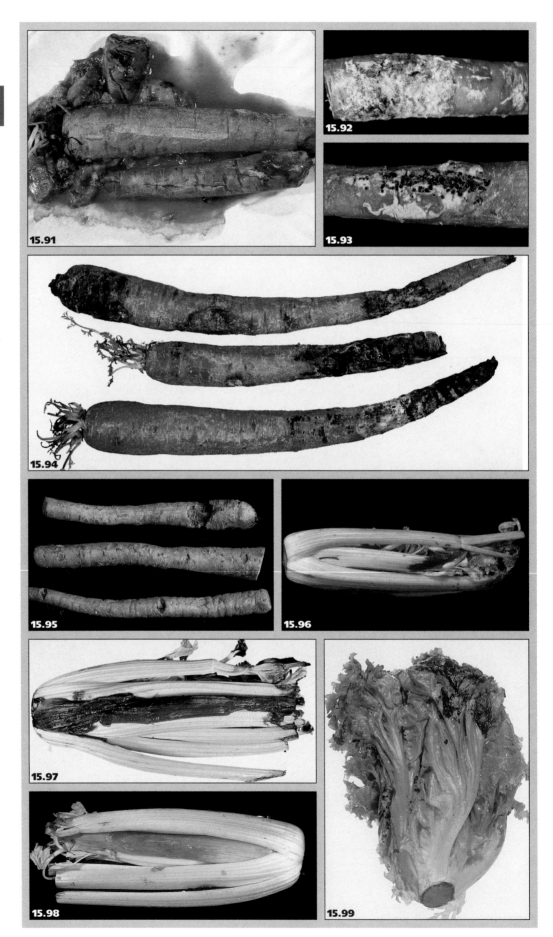

# Postharvest Disease Symptoms:

## Vegetables

Fig. 15.91. Bacterial soft rot (*Erwinia carotovora*) of carrot.

Fig. 15.92. Botrytis rot (*Botrytis cinerea*) of carrot.

Fig. 15.93. White rot (*Sclerotinia minor*) of carrot.

Fig. 15.94. Black mold (*Thielaviopsis basicola*) of carrot.

Fig. 15.95. Crater rot (*Rhizoctonia carotae*) of carrot.

Fig. 15.96. Bacterial soft rot (*Erwinia carotovora*) of celery leaflets.

Fig. 15.97. Bacterial soft rot (*Erwinia carotovora*) of celery petiole.

Fig. 15.98. Pink rot (*Sclerotinia sclerotiorum*) of celery.

Fig. 15.99. Bacterial soft rot (*Erwinia carotovora*) of lettuce.

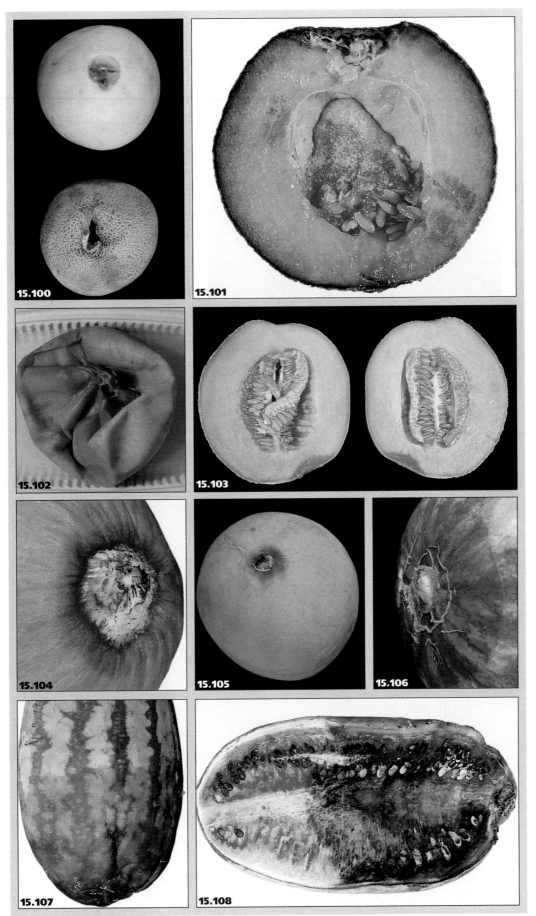

Fig. 15.100. Sour rot (*Geotrichum candidum*) of melons.

Fig. 15.101. Sour rot (*Geotrichum candidum*) development of melons.

Fig. 15.102. Sour rot (*Geotrichum candidum*) with honeydew collapsed.

Fig. 15.103. Rhizopus rot (*Rhizopus stonolonifer*) of cantaloupe.

Fig. 15.104. Fusarium rot of melons.

Fig. 15.105. Trichothecium rot (*Trichothecium roseum*) of honeydew melon.

Fig. 15.106. Botrytis stem-end rot (*Botrytis cinerea*) of watermelon.

Fig. 15.107. Lasiodiplodia rot (*Lasiodiplodia theobromae*) of watermelon.

Fig. 15.108. Lasiodiplodia rot (*Lasiodiplodia theobromae*) of watermelon after 24-hour exposure to air.

134

# Postharvest Disease Symptoms:

## Vegetables

Fig. 15.109. Neck rot (*Botrytis* spp.) of onion.

Fig. 15.110. Bacterial soft rot (*Erwinia carotovora*) of onion.

Fig. 15.111. Smudge (*Colletotrichum circinans*) of onion.

Fig. 15.112. Fusarium basal rot (*Fusarium* spp.) of onion.

Fig. 15.113. Fusarium bulb rot (*Fusarium* spp.) of onion.

Fig. 15.114. Bacterial soft rot of potato.

Fig. 15.115. Bacterial ring rot of potato. *Courtesy of H. Moline, USDA, Beltsville, Maryland.*

Fig. 15.116. Late blight (*Phytophthora infestans*) of potato (external). *Courtesy of H. Moline, USDA, Beltsville, Maryland.*

Fig. 15.117. Late blight (*Phytophthora infestans*) of potato (internal). *Courtesy of H. Moline, USDA, Beltsville, Maryland.*

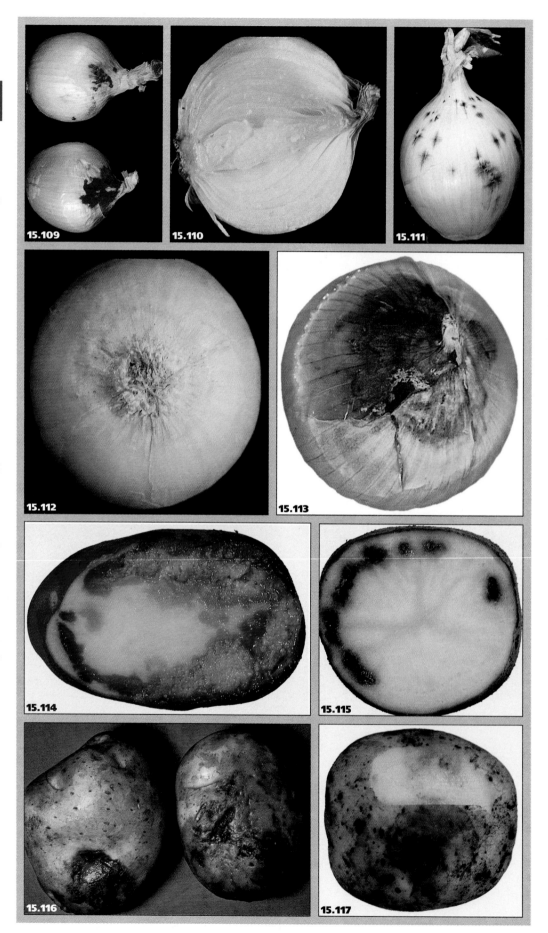

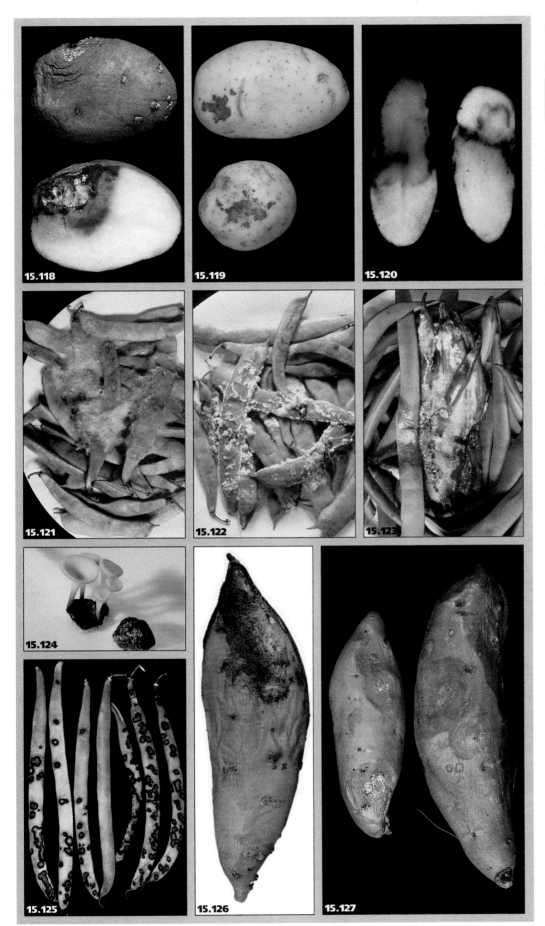

# Postharvest Disease Symptoms:

## Vegetables

Fig. 15.118. Fusarium dry rot (*Fusarium solani* or *F. roseum*) of potato. *Courtesy of H. Moline, USDA, Beltsville, Maryland.*

Fig. 15.119. Black scurf (*Rhizoctonia solani*) of potato.

Fig. 15.120. Pink rot (*Phytophthora erythroseptica*) of potato. *Courtesy of H. Moline, USDA, Beltsville, Maryland.*

Fig. 15.121. Botrytis rot (*Botrytis cinerea*) of snap bean.

Fig. 15.122. White rot (*Sclerotinia minor*) of snap bean.

Fig. 15.123. White rot (*Sclerotinia minor*) of snap bean.

Fig. 15.124. Apothecium of *Sclerotinia sclerotiorum*. *Courtesy of J. C. Tu, Harrow Research Station, Ontario, Canada.*

Fig. 15.125. Anthracnose (*Colletotrichum lindemuthianum*) of snap bean. *Courtesy of H. Moline, USDA, Beltsville, Maryland.*

Fig. 15.126. Rhizopus rot (*Rhizopus stolonifer*) of sweet potato.

Fig. 15.127. Fusarium rot (*Fusarium oxysporum*) of sweet potato.

# Postharvest Disease Symptoms:

## Vegetables

Fig. 15.128. Alternaria rot (*Alternaria alternata*) of tomato.

Fig. 15.129. Buckeye rot (*Phytophthora* spp.) of tomato.

Fig. 15.130. Ghost spot (*Botrytis cinerea*) of tomato.

Fig. 15.131. Sour rot (*Geotrichum candidum*) of tomato, early symptoms.

Fig. 15.132. Sour rot (*Geotrichum candidum*) of tomato.

Fig. 15.133. Rhizopus rot (*Rhizopus stolonifer*) of tomato, early stage.

Fig. 15.134. Rhizopus rot (*Rhizopus stolonifer*) of tomato, late stage.

Fig. 15.135. Anthracnose (*Colletotrichum* spp.) of tomato.

Fig. 15.136. Fusarium rot (*Fusarium* spp.) of tomato.

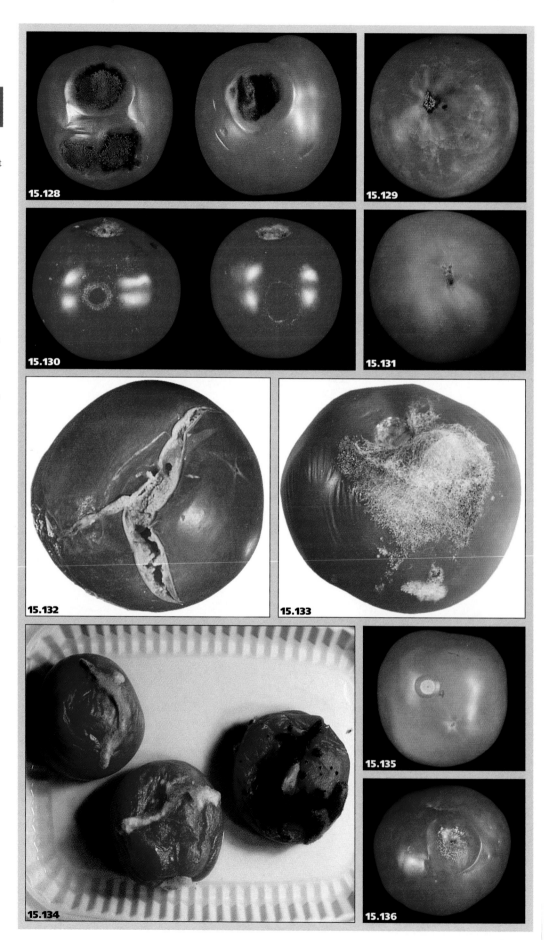

Bitter rot is primarily observed in the warm, humid growing areas of the Southeastern United States. It probably occurs wherever apple or pear is grown near the southern limits of production. The disease is rare on western U.S. apple and pear.

The fungus has a large host range in fruit or plants of temperate, subtropical, or tropical zones. Literally hundreds of names had been published, according to the region or host upon which it was first found. A 1970 study reduced several hundred species to synonymy. The minimum temperature for development of *Colletotrichum gloeosporioides* is 3° to 5°C (37° to 41°F), but some isolates from tropical fruits have much higher minimum temperatures.

The disease may occasionally occur in immature fruit, but characteristic disease symptoms are limited to fruit that are nearly or fully grown. Disease lesions first appear as light brown, circular spots that enlarge rapidly and become sunken (fig. 15.14). Under humid conditions, conidia are produced in a semigelatinous mass, often in pinkish concentric circles. As lesions age, sporulation ceases and lesions become dark brown to black.

The fungus may sometimes produce cankers in limbs of pome fruit trees. It may grow saprophytically on lesions of other diseases, producing conidia and sometimes ascospores. Diseased fruit may overwinter as mummies in the tree or on the ground and produce conidia. The fungus can penetrate unwounded fruit, forming latent infections that become active when fruit ripen.

Elimination of cankers in the orchard and rotting fruit on the orchard floor reduces the sources of inoculum. Benzimidazole fungicides applied in the orchard generally eliminate much of the inoculum and protect the fruit from infections. Postharvest handling at 0° to 3°C (32° to 37°F) should prevent disease development.

### Black rot

Sexual state:  *Botryosphaeria obtusa* (Schw.) Shoem.
         = *Physalospora obtusa* (Schw.) Cooke
         = *Physalospora everhartii* Sacc.
Asexual state: *Sphaeropsis malorum* Berk.
         = *Sphaeropsis biformis* Peck
         = *Sphaeropsis cerasina* Peck
         = *Sphaeropsis druparum* (Schwein.) Cooke
         = *Sphaeropsis fertilis* Peck
         = *Sphaeropsis maclurae* Cooke
         = *Sphaeropsis pennsylvanica* Berk. & M.A. Curtis
         = *Sphaeropsis phlei* Ellis & Everh.
         = *Sphaeropsis rosarum* Cooke & Ellis

The black rot organism may infect leaves (frogeye leaf spot), produce cankers in trees, and rot fruit. Injury to trees can be highly destructive, particularly in winter-injured or neglected orchards.

*Sphaeropsis malorum* infects fruit usually via wounds. The rotted area is at first brown but in time usually turns dark brown to black (fig. 15.15). Often concentric bands alternating brown with very dark brown or black appear in the lesion. Black, conidia-producing bodies (pycnidia) usually form on cankers or on mummified fruit. Sometimes the spores are forced out in the form of coils of mucilaginous material in which they are embedded. The spores are disseminated largely by rain, splashing rain, or wind-driven mist.

Conidia of *Sphaeropsis malorum* germinate optimally at about 25° to 27°C (77° to 81°F) and slowly at 15°C (59°F). The fungus cannot grow and develop disease under usual refrigerated storage or transport conditions.

### White rot

Sexual state:  *Botryosphaeria dothidea* (Moug.:Fr.) Ces. & De Not.
         = *Botryosphaeria berengeriana* De Not.
         = *Physalospora suberumpens* Ellis & Everh.
Asexual state: *Dothiorella gregaria* Sacc. Some taxonomists consider the asexual state to be *Fusicoccum aesouli* Corda.

White rot is a relatively minor fruit rot found in the humid eastern United States on apple, pear, and peach. In California, the disease also causes a stem-end rot of avocado and citrus fruits. The disease has been found on apple in South Africa.

Rotting apples appear bleached (fig. 15.16). Cankers formed in trees may produce spores that are distributed by air, water, or wind-driven mist.

Further information on this fungus can be found in the section on Botryosphaeria rot of avocado.

### Mucor rot

Pathogen:  *Mucor piriformis* Fisher

Mucor rot fungus appears widely distributed in apple- and pear-growing regions of North America, Europe, Australia, and South Africa. Losses are generally not serious except in the U.S. Northwest, where the pear cultivar 'Beurre d'Anjou' has suffered serious storage losses. The disease is not a problem in California apple or pear, although *M. piriformis* appears to be widely distributed in the state and has been observed in strawberry and feijoa. Stone fruits imported from Chile occasionally suffer losses.

At first glance, mucor rot looks like Rhizopus rot. A major difference is that *M. piriformis*, growing in high humidity, forms white, tall sporangiophores that are topped with a single black sporangium; sporangiophores of *Rhizopus* spp. are much shorter. *Mucor piriformis* rots fruit at 0°C (32°F); *Rhizopus* spp. seldom cause fruit rots at 5°C (41°F) or less. Disease symptoms on 'Buerre d'Anjou' pear are seen as watersoaked areas of the skin, usually near the stem end or in wounds. Sporangiophores may emerge from the fruit through breaks in the skin, but often there are few sporangiophores in pome fruits until or unless the rotted fruit is nearing collapse.

Sporangiospores of *M. piriformis* may become abundant in the orchard soil near the surface. Any contact of fruit with soil on the ground or in bins risks infections in storage. Chlorine or sodium o-phenylphenate may be used in dump or flume water to minimize the buildup of sporangiospore populations. Rotted fruit should be picked up from the ground after harvest to reduce inoculum the following year.

## Alternaria rot

Sexual state:   Unknown
Asexual state: *Alternaria alternata* (Fr.:Fr.) Keissl.
= *Alternaria tenuis* Nees
= *Alternaria fasciculata* (Cooke & Ellis) Jones &Grout
= *Macrosporium fasciculatum* Cooke & Ellis
= *Macrosporium maydis* Cooke & Ellis

*Alternaria alternata* is not a vigorous pathogen but can grow at $-2°$ to $-3°C$ (28° to 26°F). The fungus may attack wounds and fleshy stems of cultivars such as 'Beurre d'Anjou' pear or 'Yellow Newtown' apple. It is one of several fungi that invade delayed sunscalded fruit in storage (fig. 15.17). Chilling injury in chilling-susceptible apple cultivars, such as 'Yellow Newtown,' increases the susceptibility of the fruit to Alternaria rot (fig. 15.18). The fungus may contribute to "moldy core" or to "core rot."

## Cladosporium rot

Sexual state:   *Mycosphaerella tassiana* (De Not.) Johans.
Asexual state: *Cladosporium herbarum* (Pers.:Fr.) Link
= *Cladosporium caricicola* Corda
= *Cladosporium epiphyllum* (Pers.:Fr.) Fr.
= *Cladosporium fasciculatum* Corda
= *Cladosporium fuscatum* Link
= *Cladosporium graminum* (Pers.:Fr.) Link
= *Helminthosporium flexuosum* Corda

Cladosporium rot fungus is primarily a saprophyte, but it is a wound pathogen of mostly overripe and senescent fruit (fig. 15.19). It colonizes delayed sunburn-damaged areas in 'Yellow Newtown' apple. Bruised or wounded tissues may be colonized, particularly in storage near the end of the fruit's life. The fungus may also be present in "moldy cores" and in "dry core rots."

## Pleospora rot

Sexual state:   *Pleospora* sp.
Asexual state: *Stemphylium botryosum* Wallr.

*Pleospora* is a weak parasite that may attack fruit that are overripe or suffering from sunburn or chilling injury. It is commonly found colonizing "delayed sunscald" on 'Yellow Newtown' apple (fig. 15.20). This fungus may also be involved with "moldy core" or "dry core rot." Symptoms are usually nondescript, and identification requires culturing the fungus.

## Phomopsis rot

Sexual state:   *Diaporthe perniciosa* Marchal
Asexual state: *Phomopsis mali* Roberts

Phomopsis rot was observed as a stem-end rot in 1975-76 in controlled atmosphere stored 'Yellow Newtown' apples in Santa Cruz County, California (fig. 15.21). The fruit were believed to have suffered minor chilling injury. Symptoms are a brown skin and flesh discoloration along with sparse or no aerial mycelium. Differentiation from other rots without distinctive symptoms requires culturing the fungus. It is likely that orchard inoculum may come from leaf spots or pycnidia in trees or from windfall fruit that have mummified.

## Pink rot

Sexual state:   Unknown
Asexual state: *Trichothecium roseum* (Pers.:Fr.) Link
= *Cephalothecium roseum* Corda

*Triochothecium roseum* occurred in apple and pear before the advent of modern temperature management and before effective control of scab. Often the main development was on scab lesions caused by the fungus *Venturia inaequalis* (Cke.) Aderh. It is now rare on apple and pear in this country. However, in the Himalayan foothills of northern India, which is subject to monsoon rains, *T. roseum* is common before and after harvest.

## Side rot

Sexual state:   Unknown
Asexual state: *Phialophora malorum* (M. N. Kidd & A. Beaumont) McColloch
*Sporotrichum malorum* M. N. Kidd & A. Beaumont
= *Sporotrichum carpogenum* Ruhle

Incidence of side rot is evidently low except in the 'Buerre Bosc' pear in which significant losses have been experienced in the Pacific Northwest. Lesions are characteristically round or nearly so. There is usually no sporulation on the lesion surface (fig. 15.22). Similar lesions are sometimes produced by *Cladosporium herbarum*, which also commonly does not sporulate on the lesion surface. Isolation in pure culture is the usual means of identifying the causal fungus. *Phialophora malorum* is susceptible to sodium orthophenylphenate in a flotation fluid with sodium silicate.

## Sooty blotch and fly speck

Sooty blotch and fly speck are caused by two separate organisms that are commonly found growing together superficially on apple surfaces. Sooty blotch, produced by the asexual fungus *Gloeodes pomigena* (Schw.) Colby, is observed as a thin coating of gray mycelium growing in irregular patches.

Fly speck is caused by the asexual fungus *Zygophiala jamaicensis* Mason. It is observed as small black bodies on the fruit surface reminiscent of fly crottes.

Fly speck is commonly associated with blotch, which is why the two organisms are generally considered together as one condition.

Sooty blotch and fly speck are limited to warmer areas with high rainfall, but an unusual instance occurred in California's hot, dry San Joaquin Valley. When a grower sprinkled water on fruit to obtain evaporative cooling to reduce sunburn injury of 'Granny Smith' apple, blotch and fly speck appeared.

Sooty blotch and fly speck organisms do not establish infections in the fruit, but rather they grow on the surface. Nevertheless, the surface mycelium is difficult to remove and may result in unattractive fruit.

### Brown rot

Sexual state: *Monilinia fructigena* Honey
        = *Sclerotinia fructigena* Aderhold ex. Sacc.
Asexual state: *Monilia fructigena* Pers. ex Fr.

Brown rot of apple, rare in North America, is common in Europe. The organism is sufficiently similar to the stone fruit brown rot fungi, *Monilinia fructicola* and *M. laxa*, to cause considerable trouble with identification. The rare occurrence of brown rot on ripe 'Bartlett' pear is generally attributed to *M. laxa* or *M. fructicola*. The life cycle of *M. fructigena* is similar to the life cycle of the stone fruit organisms (see stone fruit section).

### Phytophthora rot

Pathogens:    (1) *Phytophthora cactorum* (Lebert & Cohn) Schrot.
               (2) *Phytophthora syringae* (Kleb.) Kleb.

Phytophthora fruit rot is sometimes called "sprinkler rot" because rotted fruit are often found in the lower part of the tree, where irrigation water from sprinklers contacts the trees. Irrigation water can deposit sporangia or zoospores on the surface of fruit if it comes from a stream contaminated with the fungus, or if it splashes contaminated particles of soil onto the fruit. Spores readily penetrate the fruit surface to produce a rapidly developing rot (fig. 15.23). Healthy fruit in contact with diseased fruit frequently become diseased. Diseased areas are dark brown in 'Bartlett' pear.

In the western United States, Phytophthora rot of fruit seldom warrants control measures. Presumably, metering a copper fungicide into the sprinkler irrigation water will prevent infection, if needed. These fungi may produce a collar or root rot among trees in the orchard.

## Apricot, cherry, nectarine, peach, plum and prune (stone fruits)

### Brown rot

Sexual state:  (1) *Monilinia fructicola* (Wint.) Honey
            = *Sclerotinia fructicola* (Wint.) Rehm
Asexual state: (1) *Monilia* sp.
Sexual state:  (2) *Monilinia laxa* (Alderh. & Ruhl.) Honey
            = *Sclerotinia laxa* Aderh. & Ruhl.
Asexual state: (2) *Monilia laxa* (Ehrenb.) Sacc. & Voglino
            = *Monilia cineria* Bonord.

In California, the two species of brown rot organisms above are responsible for the most serious disease of most stone fruits. A third organism, *Monilinia fructigena* Aderh. & Ruhl., attacks and rots apple and pear as well as stone fruits in Europe.

Most active in early spring during blossoming or after fruit have started to ripen, these fungi may grow from the rotted blossom into the pedicel and into the smaller twigs or small branches, which may be girdled, causing young leaves to die. Fruit rot may occur before harvest but often occurs postharvest.

*Monilinia laxa* is more likely to cause blossom blight and twig blight; *M. fructicola* is more likely to cause fruit rot. The two organisms are so similar morphologically that they cannot be easily distinguished by usual microscopic methods. Similarly, *M. fructigena* is difficult to distinguish from the other two. Consequently, considerable confusion about the true identities of the brown rot organisms exists throughout the world's stone fruit-growing regions.

Airborne conidia, landing on green fruit in the orchard, may germinate and penetrate fruit skin, but they do not proceed further. These quiescent infections only become active as the fruit ripens (fig. 15.24) and provide abundant conidia for infections. These fungi nest from diseased to healthy fruit (fig. 15.25). Infections commonly develop in mechanical wounds (fig. 15.26). Stem-end rots occur in peach when fruit skin is slightly torn during harvest (fig. 15.27).

A sexual (mushroom) state may be essential to overwinter the fungus in the northern climates. It is not essential in California, where the fungus overwinters in rotten fruit mummies and diseased twigs. Conidia are produced when temperature and humidity are favorable.

*Monilinia fructicola* grows slowly at temperatures near 0°C (32°F) in culture. In fruit, where the fungus must overcome resistance to infection, minimum temperature for growth is higher. During transit at 2° to 5°C (36° to 41°F) the fungus grows slowly. Therefore, brown rot of fruit in transit usually does not occur unless fruit were not adequately cooled, transit temperature exceeded 5°C (41°F), or the fruit were overripe when loaded.

### Gray mold

Sexual state:  *Botryotinia fuckeliana* (de Bary) Whetzel
Asexual state: *Botrytis cinerea* Pers.:Fr.

Gray mold affects all stone fruits and many other fruits, vegetables, and ornamental species. Serious blossom blighting by *Botrytis cinerea* is not common in California on peach, plum, nectarine, and cherry but it is occasionally serious in apricot during wet spring weather, when apricot may suffer blossom blighting and rotting of young fruit in which *B.*

140

*cinerea* is joined by a close fungus relative, *Sclerotinia sclerotiorum* (Lib.) de Bary.

Gray mold causes large economic losses because it commonly develops in fruit stored at temperatures as low as 0°C (32°F). Its symptoms are similar to those of brown rot, and inexperienced observers can confuse the two.

Germinating spores of *B. cinerea* can penetrate the unbroken cuticle and epidermis of blossoms, very young fruit, and young leaves of many plants. However, most infections of stone fruits by *B. cinerea* are likely to result from contamination of harvesting and handling wounds. Contact infections occur, however, when fungal mycelia grow from a rotting fruit to nearby healthy fruit, resulting in an ever-enlarging nest of rotting fruit in the container.

### Rhizopus or "whiskers" rot

Pathogens:     *Rhizopus stolonifer* (Ehr.:Fr.) Vuill.
        = *Rhizopus nigricans* Ehrenb.
        *Rhizopus arrhizus* Fischer
        = *Rhizopus nodosum* Namyslowski
        = *Rhizopus oxyzae* Went & Prinsen Geerligs
        = *Rhizopus tritici* K. Saito
        *Gilbertella persicaria* (Eddy) Hesseltine
        = *Choanephore persicaria* Eddy

At 20° to 25°C (68° to 77°F), Rhizopus rot lesions grow rapidly in ripe or near-ripe fruits. If fruit are promptly cooled to 5°C (41°F) or lower, growth of the fungus nearly stops. At lower temperatures, germinating spores and young mycelia fail to grow and the fungus is killed. It is commonly observed that peaches packed unrefrigerated quickly rot. If similar peaches are promptly cooled to near 0°C (32°F) and shipped at about 1° to 3°C (34° to 37°F), they seldom develop Rhizopus rot when they are removed from refrigeration.

Germinating spores of Rhizopus rot fungi initially attack fruit via wounds but spread from rotting fruit to nearby sound fruit by "nesting" (fig. 15.28). Where moist juice is on the fruit surface, spores are capable of penetrating the fruit's cuticle and epidermis. If an infected fruit is packed in a shipping container and held at elevated temperatures, the mycelium quickly grows from it to attack nearby fruit (fig. 15.29) and eventually occupies the entire container. Mycelial growth at lesions is initially white, then it changes as black sporangia are produced.

A serious problem in canned apricots due to Rhizopus rot has occurred in South Africa, Europe, and Australia, as well as in California. Severe softening of the canned fruit was usually observed 9 or more months after canning. It is now known that juice from rotting fruit contains pectolytic enzymes that remain active despite washing and the heat of canning, softening all fruit within a can. A wash containing sodium hydroxide inactivates the enzyme (fig. 15.30).

### Mucor rot

Pathogens:     *Mucor piriformis* Fischer
        *Mucor hiemalis* Wehmer

The *Mucor* spp. resemble *Rhizopus* spp. in overall characteristics in culture and in rotting fruit, except that the white sporangiophores of the *Mucor* spp. often grow very long under high humidity conditions. Mucor rot has been observed in shipping containers of peaches and nectarines imported from Chile (fig. 15.31). It is not known why this fungus has not been prevalent in California, even though it has long been present and occasionally causes fruit rot of strawberry. *Mucor piriformis* is capable of rotting fruit held at −1°C (30°F).

Inoculum is produced by the fungus growing in discarded fruit in the field or on various types of organic matter in the soil. The germinating spores are believed incapable of penetrating the unbroken skin of fruit, but they may colonize harvest and handling wounds. Nesting, by growth of the fungus from a rotting fruit to nearby sound fruit in containers, occurs in a manner similar to nesting by *Rhizopus stolonifer*.

### Blue mold

Sexual state:   Unknown
Asexual state: *Penicillium expansum* (Lk.) Thom.

Blue mold rot is commonly found in peach, plum, and nectarine only after a week or more in storage at 0°C (32°F). Even after several weeks of storage, incidence of blue mold is usually much less than that of Botrytis rot. Blue mold is also common in refrigerated sweet cherries, particularly if cold, moist weather precedes harvest. It is likely that other *Penicillium* spp. occasionally are involved.

*Penicillium expansum* is capable of contact infection from rotted to healthy fruit to form nests in containers. However, the relatively slow growth, compared with *Monilinia fructicola* or *Botrytis cinerea*, and the short storage period for stone fruits limit nesting.

Spores of *P. expansum* are produced on fruit or on various types of organic matter on or in the soil. The readily airborne spores may be produced in quantities large enough to cover surfaces of fruit-handling equipment. To establish lesions, spores must contaminate wounds. The lesions produced are soft and watery but less so than in Rhizopus rot. At first, mycelium growing from lesions is white but turns an iridescent blue as the fungus starts to sporulate (fig. 15.32). *Penicillium expansum* readily rots fruits at 0°C (32°F), having a minimum temperature for growth of about −3.3° to −2.2°C (26° to 28°F).

### Alternaria rot

Sexual state:   Unknown
Asexual state: *Alternaria alternata* (Fr.) Keissler

Alternaria rot is commonly found in dark plums, where it colonizes sunburn-injured tissues (fig. 15.33), and in California sweet cherry, often colonizing an aborted fruit of a double. Alternaria rot is also common in apricot if unseasonable rain has caused the fruit to split.

Much of the Alternaria rot of sunburned fruit develops before harvest, but the lesions may enlarge after harvest (fig. 15.34). Harvest wounds are occasionally found colonized by the Alternaria rot organism. It is possible that Alternaria rot would be more common in wounds if highly vigorous competing organisms were not present.

Fruit tissues in lesions of *A. alternata* are firm and relatively dry in contrast with the soft, watery tissue found with Rhizopus or Mucor rots and, to a lesser extent, with brown rot, gray mold, or blue mold rot. *Alternaria alternata* grows at temperatures as low as 0°C (32°F), but the slow rate of growth generally limits damage unless fruit are held for excessive periods in low-temperature storage.

### Cladosporium rot

Sexual state:  *Mycosphaerella tassiana* (De Not.) Johans.
Asexual state: *Cladosporium herbarum* (Pers.:Fr.) Link
Note: See Cladosporium rot of apple and pear for synonyms.

Examinations of the spore flora in the air in refrigerated storage have sometimes shown *Cladosporium* to be the most prevalent fungus. Its presence is facilitated by its capability to grow at temperatures as low as 0°C (32°F). Nevertheless, stone fruits are seldom attacked unless they have been in storage more than one month.

## Kiwifruit

### Botrytis rot

Sexual state:  *Botryotinia fuckeliana* (de Bary) Whetzel
Asexual state: *Botrytis cinerea* Pers:Fr.
Note: See Botrytis rot of apple and pear for synonyms.

Botrytis rot in kiwifruit is a major postharvest disease in California, especially after fruit are more than 4 months in storage at 0°C (32°F). The first indication of *B. cinerea* activity is extreme softness localized at the stylar or, more commonly, at the stem end (fig. 15.35). Occasionally the rot is centered at a wound on the side of the fruit. The fungus, generally first visible on the fruit surface as white tufts of mycelium, may spread until the lesion surface is covered. As spore production occurs, the mycelium turns gray. Occasionally, irregular-shaped sclerotia, at first gray but black at maturity, form on fruit surfaces in place of normal sporulating mycelium (fig. 15.36).

The conidia produced in storage have limited effect because the spores are unable to penetrate the fruit. However, mycelium readily penetrates sound fruit. Mycelium growing from a diseased fruit contacts surrounding fruit, enlarging a nest of rotting fruit (fig. 15.37).

The success of *B. cinerea* in rotting fruit in refrigerated storage is due to its ability to grow, albeit slowly, at −2°C (28°F). Nevertheless, fungal growth rate at 0°C (32°F) is minuscule compared with that at higher temperatures. Thus, fruit should be cooled to near 0°C (32°F) immediately after harvest.

Kiwifruit in California are relatively resistant to disease until nearly ripe. Ripening progress is followed by measuring flesh firmness with a penetrometer 8-mm in diameter. Firmness of kiwifruit is reduced by about half every 40 to 50 days at 0°C (32°F). Fruit commonly become susceptible when they have ripened to a firmness of 6.6 to 9 newtons (1.5 to 2 lb-force). Exposure to ethylene during storage dramatically hastens softening and shortens fruit life. Fruit injury in handling also hastens incidence of rot. Widely used measures are prompt cooling and storage at 0°C (32°F); monitoring storage rooms to detect ethylene; and removing ethylene.

### Surface mold, Alternaria rot, juice blotch

Sexual state:  Unknown
Asexual state: *Alternaria alternata* (Fr.) Keissler

*Surface mold.* *Alternaria alternata* grows in storage primarily on senescing fruit calyces that have not been removed by brushing (fig. 15.38). Although the fruit proper is not affected, the mycelial growth covering the calyces, or other dead organic matter, is unsightly. It sometimes causes the purchaser to believe, erroneously, that the fruit is rotting. Removing molding fruit from the high humidity of the transit vehicle causes the mycelium to dry and collapse, leaving little or no indication of its presence.

*Alternaria rot.* Fruit frequently are damaged or their disease resistance weakened by sunburning, particularly if they are produced on young, poorly shaded vines. *Alternaria alternata* frequently colonizes sunburned fruit (fig. 15.39), necessitating its elimination by sorting.

*Juice blotch.* Juice blotch may result when fruit are crushed in handling. The juice contaminates other fruit and packing equipment. Juice on sound fruit provide a medium for *A. alternata* and other fungi to grow and create unsightly black blotches.

Juice blotch is most likely to occur if fruit are moved on packing lines after more than a month in storage. By then fruit are less firm and susceptible to crushing.

### Dothiorella rot

Sexual state:  *Botryosphaeria dothidea* (Moug.:Fr.) Ces. & de Not.
Asexual state: *Dothiorella gregaria* Sacc.
Some taxonomists believe the binomial for the asexual state should be *Fusicoccum aesculi* Corda.

*Dothiorella gregaria*, believed to be the same fungus that attacks citrus and avocado fruit in California, attacks peach fruit and causes a "white rot" of apple in southeastern states. In addition to fruit rotting, the fungus causes cankers in peach trees. There is little indication that this fungus will seriously affect California-grown kiwifruit, either as a fruit rot or as a disease in the vineyard. In storage the disease is found in overripe fruit. Generally, a portion of the fruit surface collapses (fig. 15.40). It is usually necessary to culture the organism to distinguish it from Phoma rot.

## Phoma rot

Sexual state:   Unknown
Asexual state: *Phoma* spp.

Phoma rot is occasionally observed on kiwifruit, especially near the end of the storage season. The rots are on the side of the fruit and are probably associated with wounds. The fruit surface characteristically becomes depressed, often like a crater (fig. 15.41), without superficial mycelium. The flesh in the area beneath the crater may be colorless but is frequently pink or purple.

## Phomopsis rot

Sexual state:   *Diaporthe actinidiae* Som. & Ber.
Asexual state: *Phomopsis* sp.

Phomopsis rot is occasionally observed in both California-grown and New Zealand-grown kiwifruit (fig. 15.42). It appears to be found mostly in overripe fruit and does not appear to develop into a serious postharvest disease. Some leaf spots and wilted canes in the vineyard yield a *Phomopsis* spp. Affected fruit commonly are colonized at the stem end, where frothy juice often appears. That and the swarms of vinegar flies that are attracted suggest that yeasts may also be present within the lesion. Gaps or spaces in the lesion are common.

## Sclerotinia rot

Sexual state:   (1) *Sclerotinia sclerotiorum* (Lib.) de Bary
= *Whetzelinia sclerotiorum* (Lib.) Korf & Dumont
= *Sclerotinia libertiana* Fuckel
(2) *Sclerotinia minor* Jagger
= *Sclerotinia intermedia* Ramsey
= *Sclerotinia sativa* Drayton & Groves
Asexual state: Unknown

*Sclerotinia sclerotiorum* and *S. minor* grow at below 0°C (32°F) and cause serious disease losses in certain stored vegetables such as carrot and cabbage. The organisms have been observed only rarely in California kiwifruit storages.

## Mucor rot

Pathogen:   *Mucor piriformis* Fischer

*Mucor piriformis* grows below 0°C (32°F) and could become important in kiwifruit storages, but so far it has only been observed occasionally.

## Blue mold

Sexual state:   Unknown
Asexual state: *Penicillium expansum* (Lk.) Thom.

*Penicillium expansum* produces a very soft, wet rot in the commodities it attacks. Its occurrence in kiwifruit is mostly limited to overripe fruit (fig. 15.43). It is believed to infect fruit only through injuries to the skin or by acting as a secondary invader of lesions made by another fungus, such as *Botrytis cinerea*. It typically occurs as a side rot in kiwifruit. Initially white, the fungus colony turns blue as it starts to produce conidia. The fungus grows at below 0°C (32°F).

## Buckshot rot

Pathogen:   *Typhula* spp.

*Typhula* spp. appeared recently in California kiwifruit storages. The disease, buckshot rot, was named because of the nearly round, black sclerotia produced on the fruit surface. The occasional presence of the fungus in kiwifruit storage is evidently the first instance that the disease has been reported attacking a fruit in storage (fig. 15.44). The fungus grows very slowly at 0°C (32°F), and it grows much better at 15°C (59°F) than at 20°C (68°F).

## Strawberry

## Gray mold

Sexual state:   *Botryotinia fuckeliana* (de Bary) Whetzel
Asexual state: *Botrytis cinerea* Pers.:Fr.

Gray mold is the most serious postharvest disease commonly found in strawberry under modern refrigerated storage and transit conditions. The minimum temperature for growth is about −2°C (28°F). The disease is common on ripening and ripe strawberry fruit. Nesting caused by growth of the disease from rotting to sound fruit is characteristic. Infection occurs on immature, green fruit particularly under wet conditions. The fungus may attack strawberry planting stock in cold storage.

Rot may be initiated anywhere on the fruit. Affected tissue turns dull pink to brown. Fruit may become completely rotted without disintegration, and very little juice is exuded. In time, the lesion exhibits on the fruit surface white mycelium that turns gray as the fungus sporulates (figs. 15.45 and 15.46). The fungus "nests" when mycelium from a rotting fruit penetrates and colonizes adjacent fruit. Occasionally, irregularly shaped, black sclerotia, 1 to 7 mm in diameter, may form on the fruit surface if fruit are held more than a month at 0°C (32°F).

Infections are initiated in various ways. (1) During blossoming, stamens and petals become infected when conidia land and germinate on them. From infected stamens and petals the fungus may grow into the sub-calyx area of the fruit. (2) Infected petals may be shed and land on the fruit, and the fungus grows from them into the fruit. (3) Sound fruit may contact the ground and become infected by fungus conidia, mycelia, or sclerotia in the soil. (4) Conidia may be dispersed by air currents, splashing water, or wind-driven mist.

*Botrytis cinerea* mycelia do not have well-developed and obvious appressoria to penetrate a sound fruit as contrasted with *Colletotrichum* spp. Nevertheless, apparently undifferentiated mycelial strands grow from a rotting fruit to contact a sound fruit, which is penetrated without apparent pressure.

Use of plastic sheeting with slits or holes through which plants are planted greatly reduces rot in the field by preventing fruit contact with the soil, which may be heavily contaminated with *B. cinerea* conidia, sclerotia, or mycelia. Gray mold may be suppressed in strawberry by fungicidal sprays in the field.

Fruits should be moved promptly from the field to a cooler, to remove field heat. Storage should be at 0°C (32°F). Use of a modified atmosphere with 12 to 20 percent carbon dioxide, depending upon the cultivar, suppresses fungal activity and also slows senescence of the fruit.

## Leak or Rhizopus rot

Pathogen:   *Rhizopus stolonifer* (Ehrenb. ex Fr.) Vuill.

Leak is found throughout the world's strawberry-growing areas. Ripe fruit in the field may be rotted, but losses occur primarily after harvest (fig. 15.47). Without good temperature management, postharvest life is shortened to as little as 1 to 3 days if *Rhizopus stolonifer* is active. A ubiquitous fungus that grows as a saprophyte on decaying organic matter, is capable of infecting (via wounds) many fruits after they are completely ripe. Sexual spores (zygospores) are produced when mated with opposite strains of the fungus.

Modern temperature management, consisting of rapid removal of field heat, and storage and transport at or near 0°C (32°F), has largely eliminated leak as a significant disease. Postharvest temperatures are usually below the minimum temperature for its growth. Furthermore, germinating sporangiospores are killed by low temperatures, and the rot after removal to ambient temperatures is usually minor.

## Mucor rot

Pathogens:   *Mucor piriformis* Fischer
*Mucor hiemalis* Wehmer

Mucor rot resembles Rhizopus rot sufficiently to sometimes cause confusion (fig. 15.48). The most striking difference is the ability of *Mucor piriformis* to grow at low temperatures (minimum temperature for growth is below 0°C [32°F]) while the *Rhizopus* sp. does not. Sporangiophores are extremely long compared to those of *R. stolonifer*. Like *R. stolonifer*, *Mucor piriformis* causes copious leakage of juice from strawberries.

Despite an ability to grow at low temperatures, the incidence of Mucor rot in California-grown strawberries in storage or transit is usually much less than that of *Botrytis cinerea*, which is also capable of growing at below 0°C (32°F).

## Anthracnose

Sexual state:    *Glomerella cingulata* (Stoneman) Spauld. & Schrenk
Asexual state: *Colletotrichum gloeosporioides* (Penz.) Penz. & Sacc.
With several hundred synonyms.

The sexual state of other asexual fungi listed below is unknown.
*Colletotrichum acutatum* Simmonds
*Colletotrichum fragariae* Brooks
*Colletotrichum dematium* (Pers.) Grove

Anthracnose disease organisms may attack strawberry in the field, causing lesions on stolons, petioles, crowns, and leaves as well as fruit. Such attacks are common in Florida and other humid growing areas. Anthracnose species attack fruit primarily in the field, but the disease may develop after harvest.

Anthracnose lesions develop as tan or light brown, circular, sunken lesions on ripe or ripening fruit. As sporulation occurs (fig. 15.49), cream to salmon or pink spore masses erupt from subepidermal acervuli.

Anthracnose fungi overwinter as dormant mycelium on and in infected plants. Conidia are produced that are spread by wind, wind-driven mist, splashing water, or insects. Conidia landing on fruit germinate and form appressoria that penetrate the unwounded epidermis. Fruit lesions produce acervuli with many conidia that further spread the disease.

## Leather rot

Pathogen:   *Phytophthora cactorum* (Leb. & Cohn.) Schroet.

*Phytophthora cactorum*, a common soil pathogen with a wide host range, causes root and crown rots of many hosts. On strawberry, the fungus attacks fruit at various stages of maturity including green fruits (fig. 15.50) on occasion.

Infected fruit that are ripe or nearly so are likely to be picked and included with sound berries. Infected fruit are likely to be noticeably lighter colored than nondiseased fruit. Diseased fruit remain quite firm and "leathery" (fig. 15.51). Diseased fruit have an unpleasant taste that permits ready identification of infected berries.

The disease is likely to be most abundant during or after excessively wet weather. In California, growing strawberry plants through plastic film prevents direct contact of fruit with the soil. For that reason, the incidence of leather rot in fruit grown on plastic is rare.

## Table grapes

### Botrytis rot

Sexual state:  *Botryotinia fuckeliana* (de Bary) Whetzel
Asexual state: *Botrytis cinerea* Pers.:Fr.

*Botrytis cinerea* is the most serious rot problem of table grapes in storage and marketing (fig. 15.52). The fungus may infect berries in the vineyard, particularly when extended periods of rainy weather before harvest occur. However, wounds are common entry points for the fungus. A frequently invaded wound is at the stem where berries are joined and may be partially loosened by harvesting and handling.

Other organisms frequently found in stored grapes include *Penicillium expansum* (fig. 15.53), *Alternaria alternata*, and *Cladosporium herbarum*.

Commercial control of the diseases is obtained by fumigating grapes in storage at 0°C (32°F), using 0.5 percent sulfur dioxide gas for about 20 minutes and then clearing the air of the gas. This initial fumigation should follow harvest within 12 hours. For stored grapes, treatment should be repeated every 7 days at a concentration of 0.05 to 0.2 percent for 30 to 45 minutes. If a high disease potential is expected, as a result of rain before harvest or other causes, sulfur dioxide concentrations may be set at the upper levels. Sulfur dioxide has little effect on fungal lesions established in the vineyard before harvest, but it prevents fruit-to-fruit contact infection and nesting of the fungus within packed containers.

Some injury is associated with fumigation when sulfur dioxide is absorbed into wounds in the berry. The injury increases with each successive fumigation. The most obvious symptom of injury is bleaching of tissue surrounding a wound or near the point of attachment of berry to stem.

## Postharvest Diseases of Subtropical Fruits

### Avocado

#### Dothiorella rot

Sexual state:  *Botryosphaeria dothidea* (Mout.:Fr.) Ces & De Not.
Asexual state: *Dothiorella gregaria* Sacc.
Some taxonomists consider the asexual state to be *Fusicoccum aesculi* Corda

Dothiorella rot develops extensively only in ripened fruit. A lesion may appear on any part of the fruit as a black spot in otherwise green skin. The spots may reach 1.3 cm (0.5 inch) in diameter within 3 or 4 days. The uniformly dark spots are circular, not sunken. The surface is somewhat softer than uninvaded skin. Most of the rot is limited to the skin, and if the flesh is at all affected, only the outermost part appears slightly watery. From this stage the spot spreads more rapidly, becoming soft and somewhat sunken and uneven, while a watery rot spreads slowly into the flesh. Frequently, lesions develop as a stem-end infection. These may appear earlier and penetrate more rapidly and deeply than spots in the skin (fig. 15.54).

### Stem-end rot

Sexual state:  *Botryosphaeria rhodina* (Cooke) Arx
Asexual state: *Lasiodiplodia theobromae* (Pat.) Griffon & Maubl.
Note: See stem-end rot of citrus for synonyms of *Lasiodiplodia theobromae*.

*Lasiodiplodia theobromae* has a large host range and causes serious losses in warm, humid growing areas. The fungus attacks cultivated crops as a stem-end rot (such as stem-end rot of citrus, Java black rot of sweet potato, and stem-end rot of watermelon). Symptoms of Lasiodiploidia rot in avocado (fig. 15.55) are similar to those of Dothiorella rot.

### Fusarium rot

Pathogen:  *Fusarium* spp.

Studies in Israel showed that among 36 isolates of *Fusarium* spp. causing avocado fruit rots, 19 were *Fusarium roseum* (Lk.) Snyder & Hansen. *Fusarium moniliforme* Sheld. accounted for 8, *Fusarium solani* (Mart.) Sacc. for 5, and *Fusarium oxysporum* Schlechtend.:Fr. for 4. The optimum temperature range for growth of these species is 20° to 30°C (68° to 86°F).

Lesions (often near the stem end) darken the skin; the invaded flesh is discolored and may become light brown to nearly black. Typically, the flesh remains firm or only slightly softer than uninvaded tissue, and there is little or no leakage of moisture from invaded areas. Often tissue shrinkage results in voids or gas pockets within invaded tissue. Light-colored mycelia may grow into these voids.

### Anthracnose

Sexual state:  *Glomerella cingulata* (Stonem.) Spauld. & Schrenk
Asexual state: *Colletotrichum gloeosporioides* (Penz.) Sacc.
There are several hundred synonyms not listed.

Anthracnose is a serious disease of avocado where growing conditions are humid. In the orchard the fungus may grow and sporulate on dead twigs or branches or in dead spots of leaves. Conidia of the fungus are spread by wind and wind-driven rain to fruit on the tree. Infections usually remain latent until about the time fruit soften after harvest.

Anthracnose first appears on ripening avocado fruit as circular discolorations of the skin that may appear at one location or scattered over the fruit surface. Centers of disease lesions become slightly sunken. Sporulation commonly occurs at the center of lesions, and orange to pink spore masses are exuded from the usually numerous ascervuli (fig. 15.56). Fruit flesh beneath lesions is greenish-black and decayed, and may be fairly firm to very soft.

Control generally includes preharvest spraying with benzimidazole fungicides and maintaining optimum temperature in storage, 7°C (45°F) to 13°C (55°F), depending on the cultivar.

**Other diseases.** In addition to diseases discussed above, those caused by *Phomopsis* spp., *Rhizopus stolonifer*, and other fungi are usually of minor extent or limited to certain growing areas.

## Citrus fruits

### Blue and green mold rots

Sexual state:  Unknown
Asexual state: Blue mold. *Penicillium italicum* Wehmer
Green mold. *Penicillium digitatum* (Pers.:Fr.) Sacc.

Blue and green mold rots, the most serious fruit diseases of citrus, occur wherever citrus fruits are grown. All species of citrus fruits are susceptible. Fruit may become diseased while still on the tree, during handling, storage, and transportation, or in the market.

The first indication of rot is usually a soft water-soaked area at the fruit surface, usually 0.6 to 1.3 cm (¼ to ½ inch) in diameter when first noticed. At favorable warm temperatures the lesion may grow to 3.8 to 5 cm (1.5 to 2 inches) within 24 to 36 hours. At about that stage, colorless mycelium can be detected on the fruit surface.

At room temperatures, blue mold develops slower than green mold, but the blue mold fungus is somewhat better adapted to low temperatures. Sporulation of *Pencillium italicum* is blue, surrounded by a narrow band of white mycelium. Green mold is olive green surrounded by a broad zone of white mycelium, which in turn may be surrounded by a narrow watersoaked area (fig. 15.57). Often the blue and green molds occur as mixed infections. The slower-growing blue mold lesion appear to be "taken over" by the more vigorously growing green mold.

Most infections are believed the result of fungal penetration of wounds in the fruit rind, although wounds may be small. *Penicillium italicum*, however, is capable of growing from a rotting fruit contacting an adjacent sound fruit. Given sufficient time, a "nest" of rotting fruit may involve several fruit.

Minimum temperature for rotting is below 7°C (45°F) for both species.

### Brown rot

Pathogen:  *Phytophthora citrophthora* (R. E. Smith & E. H. Smith) Leonian.

Brown rot, caused by any of several *Phytophthora* spp. in addition to *P. citrophthora*, is found in all citrus-growing regions of the world. It affects all citrus fruit; however, in lemon the highly acid juice causes the disease to be limited to the rind, core, and tissues between segments.

The rot, first observed as a slight surface discoloration, extends rapidly, and the lesion becomes a brownish tan or sometimes a slightly olive drab. The fruit remains firm and leathery. It has a characteristic pungent odor. Mold on the surface of fruit is unusual, but a delicate white fungus growth may be seen during extremely wet weather (fig. 15.58).

The disease is seldom seen except during or after very wet weather. Low-hanging fruit on trees become infected as motile spores from the soil are splashed on the lower part of the tree. Ordinarily, fruit within about 1.8 meters (6 feet) of the ground are infected, but strong winds may cause fruit high in the tree to become diseased. For the fungus to cause disease fruit must be wet for a long uninterrupted period. The motile spores are capable of penetrating the fruit rind without a wound.

Harvested fruit believed to be infected by the brown rot fungus are commonly heat treated. Submersion in water at 46° to 48°C (115° to 120°F) for 2 to 4 minutes kills the fungus providing it is confined to external layers of the rind. Fungus that has penetrated well below the rind survives heat treatment. Heat is especially injurious to lemons, and turgid fruit are particularly subject to injury. Thus, turgid fruit, particularly lemons picked in cool or humid weather, are allowed to wilt for 1 or 2 days before heat treatment.

### Phomopsis stem-end rot

Sexual state:  *Diaporthe citri* Wolf
Asexual state: *Phomopsis citri* Fawc.

Phomopsis stem-end rot is widely prevalent and results in serious losses in such humid growing areas as Florida. Optimum temperature for rotting is 23° to 24°C (73° to 75°F) and minimum temperature is 10°C (50°F). Infections appear as a softening of the rind and a slight watersoaked appearance generally turning light brown. The decayed flesh is not discolored. Mycelium sometimes occurs on the fruit surface (fig. 15.59).

Infections may develop on the calyx (button) during the growing season and remain latent until the fruit is harvested. After harvest, lesions may develop at the calyx end and grow at the sides of fruit or at the stylar end. Susceptibility to the disease increases with increasing age of the fruit at harvest.

*Phomopsis citri* also causes the orchard disease melanose. In wet weather the spores may infect a few epidermal cells of the rind of very young fruit. Infections may coalesce, forming various patterns on the rind as water containing fungal spores runs over the fruit surface. The fungus is usually dead by the time the fruit mature.

### Stem-end rot

Sexual state:  *Botryosphaeria rhodina* (Cooke) Arx
Asexual state: *Lasiodiplodia theobromae* (Pat.) Griffon & Maubl.
Note: The connection of *Botryosphaeria rhodina* as the sexual state of *Lasiodiplodia theobromae* is tenuous. See Lasiodiplodia rot of banana for synonyms.

Stem-end rot was long known as diplodia stem-end rot. Symptoms are similar to those caused by *Phomopsis citri* in that the disease normally starts at the calyx end of the fruit, but can develop from wounds at any point on the fruit. However, Lasio-diplodia rot lesions are usually a darker brown than those of Phomopsis rot. Also, the advancing margin of the Lasiodiplodia rot lesion progresses in lobes or fingers (fig. 15.60) whereas the advancing margin of Phomopsis lesions develops evenly. Lasiodiplodia rot is rapid and may decay the fruit completely in 3 or 4 days at its optimum temperature for growth, 28° to 30°C (82° to 86°F).

### Alternaria stem-end rot

Sexual state:  Unknown
Asexual state: *Alternaria citri* Ellis & N. Pierce.

Alternaria stem-end rot, usually a minor disease of overripe or badly stressed fruit, is found in all citrus-growing areas. Its market importance, however, increases when the rot is internal and hidden from the consumer's view.

Usually starting on the senescent calyx button after long storage on trees or in storage rooms (fig. 15.61) the fungus develops in the vascular tissues of the core and the inner tissues of the rind. Rot is not always apparent at the surface, although the central core and inner rind may be seriously rotted (fig. 15.62).

Onset depends upon the physiological condition of the buttons. Therefore, treatment with 2,4-dichlorophenoxy acetic acid (2,4-D) has been used to delay senescence of the buttons.

### Anthracnose

Sexual state:  *Glomerella cingulata* (Ston.) Spauld. & Shrenk
Asexual state: *Colletotrichum gloeosporioides* (Penz.) Sacc.

Anthracnose occurs on citrus fruit in the orchard, storage, or market. Conidia, produced in ascervuli on the fruit rind, are at first pinkish but soon darken (fig. 15.63).

The fungus enters the fruit at mechanical injuries or senescent buttons. Drought, freeze damage, and other factors that weaken trees may increase susceptibility. Fruit, such as tangerines, that may be picked green and subjected to lengthy degreening treatments with ethylene are often susceptible to attack by *C. gloeosporioides*.

Harvesting at proper maturity, careful handling, avoidance of long storage or excessive degreening, and maintaining temperatures below 10°C (50°F) during postharvest handling reduce anthracnose losses.

### Sour rot

Sexual state:  *Galactomyces geotrichum* (E.E. Butler & L. J. Petersen) Redhead & Malloch.
Asexual state: *Geotrichum candidum* Link
    = *Oospora lactis* (Fresen.) Sacc.
    = *Oospora lactis* (Fresen.) Sacc. var. *parasitica*
    Prit. & Port.

Sour rot disease is widely distributed throughout most citrus-growing areas, and occurs most often in storage and transit and in ripe or overripe fruit. The rot is a soft, messy, putrid, sour smelling, disgusting mass. Fruit flies are attracted to diseased fruit.

*Geotrichum candidum* is common in orchard soils where diseased fruit have rotted in previous years. Fruit that drop to the ground should be discarded because wounds can be contaminated by the sour rot fungus. The fungus can spread by contact infection from rotting to adjacent sound fruit, creating a nest of rotting fruit (fig. 15.64).

Sour rot does not develop at temperatures below about 5°C (41°F), which are appropriate for orange and mandarin. Lemon, lime, and grapefruit may be injured at temperatures below about 10°C (50°F).

### Sclerotinia rot

Sexual state:  *Sclerotinia sclerotiorum* (Lib.) de Bary
Asexual state: Unknown

*Sclerotinia sclerotiorum* produces an apothecium (a small mushroom) in the orchard from which sexual spores are forcibly ejected. The spores are readily spread by wind and may become established in fruit at the calyx or in wounds. Decaying fruit appears leathery, but mycelial growth results in a cottony look (fig. 15.65). The fungus nests by contact infection from an infected fruit to all surrounding fruit in a container.

The fungus grows in fruit that have fallen to the ground or on one of many host plants. Sclerotia form up to 1 cm (0.4 inch) long from which the apothecia are produced. Alternatively, the sclerotia produce mycelium that attacks susceptible plants. Elimination of weeds and clean cultivation of the orchard floor considerably reduce incidence of the disease.

### Trichoderma rot

Sexual state:  *Hypocrea* sp.
Asexual state: *Trichoderma viride* Pers.:Fr.
    = *Trichoderma lignorum* Tode.
    Note: The relation between *Trichoderma viride* and *Hypocrea* sp. is tenuous.

Trichoderma rot starts in injuries contaminated with soil. The fungus may also be spread by contact between infected and sound fruit. Lemons are most likely to become diseased, particularly after long-term storage. Grapefruit and oranges are also commonly diseased. Rotting citrus fruit turn brown (fig. 15.66) and have a firm, pliable texture. A coconutlike odor is typical. Control measures involve good sanitation and avoidance of contact with soil.

**Botrytis rot**

Sexual state: *Botryotinia fuckeliana* (de Bary) Whetzel
Asexual state: *Botrytis cinerea* Pers.:Fr.

In cool growing areas with moist, foggy weather during spring, *Botrytis cinerea* may grow abundantly in old flower petals, which become attached to fruit surfaces and cause mycelial infection. Quiescent infections may be established in the fruit's stem end. In storage, nesting occurs in packed fruit (fig. 15.67).

# Postharvest Diseases of Tropical Fruits

## Banana

### Crown rot disease complex

Pathogens: The complex of organisms responsible for crown rot generally includes *Fusarium roseum*, *Lasiodiplodia theobromae*, *Thielaviopsis paradoxa*, *Verticillium theobromae*, *Nigrospora sphaerica*, *Deightoniella torulosa*, and *Colletotrichum musae*. Many other fungi are sometimes present. Inoculations with various combinations of fungi show that the greatest damage results from combinations of *T. paradoxa*, *L. theobromae*, *C. musae* and *D. torulosa*. *Fusarium roseum* makes a less serious contribution.

Crown rot in banana is largely a consequence of a technological change in which the hands are cut from the stems and packaged in fiberboard cartons rather than shipped as stems. The same organisms that now attack the hand tissue formerly attacked the cut surface of the stem in transport. The disease is characterized by the darkening of the hand and adjacent peduncle and loss of the ability of the hand to support the fruit (fig. 15.68). From the rotting hand tissue the fungi grow into the finger neck and, with time, down into the fruit proper. Creasing of the finger stem in handling is a common injury that, along with cut surfaces of hands, permits initiation of the disease (fig. 15.69).

### Anthracnose

Asexual state: *Colletotrichum musae* (Berk. & Curt.) Arx
= *Gloeosporium musarum* Cooke & Mass.
= *Myxosporium musae* Berk. & Mass.

Anthracnose disease becomes evident as fruit ripen. Symptoms include more or less circular and somewhat sunken spots in which the skin becomes black and cracks in the peel sometimes occur (fig. 15.70). Salmon-colored masses of spores, produced by acervuli within the lesion, can be seen. Disease lesions on the fruit seldom extend into the flesh, but they render bananas and plantains unmarketable. *Colletotrichum musae* is a participant, with other fungi, in crown rot disease.

In the field, *C. musae* is found growing on all aboveground parts of the banana plant, particularly on the persistent bracts, during the wet season. Spores landing on green fruit germinate if free water is present or in very high humidity. On the germ tube is formed an appressorium from which an infection hypha penetrates the fruit cuticle. Green fruit are readily penetrated, but the infection remains latent until the fruit ripens and conditions become favorable for lesion development. Infection usually takes place between adjacent epidermal cells. Only a small proportion of the appressoria give rise to disease lesions. Infections also readily occur in handling wounds or where the skin splits between carpels.

### Lasiodiplodia rot

Sexual state: *Botryosphaeria rhodina* (Cooke) Arx
= *Physalospora rhodina* (Berk. & Curt.) Cooke
Asexual state: *Lasiodiplodia theobromae* (Pat.) Griffon & Maubl.
= *Botryodiplodia theobromae* Pat.
= *Diplodia theobromae* (Pat.) W. Nowell
= *Botryodiplodia gossypii* Ellis & Barth.
= *Diplodia gossypina* Cooke
= *Diplodia natalensis* Pole-Evans
= *Lasiodiplodia triflorae* Higgins
= *Diplodia tubericola* (Ellis & Everh.) Tauben.
= *Lasiodiplodia tubericola* Ellis & Everh.
Note: The connection between *Botryosphaeria rhodina* and *Lasiodiplodia* is tenuous.

In Lasiodiplodia rot, the fungus produces pycnidia on dead or dying leaves, bracts, and decaying fruit of the banana in the field. Similar dead or senescent plant materials of many other hosts also provide a substrate for colonization of the fungus and development of pycnidia. Conidia of the fungus are readily airborne. Rain, wind-driven rain, or splashing rain contribute to dispersal.

Freshly cut surfaces of banana stems, hands, and fruit provide areas moist with the juice and nutrients required for prompt germination and growth of conidia (fig. 15.71). The optimum temperature for growth of *L. theobromae* is about 27° to 30°C (81° to 86°F) and most isolates produce some growth over the range of 15° to 37°C (59° to 97°F). No growth is detected in culture after 4 weeks at 10°C (50°F).

### Thielaviopsis or Ceratocystis rot

Sexual state: *Ceratocystis paradoxa* (Dade) Moreau
*Ceratostomella paradoxa* Dade
*Endoconidiophora paradoxa* (de Seyn.) Davidson
*Ophiostoma paradoxum* (Dade) Nannf.
Asexual state: *Thielaviopsis paradoxa* (de Seyn.) Hohnel
Note: Asexual state is considered by some to be *Chalara paradoxa* (De Seyn.) Sacc.

*Thielaviopsis paradoxa* has a higher growth rate in culture than most common tropical fruit rotting fungi at optimum temperatures of 25° to 30°C (77° to 86°F). Growth occurs, but at a much reduced rate, at normal transit temperatures of 12° to 14°C (54° to 57°F) for banana.

Because of its high growth rate, *Thielaviopsis paradoxa* is a wound pathogen of considerable potential for damage. Although the growth and the extent of rotting are greater in mature fruit, *T. paradoxa* invades stem tissue and green fruit as well. A common place of entry at harvest is the cut stem or hand. Although commonly a stem-end rot (fig. 15.72), the fungus can enter and colonize wherever there is an injury. When the rot occurs in market areas on ripe or near-ripe fruit, the invaded flesh becomes soft and watersoaked.

## Cigar-end rot

Two fungi, *Verticillium theobromae* and *Trachysphaera fructigena*, may cause the disease singly or in combination.

Sexual state:  Unknown

Asexual states: *Verticillium theobromae* (Turc.) Mason & Hughes
= *Stachylidium theobromae* Turc.
*Trachysphaera fructigena* Tabor & Bunt.

Cigar-end rot causes serious losses in the West Indies, the Canary Islands, and east to Africa, Iran, and India. The disease appears to be absent from banana-growing areas of Central and South America. *Verticillium theobromae* is present in American growing areas, but *Trachysphaera fructigena* evidently is not.

Infection occurs in senescent floral parts from which the disease spreads to the fruit finger. Cigar-end rot in Egypt, for example, may involve half of the finger or more. The rotted portion of the banana finger is dry and tends to adhere to the fruit. The appearance of the rotted portion is like the ash of a fine cigar (fig. 15.73).

## Squirter disease

Sexual state:  Unknown

Asexual state: *Nigrospora sphaerica* (Sacc.) Mason
= *Trichosporum sphaerica* Sacc.

Found in Australia, primarily Queensland, and in the Cook Islands and Norfolk Island, squirter disease is associated with the Australian practice of shipping separate fruit (singles), often without refrigeration during transport. Infected green fruit are not detected at the time of packing but rot during transport and ripening.

The disease's name is descriptive of the condition of fruit in advanced stages of the disease. Fruit flesh darkens and tends to liquefy, and under pressure the liquefied flesh squirts out of the fruit.

The earliest recognizable symptom is the development of a dark center or core. Sometimes a line of dark red gumlike substance is distributed along the center. The disease may be confined to either end or may affect the entire fruit.

The fungus usually enters the freshly cut peduncle in fruit packed as singles. Spores germinate in the moist cut surface or may be drawn slightly into xylem vessels of the peduncle. Hyphae may grow in the xylem vessels through the peduncle into the fruit proper with little visible effect on the peduncle. Consequently, initial symptoms may develop in the fruit's interior with no obvious connection to the outside.

Optimum temperature for growth of *N. sphaerica* on malt agar is 22° to 25°C (72° to 77°F); maximum and minimum temperatures permitting growth are 32.5°C (90.5°F) and 5°C (41°F), respectively.

*Nigrospora sphaerica* is commonly found on bananas in the Western Hemisphere, but squirter dis-

ease symptoms are not. Some have proposed that the squirter disease organism is a separate species, and the name *Nigrospora musae* was suggested as a name for it. That suggestion has not generally been accepted. Instead, it is believed that the Australian organism is an especially vigorous pathogenic strain of *N. sphaerica*.

## Fusarium rot

Sexual state:  Unknown

Asexual state: *Fusarium roseum* Link 'gibbosum' Snyder & Hans.

The fusarium rot fungus is a major contributor to the crown rot complex. It also attacks fruit peduncles and fruit. Occasionally, infection begins in adhering floral parts and extends up the finger from the blossom end. These blossom-end rots resemble cigar-end disease (fig. 15.74), but Fusarium disease development is usually comparatively slow. Also, the disease may extend from the peduncle or from the stylar end, but it is not likely to extensively invade the finger.

Injuries to fruit skin are sometimes colonized. The invaded area of the skin darkens and becomes dry, and the dead skin sometimes splits to reveal the flesh below. The flesh, colonized while the skin was still green, often appears slightly brown or pink and is invariably dry and pithy. When the fruit ripens, those portions of the flesh that have been invaded remain hard and dry.

*Fusarium roseum* 'gibbosum' is prevalent in plantations where it colonizes organic matter such as stigmas and styles of immature fruit, dead floral parts, and rotting fruit on the ground. Spores are effectively dispersed by splashing raindrops or wind-driven rain.

# Mango

## Anthracnose

Sexual state:   *Glomerella cingulata* (Stonem.) Spauld. et Schrenk

Asexual state: *Colletotrichum gloeosporiodes* Penz.
Note: There are several hundred synonyms.

Fruit are infected with anthracnose rot in the young stage as well as mature, but infections in unripe fruit are mostly latent and do not start to appear until the fruit begin to ripen. Some infections enlarge and darken. Acervuli develop beneath the fruit cuticle, sometimes situated in more or less concentric circles. When acervuli development causes the cuticle to break, salmon-colored spores in a mucilaginous liquid can be observed (fig. 15.75). Under conditions of high humidity, the lesion may largely be covered with salmon-colored spores. Sometimes, however, spore production is less obvious and lesions appear as black spots (fig. 15.76). Lesions may remain limited to the fruit skin or may invade and darken the flesh to the stone. The bitter taste associated with this organism on apples and pears is less obvious in mangos.

Isolates of *Colletotrichum gloeosporioides* from tropical sources grow at 10°C (50°F) but not at 5°C (41°F).

### Stem-end rot

Sexual state: *Botryosphaeria rhodina* (Cooke) Arx
Asexual state: *Lasiodiplodia theobromae* (Pat.) Griffin & Maulb.
Note: See Lasiodiplodia rot of banana for synonyms. The connection between *Botryosphaeria rhodina* and *Lasiodiplodia theobromae* is tenuous.

Under high humidity and at 30°C (86°F), spores of the stem-end rot fungus readily infect fruit when placed on the pedicel scar or in injuries to the pedicel or injuries to the fruit peel. Sections of inoculated pedicels contain mycelium, largely in the vascular tissue and sparingly in the cortex. The fungus grows from the pedicel into a circular black lesion around the pedicel. Later the fungus commonly grows unevenly, with the advancing margin of the lesion growing much faster at certain places than at others (fig. 15.77), a characteristic of growth also observed in citrus fruits.

**Other diseases.** Several other diseases occasionally cause postharvest diseases of mango. In areas where refrigeration hardly exists, rots caused by *Rhizopus* spp., *Aspergillus niger*, *Macrophomina* spp., and other fungi may be prevalent. Severe cases of sooty mold occur in high-rainfall areas of India. Most of the rot in American-grown mangos imported into the United States is due to stem-end or side rots caused by *Lasiodiplodia theobromae* or due to anthracnose caused by *Colletotrichum gloeosporioides*.

## Papaya

### Anthracnose

Sexual state: *Glomerella cingulata* (Stonem.) Spauld. & Schr.
Asexual state: *Colletotrichum gloeosporioides* (Penz.) Arx

Anthracnose is found almost everywhere papaya is grown and may be the most serious cause of loss of harvested fruit. Infections occur in the green fruit in the orchard but are latent until the fruit ripen after harvest.

The highly variable nature of *Colletotrichum gloeosporioides* is evidenced by the synonym list of around 600 binomials. Further, its host range is wide, including leaves, young twigs, and fruit of many species. Besides the destructive fruit rot of papaya, anthracnose causes rots of many other commodities. Its optimum temperature for growth on papaya fruit is between 26° and 29°C (79° and 84°F), and its minimum temperature for growth is about 9°C (48°F). Isolates from Hawaiian papaya fruit grow at 10°C (50°F) but not at 5°C (41°F).

The anthracnose disease organism is present in the field, growing on senescent leaves and petioles, fallen fruit, and other types of organic matter. Conidia are produced in an acervulus in a water-soluble matrix. Evidently conidia are not readily transported in dry wind, but spores are readily dispersed by rain and wind-driven rain. Upon landing on a fruit, the conidium may germinate to produce a germ tube, which forms an appressorium. From the appressorium a fine infection hypha penetrates the fruit cuticle. If fruit are green, the fungus usually remains quiescent. With the onset of ripening, conditions favor growth of the fungus and lesions develop. Certain stress conditions also render the fruit more susceptible, as shown by the enhanced development of anthracnose lesions following fumigation of fruit with methyl bromide for fruit fly control.

Lesions are first detectable as tiny, brown, superficial, watersoaked lesions that may enlarge to 2.5 cm (1 inch) or more in diameter. Several lesions may grow together to produce a large, irregular compound lesion. Lesions become sunken but usually do not extend deep into the fruit flesh (fig. 15.78). Invaded flesh may taste bitter. On the surface, salmon-colored spore masses may form, sometimes giving the lesion a target or bull's-eye appearance. The lesions may remain a light brown to salmon color, but eventually they darken to dark brown or black. Sometimes the spore masses remain inconspicuous.

Conditions for controlling the disease by a postharvest heat treatment are almost ideal: infections are present in the fruit but have caused no damage at time of harvest; control of the disease in the orchard has been unsatisfactory; and the fruit tolerate a heat treatment that effectively inactivates latent infections in the fruit. The heat treatment consists of a preheating step of 30 minutes at 42°C (108°F) followed by 20 minutes at 49°C (120°F).

Central to the control of anthracnose and other postharvest fruit rots, as well as maintenance of the physiological condition of the fruit, is good temperature management during transportation and marketing. If fruit are quickly cooled to about 13°C (55°F), growth of the pathogen and ripening changes in the fruit are slow. If fruit are ripened quickly at 20°C (68°F) following transportation, there is little opportunity for disease to develop during normal marketing. To be avoided is slow ripening at 15° to 17°C (59° to 63°F) because the fungus can grow appreciably during the time required for ripening at those temperatures.

### Phoma (Ascochyta) rot

Sexual state: *Mycosphaerella caricae* H. & P. Sydow.
Asexual state: *Phoma caricae-papayae* (Tarr.) Punith.
= *Ascochyta caricae-papayae* Tarr

*Phoma caricae-papayae* attacks the trunk, leaves, flowers, leaf and fruit pedicels, and green and ripe fruit in the orchard, particularly during winter and early spring. Fruit still attached to trees may show a black rot extending into the fruit from the stem end or from a point of contact with a dead leaf or infected leaf stalk. An isolated lesion first appears as a small watersoaked spot that develops into a shrunken, black, circular lesion from 2.5 to 7.5 cm (1 to 3 inches) in diameter. Multiple infections may coalesce to form a large rotting area. Fruit attacked in the field may wither and drop, particularly if still immature.

In the field, *Phoma caricae-papayae* frequently colonizes pedicels of senescent leaves from which the trunk of the plant is invaded. Fruit may become infected in the field during harvesting when they are injured.

After harvest, the disease lesion on a fruit is frequently observed first in the region of the stem, where rotting tissue becomes dark brown to coal black, while the invaded tissue remains relatively firm and dry (figs. 15.79 and 15.80). The surface of the lesion eventually appears roughened or pebbly from the development of pycnidia, which rupture the host epidermis.

### Phomopsis rot

Sexual state:   Unknown
Asexual state: *Phomopsis caricae-papayae* Petr. & Cif.

*Phomopsis caricae-papayae* invades senescent leaf petioles and other papaya trash in the field where abundant pycnidia are produced. During rainy weather, spores are carried by splashing rain, wind, or wind-driven rain.

Phomopsis rot commonly originates in the area of the peduncle or a fruit skin wound. Presumably, spores of the fungus cannot easily enter the fruit except through wounds or the freshly ruptured peduncle.

The disease can develop rapidly in ripe fruit. The invaded flesh and skin tissue is soft, watersoaked, and slightly darker than uninvaded tissue (figs. 15.81 and 15.82). Surface mycelium may be almost absent. Sometimes, however, other fungi are also present. Pycnidia of the fungus are usually produced only after the fruit is nearly completely rotten.

### Lasiodiplodia rot

Sexual state:   *Botryosphaeria rhodina* (Cooke) Arx
Asexual state: *Lasiodiplodia theobromae* (Pat.) Griffin & Maulb.
                Note: The connection between *Botryosphaeria rhodina* and *Lasiodiplodia theobromae* is tenuous.

Lasiodiplodia rot, commonly initiated near the fruit peduncle, causes a stem-end rot that may be confused with Phoma stem-end rot. Lasiodiplodia rot may usually occur at injuries to the fruit skin. The conidia ordinarily cannot penetrate uninjured skin. Fresh mechanical injuries or the fresh surface of the peduncle at the point of separation are ideal infection sites.

*Lasiodiplodia theobromae* is found in the orchard in dropped fruit or in senescent or decaying organic matter of various types. Spores are disseminated by wind during wet weather and especially by splashing

rain. The minimum temperature for growth is 10°C (50°F) or slightly above. Growth is rapid at the optimum temperatures of 25° to 30°C (77° to 86°F).

### Phytophthora rot

Pathogen:       *Phytophthora nicotianae* Breda de Haan var. *parasitica* (Dast.)Waterh.

Phytophthora rot in the field can be confused with other *Phytophthora* spp. or *Pythium* spp. Fruit of any age may be infected on the tree. As the disease progresses, the fruit starts shriveling, turns dark brown, and falls to the ground where more shriveling and mummification takes place.

Postharvest fruit rot is typically a stem-end rot. After starting near the abscission layer, the fungus progressively invades the tissue inter- and intra-cellularly until the entire fruit decays. The epidermis and fruit flesh, although watersoaked, remain near normal in color. Fruit surfaces are commonly covered with white mycelium that becomes encrusted (fig. 15.83).

### Alternaria rot

Sexual state:   Unknown
Asexual state: *Alternaria alternata* (Fr.) Keissler
                Note: See apple and pear for synonymy.

*Alternaria alternata* seriously affect papayas that have been stored or transported at chilling temperatures below 12°C (54°F). Such chilled fruit typically appear sound for a time, but the chilling injury increases their susceptibility to *A. alternata* and certain other fungi (fig. 15.84).

## Pineapple

### Thielaviopsis rot, black rot, water blister

Sexual state:   *Ceratocystis paradoxa* (Dade) C. Moreau
                = *Ceratostomella paradoxa* Dade
                = *Ophiostoma paradoxa* (Dade) Nannf.
Asexual state: *Thielaviopsis paradoxa* (de Seyn.) Hohnel
                = *Thielaviopsis paradoxa* (de Seyn.)
                Note: Some consider *Chalera paradoxa* (de Seyn.) to be the preferable name for the asexual state.

Thielaviopsis rot, well known throughout the world's pineapple-growing areas and markets, poses a major problem almost everywhere to the orderly production and marketing of fresh fruit. For example, 3 to 5 percent of Puerto Rican fruit on the New York market was reported rotted in usual shipments, but in fruit harvested during or just after a rainy period, rot affected 25 percent.

The disease organism is found on a wide range of hosts throughout the world's temperate and tropical regions. Although the same organism attacks both pineapple and banana fruit, isolates of the organism from rotting bananas, in our experience, have not readily decayed pineapples. Similarly, isolates from rotting pineapple do not rot bananas.

The optimum temperature for growth of the fungus in culture is about 27°C (81°F); the maximum is about 37°C (99°F) and the minimum is between 0° and 5°C (32° and 41°F). However, fruit rotting is almost arrested at 10°C (50°F).

Thielaviopsis rot is usually not readily noticeable until its invasion is advanced. Rot may start at the stem end and advance through most of the flesh with little external evidence of decay (figs. 15.85, 15.86, and 15.87). The only external indication of damage is a slight darkening due to watersoaking of the skin over rotted portions of the fruit. As the flesh softens, the skin above the affected tissue readily breaks under slight pressure.

When fruit are cut, areas invaded by the fungus are soft and watery as well as deeper in color than adjacent healthy tissue. In the disease's final stages the core disintegrates with the flesh. A sweetish odor distinct from the sour odor of rot by yeasts accompanies the decay. Yeasts often cohabit the rot lesion and attract vinager flies.

### Fruitlet core rot

Sexual state:   (1) *Gibberella fujikori* Sawada & Ito
Asexual state: (1) *Fusarium moniliforme* Sheld.
               (2) *Penicillium funiculosum* Thom

Fruitlet core rot, evident throughout the pineapple-growing world, is characterized by rot of individual fruitlets, and is usually visible only when the fruit is cut to reveal a brown to black discoloration of the flesh. Inoculations of fruitlets with *Penicillium funiculosum* or *Fusarium moniliforme* or with a mixture of conidia of the two fungi result in typical fruitlet core rot (fig. 15.88).

The two types of symptoms are: a wet rot resulting when mature, juicy fruit are infected, and dry spots that develop when fruit are inoculated before flower buds are visible and before the fruit mature and become juicy.

The two fungi appear incapable of infection unless the epidermis of the floral cavity has been broken. Possibly insects that enter the floral cavity in search of nectar carry the fungus spores into the cavity with them, and cause the damage. Also, growth cracks in the floral cavity may provide the needed infection court.

### Yeasty fermentation

Asexual state: *Saccharomyces* spp.(possibly other yeasts).

Yeasty fermentation, usually associated with overripe fruit, may start while the fruit is still on the plant or after harvesting. Evidently the organisms enter the fruit through wounds.

Fruit flesh becomes soft and bright yellow and is ruptured by large gas cavities. Production of gas forces juice from dead and dying cells out through the cracks and wounds in the shell as a frothy, sticky liquid (fig. 15.89). In time, the juice within the tissues may largely be fermented, leaving spaces within the tissues (fig. 15.90).

### Marbling

Pathogen:     *Erwinia carotovora* (Jones) Bergey et al.

Marbling is a bacterial disease caused by *Erwinia carotovora*, but other, unrelated bacteria may also cause nearly identical symptoms. It has been associated with fruit of low acidity such as summer fruit from warmer pineapple-growing areas.

Disease symptoms include speckled browning and abnormal hardening of internal tissues. Browning varies from bright yellowish or reddish-brown to a very dark dull brown. Hardening is most pronounced in tissues that are brown, but adjacent tissues of normal color may be abnormally crisp. These symptoms may affect flesh of the entire fruit or be limited to a single fruitlet.

Infection is thought to take place at flowering or shortly thereafter, but the disease develops only during ripening. Large fruit seem more susceptible than small, possibly because they are generally less acid.

## Postharvest Diseases of Vegetables

### Carrot

#### Bacterial soft rot

Pathogen:     *Erwinia carotovora* (Jones) Bergey et al.

The bacteria causing soft rot may affect carrot foliage in the field and may be present in soil adhering to carrot roots. Harvest bruises and insect wounds offer entrance into the root.

The soft rot bacteria decay carrot tissues rapidly at room temperature. Carrot tissues become watersoaked and the middle lamella is solubilized by the action of pectolytic enzymes of the fungus. The result is a soft, slimy, often foul-smelling, wet mass of carrot cells (fig. 15.91).

Bacterial soft rot seldom occurs at temperatures below 5°C (41°F), although the minimum temperature for growth of the bacterium is as low as 0° to 2°C (32° to 36°F). Thus, serious loss from bacterial soft rot usually occurs when carrots have not been adequately cooled or cooling has been delayed.

#### Gray mold rot

Sexual state:   *Botryotinia fuckeliana* (de Bary) Whetzel
Asexual state: *Botrytis cinerea* Pers.:Fr
               Note: See gray mold rot of apple and pear for synonymy.

Gray mold rot is common in carrot, particularly during storage periods of 5 months or longer. Lesions are typically covered with mycelium of the fungus, which is initially white. In time, the fungus sporulates abundantly, turning gray (fig. 15.92). Later, black, irregular sclerotia appear on the host's surface.

Conidia of *B. cinerea* are capable of directly penetrating many tissues. It appears likely, however, that most infections in storage occur from mycelium or sclerotia in adhering soil particles entering wounds made during harvesting and handling. Mycelia grow from diseased roots to nearby sound roots, like *S. sclerotiorum* or *S. minor*, to form a nest of rotting carrots.

### Carrot white rot

Sexual state: (1) *Sclerotinia sclerotiorum* (Lib.) de Bary
(2) *Sclerotinia minor* Jagger

Asexual state: Unknown
See white rot of snap beans for synonymy of the organisms and figure 15.124 for illustration of apothecia of *Sclerotinia sclerotiorum*.

White rot organisms cause a soft, but not slimy, rot of carrot roots in storage at 0°C (32°F). Disease is likely to become serious if the organisms were active before harvest, when a cottony white mycelium develops on the root (fig. 15.93). In time, distinctive black, rounded sclerotia form.

The minute microconidia that are produced appear to serve as spermatia (male cells), permitting development of the perfect (ascospore) state. Ascospores play little or no part in infecting mature carrot roots in storage. Instead, mycelia in soil may produce infection hyphae that penetrate the host cuticle. More often, the mycelium in soil invades wounds caused by harvesting or handling.

Mycelia may grow from a colonized root to penetrate and infect nearby sound roots. The result is an ever-enlarging nest of rotting roots bound together by fungus mycelium.

### Fusarium dry rot

Several *Fusarium* spp. attack carrot, causing a relatively dry, spongy rot in storage. The disease usually is observed only after several months in storage, generally above 8° to 10°C (46° to 50°F).

### Black mold

Sexual state: Unknown

Asexual state: *Thielaviopsis basicola* (Berk. & Br.) Ferr.
= *Trichocladium basicola* (Berk. & Br.) Carmichael

Black mold is occasionally found on carrots (fig. 15.94). *Thielaviopsis basicola* attacks the root systems of many growing plants, particularly cucurbits, legumes, and members of the Solanaceae. This fungus produces two types of conidia. Hyaline conidia are produced internally within a conidiophore and are expelled in chains of several conidia that soon break apart as they age, and brown-walled chlamydospores are produced on hyaline conidiophores.

Black mold develops only if temperatures exceed 5°C (41°F), and possibly after the root has become senescent.

### Crater rot

Sexual state: Unknown

Asexual state: *Rhizoctonia carotae* Rader

High losses due to crater rot have been experienced in carrot storages in the northeastern United States and in northern Europe. *Rhizoctonia carotae* resembles *Rhizoctonia solani*, but it is known to attack only carrot. Clamp connections on the mycelium indicate that these two fungi are basidiomycetes and may be closely related.

The pathogen grows at temperatures from below 0° to 24°C (<32° to 75°F) with an optimum of 21°C (70°F). Humidities near saturation are associated with lesion development. Inoculum is present in carrot fields, and the root may be infected at or soon after harvest.

Pits develop in the root and become large, sunken lesions resembling a crater (fig. 15.95). Avoidance of injuries to roots during harvesting and handling reduces crater rot. Losses can be further prevented by reducing storage relative humidity below 95 percent and the temperature to 0°C (32°F).

## Celery

### Bacterial soft rot

Pathogen: *Erwinia carotovora* var. *carotovora* (Jones) Dye

Bacterial soft rot advances down the petioles from the leaflet. Sometimes a single petiole is involved with little or no tendency for the diseased petiole to infect others (figs. 15.96 and 15.97). Contributing to bacterial soft rot are such problems as inadequate or delayed cooling or warming in transit.

### Sclerotinia pink rot

Sexual state: *Sclerotinia sclerotiorum* (Lib.) de Bary

Asexual state: Unknown
Note: See Sclerotinia rot of kiwifruit for synonymy.

Pink rot results from the fungus in soil, generally present where celery has been grown. Mycelium in the soil may penetrate healthy plants at the base of petioles. Affected petioles become pink and soft (fig. 15.98), either in the field or during storage and transit. For more information on the disease, see the section on white rot in lettuce, snap bean, and carrot.

## Lettuce

### Bacterial soft rot

Pathogen: *Erwinia carotovora* var. *carotovora* (Jones) Dye

Bacterial soft rot may start in the field, where *Erwinia carotovora* is normally present. Leaves die as the infection progresses, often with the production of slime. The greatest potential for loss is as a postharvest disease (fig. 15.99). Commonly the disease follows delays in cooling or inadequate heat removal, giving the bacterium ideal conditions for growth. Often lettuce leaves appear to have become senescent and highly susceptible to rot before soft rot develops.

### Lettuce white rot

Sexual state: (1) *Sclerotinia sclerotiorum* (Lib.) de Bary
(2) *Sclerotinia minor* Jagger

Asexual state: Unknown

Note: See Sclerotinia rot of kiwifruit for synonyms.

These fungi cause the disease "lettuce drop" in the field. The unharvested head yellows and collapses as a consequence of infection. "White rot" is the name commonly given to postharvest rot of lettuce and many other commodities. The white mycelium grows abundantly on the rotting host and the fungus extends mycelial strands to sound lettuce heads which become infected, forming a nest of rotting heads.

The two causal organisms differ primarily in the size of the sclerotia they produce. *Sclerotinia sclerotiorum* produces relatively large sclerotia; those of *S. minor* are much smaller. Neither species produces asexual spores. Both species produce small flesh-colored flat or saucer-shaped apothecia (mushrooms), which eject ascospores into the air during a short period in spring (fig. 15.124).

The fungus, growing saprophytically on organic matter in the soil, may contact a plant and the mycelium penetrates directly to establish infection. Ascospores landing on lettuce leaves may also establish infections, depending upon local growing and climatic conditions.

## Gray mold rot

Sexual state: *Botryotinia fuckeliana* (de Bary) Whetzel
Asexual state: *Botrytis cinerea* Pers.:Fr.
Note: See gray mold rot of apple and pear for synonymy.

Gray mold rot resembles white rot in many respects. In the field or greenhouse the fungus reduces a lettuce head to a slimy rotten mass. As with white mold, the gray mold fungus has a wide and somewhat similar range of host plants.

*Botrytis cinerea* produces abundant conidia, causing a gray color, in contrast to *Sclerotinia* spp., which do not produce conidia. Sclerotia of *B. cinerea* function like those of *Sclerotinia* spp., but their shape is very irregular. Sexual spores are rare and appear to have little or no role in rotting host products.

*Botrytis cinerea* spreads by contact infections, like *Sclerotinia* spp., if diseased heads are mixed with healthy heads. Prompt cooling to near 0°C (32°F) and maintaining low temperature during transport and marketing minimizes losses.

## Melons

### Sour rot

Sexual state: *Galactomyces geotrichum* (Butl. & Peter.) Redh. & Mall.
Asexual state: *Geotrichum candidum* Lk.
Note: See sour rot of citrus for synonymy.

The sour rot fungus is found in the soil and in plant debris in the soil. It can colonize wounds at any point on the fruit. On muskmelons the fungus often colonizes the fresh stem scars. In advanced stages, the rot may convert the interior of the melon to a slimy mass with a sour, disagreeable odor highly attractive to vinegar flies (*Drosophila* spp.). The disease is also found, in a lesser incidence, in watermelon, usually as a stem-end rot but sometimes at the blossom end or at wounds. The flesh of the melon may eventually liquefy (figs. 15.100, 15.101, and 15.102).

*Geotrichum candidum* is widely distributed in most soils, and stems are likely contaminated by soil containing propagules of the fungus. Arthrospores of the fungus are spread by wind or wind-driven mist or splashing water. Vinegar flies attracted to rotting fruit may transport spores from fruit to fruit. Melons become more susceptible to the sour rot fungus as they mature and ripen.

### Rhizopus rot

Pathogen: *Rhizopus stolonifer* (Ehrenb.:Fr.) Vuill.
Note: See Rhizopus rot of stone fruits for synonymy.

*Rhizopus stolonifer* and possibly other *Rhizopus* spp. infect melons in injuries or at the stem scar (fig. 15.103). Stem scar infections are particularly common in cantaloupe. The resulting rot is extremely soft, and considerable liquid accumulates by the time the fungus has involved much of the fruit. At elevated temperatures the fungus colonizes the fruit very rapidly. Optimum temperature for growth of *Rhizopus stolonifer* is about 24° to 27°C (75° to 80°F); the minimum is 5°C (41°F).

### Fusarium rot

*Fusarium* spp. infect melons in contact with the soil before harvest. Development of the disease occurs as the fruit ripen. Rotted tissues may develop into voids. Disease development is slow enough that losses in California are not excessive. Cantaloupes that have been chilled may develop mold on the fruit surface. *Fusarium roseum* Lk: Fr. occasionally develops as a stem-end rot of melons (fig. 15.104).

### Trichothecium rot

Sexual state: Unknown
Asexual state: *Trichothecium roseum* (Pers.:Fr.) Link
Note: See pink rot of apple and pear for synonymy.

The fungus *Trichothecium roseum* occasionally attacks melons, particularly honeydew melon (fig. 15.105). A stem-end rot is usually produced, but all deep wounds are probably infection courts. The fungus is often found in melons nearing the end of their postharvest life. Usually recognized by the rusty-red color of the sporulating fungus, the disease is sometime confused with Fusarium stem-end rot which may appear pink or red.

Usually of little consequence, the disease does produce a potent toxin that is strictly confined to the lesion. Chances that someone may consume the rotted lesion tissue appear minuscule.

## Botrytis rot

Sexual state: *Botryotinia fuckeliana* (de Bary) Whetzel
Asexual state: *Botrytis cinerea* Pers.:Fr.
Note: See gray mold rot of apple and pear for synonymy.

Botrytis rot is found rarely as a stem-end rot of watermelon (fig. 15.106) or as invader of mechanical wounds. Botrytis rot usually follows unseasonable cool, wet weather.

## Lasiodiplodia rot

Sexual state: *Botryosphaeria rhodina* (Cooke) Arx
Asexual state: *Lasiodiplodia theobromae* (Pat.) Griff. & Maubl.
Note: The connection between *Botryosphaeria rhodina* and *Lasiodiplodia theobromae* is tenuous. See Lasiodiplodia rot of banana for synonymy.

Lasiodiplodia rot is occasionally found as a postharvest rot of California watermelon (fig. 15.107). More serious losses are found in watermelon grown in humid climates, where honeydew melon and cantaloupe are also attacked. The disease is first seen as a shriveling and drying of the stem followed by browning of the area around the stem, which progressively enlarges as the disease develops.The cut flesh is noticeably softened and lightly browned. If the cut melon is exposed to the air for a few hours, the diseased areas become black (fig. 15.108). The disease develops rapidly in the fruit at temperatures of 25° to 30°C (77° to 85°F) but slowly or not at all at 10°C (50°F).

## Gummy stem blight and black rot

Sexual state: *Didymella bryoniae* (Auersw.) Rehm.
= *Mycosphaerella citrullina* (C.O.Sm.) Gross.
= *Didymella melonis* Pass.
= *Mycosphaerella melonis* (Pass.) Chiu & Walker
Asexual state: *Phoma cucurbitacearum* (Fr.:Fr.) Sacc.

Gummy stem blight, found primarily in warm, humid growing areas, is a field disease whose infections sometimes cause postharvest losses in cucurbits. In the field, nodes of plants may appear oily green, and sap exudes that may partially dry to form drops of dark gum, giving the disease its name. On fruit first symptoms developing after harvest are dark, watersoaked spots anywhere on the fruit surface. Mature lesions are sunken, may show a pattern of concentric rings, and turn black with small pycnidia.

## Anthracnose

Sexual state: *Glomerella lagenarium* F. Stevens.
Asexual state: *Colletotrichum orbiculare* (Berk. & Mont.) Arx
= *Gloeosporium orbiculare* Berk. & Mont.
= *Colletotrichum lagenarium* (Pass.) Ellis & Halst.
= *Gloeosporium lagenarium* (Pass.) Sacc.

The anthracnose fungus can penetrate the cuticle and epidermis of leaves, stems, and fruit. With fruit, quiescent infections may be produced with no external evidence of the disease. As fruit mature and ripen the latent infections become active. Watersoaked spots appear, becoming sunken and black with the formation of acervuli. Spores are produced of a rose, brick red, or orange color as viewed en masse. Spores are distributed by water, wind-driven mist, insects, or pickers' hands.

## Onion

### Neck rot

Sexual state: (1) Unknown
(2) *Botryotinia allii* (Sawada) Yamamoto
(3) *Botryotinia squamosa* Vien.-Bourg.
= *Sclerotinia squamosa* (Vien.-Bourg.) Dennis
Asexual state: (1) *Botrytis aclada* Fresen.
= *Botrytis allii* Munn.
(2) *Botrytis byssoidea* Walker
(3) *Botrytis squamosa* Walker

Botrytis neck rot is often the most serious postharvest disease of onion. The organisms may spot leaves and cause tip dieback in the field, but the bulb is seldom affected before harvest. The disease usually progresses from the cut leaves into the inner bulb scales, which become watersoaked and light brown to dark brown. Gray fungal growth may be abundant, usually in the neck area, and sporulation may occur (fig. 15.109).

The fungus persists from one season to the next by growing on onion waste or by sclerotia in the soil. Its conidia are dispersed by wind or splashing rain.

To minimize losses from neck rot, onions should be properly cured by drying with good circulation of air around them to prevent accumulation of moisture. Foliage should have matured before leaves are cut. Bulbs should not become sunburned, and every effort should be made to avoid cuts, bruises, or abrasions. Best storage is at 0°C (32°F) and 70 to 75 percent relative humidity.

### Bacterial soft rot

Pathogen: *Erwinia carotovora* var. *carotovora* (Jones) Dye

Bacterial soft rot may occur with high humidity and temperatures. The bacterium gains entrance at cut leaves or by contamination of cuts or bruises in bulbs. Invaded fleshy scales become watersoaked and appear yellow or light brown (fig. 15.110). Scales may become thoroughly invaded. Bulbs with leaves cut before they are completely mature or bulbs that have not been well cured are likely to be affected.

### Smudge

Sexual state: Unknown
Asexual state: *Colletotrichum circinans* (Berk.) Vogl.
= *Vermicularia circinans* Berk.
= *Colletotrichum dematium* (Pers.) Grove
f. *circinans* (Berk.) Arx

Smudge is limited to the dry, outer scale leaves surrounding the bulb. Black areas on the dry scales may appear to be smeared but frequently concentric black circles are formed (fig. 15.111).

The fungus may be transported long distances on bulbs or onion sets and may persist on onion residues.

Conidia of the fungus may be disseminated by wind or splashing rain in the field. The disease may be particularly prevalent in storage, if rain has occurred during harvest.

### Fusarium bulb rot

Sexual state:  Unknown

Asexual state: *Fusarium oxysporum* Schlect.:Fr. f. sp. *cepae* (Hans.) Snyder & Hans.
*Fusarium zonatum* (Sherb.) Wr.

Fusarium bulb rot is caused by the *Fusarium* spp. listed above and probably others as well. Infection may start in the field. The fungus may directly penetrate roots or bulbs or may enter via injuries caused by soil insects.

Rot may develop at anywhere on the bulb including the basal plate (fig. 15.112). Losses are greatest in storages with high humidity or in bulbs poorly cured before storage (fig. 15.113).

### Black mold

Sexual state:  Unknown

Asexual state: *Aspergillus niger* v. Tiegh.

Black mold is most common in hot growing areas, but the fungus occurs wherever onion is grown. It can often be observed on dead plant debris where considerable moisture is present and temperatures are high. The fungus grows slowly at 13°C (55°F) but the disease occurs at temperatures of 16° to 40°C (60° to 104°F).

The most common indication of the disease is the occurrence of black, powdery spore masses on the dry outer scales or between dry scales and the outermost fleshy scales. Under hot, humid conditions the fungus may cause a slow rot of bulbs. However, the main loss is usually in appearance and marketability of the bulbs.

### Blue mold rot

Sexual state:  Unknown

Asexual state: *Penicillium expansum* Link

Blue mold rot may be present in storage causing a wet, soft lesion. Rotting can occur in storage at 0°C (32°F), but serious losses seldom occur if bulbs are properly cured and are not stressed by unfavorable temperatures or humidity levels. The disease is readily recognized by the blue to blue-green color of the sporulating fungus.

## Potato

### Bacterial soft rot

Pathogen:      *Erwinia carotovora* var. *carotovora* (Jones) Dye
*E. carotovora* var. *atroseptica* (Van Hall) Dye

In addition to the two main causal agents of bacterial soft rot, species belonging to the genera *Pseudomonas*, *Bacillus*, and *Clostridium* sometimes cause similar diseases in potato. *Erwinia carotovora* cells are motile with peritrichous flagella, are rod shaped, and are gram-negative. The bacteria enter tubers via lenticels or wounds.

Water on the surface of stored tubers reduces aeration and predisposes them to infection. Infections at lenticels appear watersoaked and circular at first. Rotting tissues become wet and cream or tan colored compared to sound tissues (fig. 15.114). The rotting tissues are soft and watery. With time infected tubers develop a foul odor.

Tubers for storage should be harvested when mature with minimal mechanical injuries. Ventilation prevents formation of water films or localized accumulation of elevated carbon dioxide. Tubers should not be washed before storage and when washed before marketing should be dried as soon as possible. Wash water should be changed frequently and should be treated with chlorine to lower the level of viable cells.

### Ring rot

Pathogen:      *Corynebacterium sepedonicum* (Spieck. & Knott.) Skapt. & Burkh.

The ring rot bacterium, *Corynebacterium sepedonicum*, is a gram- positive, nonmotile organism having predominantly wedge-shaped cells (0.4 to 0.6 × 0.8 to 1.2 μm). The bacterium is slow growing on media. The disease overwinters in infected tubers in the field or in storage. Infection occurs through tuber wounds. Contamination of knives used for cutting seed tubers or various machinery transmits the bacterial slime to tuber seed pieces. Harvesting operations in which tubers are wounded may also spread the disease (fig. 15.115).

### Late blight

Pathogen:      *Phytophthora infestans* (Mont.) de Bary.

Late blight (fig. 15.116) caused the potato famine in Europe in the mid-nineteenth century. It continues to threaten potatoes wherever they are grown, except in certain hot, dry, irrigated areas.

The disease causes serious losses in the field. Tubers that have been infected in the field may rot in storage (fig. 15.117). Tuber infections are most likely when wet weather occurs during harvest. Little spread occurs under good storage conditions.

### Fusarium dry rot

Sexual state:

Asexual state: (1) *Fusarium solani* (Mart.) Sacc.
(2) *F. roseum* (Lk.) Snyd. & Hans.

Fusarium dry rot is found wherever potatoes are grown and handled. Although the rot lesion is characteristically dry, secondary infections by soft-rot bacteria soften the lesion and make it wet. A disagreeable odor, often associated with bacterial soft rot, may develop.

Fusarium dry rot fungi are believed incapable of penetrating the tuber periderm or lenticels. Some infections are associated with insect or rodent activity,

156

and the pathogens may become secondary organisms invading lesions of other fungus diseases. Pathogens enter the tuber through mechanical injuries usually cuts and periderm-breaking bruises inflicted during harvesting and handling.

Tubers are relatively resistent to Fusarium dry rot when harvested, but susceptibility increases during storage (fig. 15.118). Disease development is most rapid at temperatures of 15° to 20°C (59° to 68°F). Cuts in tubers result in the deposition of suberin, leading to the formation of a disease-preventing periderm barrier within several days at 20°C (68°F) and somewhat longer at lower temperatures.

## Black scurf

Sexual state: *Thanatephorus cucumeris* (Frank) Donk
= *Corticium areolatum* Stahel
= *Pellicularia filamentosa* (Pat.) Rogers
= *Ceratobasidium filamentosum* (Pat.) Olive
= *Hypochnus filamentosus* Pat.
= *Corticium praticola* Kotila
= *Corticium sasakii* (Shirai) Matsumoto
= *Hypochnus sasakii* Shirai
= *Pellicularia sasakii* (Shirai) Ito
= *Botryobasidium solani* (Prill. & Delacr.) Donk
= *Corticium solani* Prill. & Delacr.
Asexual state: *Rhizoctonia solani* Kuhn
= *Moniliopsis solani* Kuhn
= *Rhizoctonia macrosclerotia* J. Matz
= *Rhizoctonia microsclerotia* J. Matz

Black scurf occurs in the field where sclerotia on tubers or mycelium in plant debris provide the inoculum. Potato stems, roots, and stolons of the growing plants may be infected. Sclerotial development on the tuber occurs in a favorable environment (low soil temperature and high moisture level).

In the harvested tuber the disease is usually observed as black or dark brown sclerotia that are tightly appressed to the periderm (fig. 15.119), often described as "dirt that won't wash off." Sclerotia are usually a few millimeters in diameter and are irregular in size and shape. Sclerotia on tubers do not ordinarily cause them to rot, but their presence detracts from their appearance.

**Other diseases.** Many diseases of tubers after harvest may be periodically or locally important; these include:

*Gangrene.* Causal agents of gangrene are *Phoma exigua* Desm. var. *foveta* (Foister) Boerema and *Phoma exigua* Desm. var. *exigua* Sutton & Water. The former is the most virulent and is found in most northern European countries. Infected seed tubers provide the initial inoculum. Spores of the fungus are washed through the soil and tuber infections may occur at that time, but more commonly infection occurs in wounds caused by harvesting or handling.

*Gray mold. Botrytis cinerea* Pers.:Fr. may attack foliage in the field and occasionally may cause significant rotting of tubers in storage if curing is not done and if storage temperature is above optimum range.

*White mold.* Causal organisms of white mold, *Sclerotinia sclerotiorum* (Lib.) de Bary and *S. minor* Jagger,

are active in the growing crop. Tubers are infected in the field and may continue rotting during storage.

*Sclerotium rot. Sclerotium rolfsii* Sacc. is primarily a field disease, but freshly infected tubers at harvest continue to rot in storage if temperatures are favorable for the fungus.

*Pink rot.* (Pathogen: *Phytophthora erythroseptica* Pethybr.). Pink rot of potato (fig. 15.120) is primarily a disease of tubers in the field. Basal stem decay and wilt of the top may occur. Affected tubers appear a dull brown with lenticels and eyes still darker. Internal decay usually begins at the stem end; affected tissue becomes rubbery. Internal tissues gradually turn pink as decay occurs. Infected tubers may continue rotting during storage.

## Snap bean

### Botrytis gray mold rot

Sexual state: *Botryotinia fuckeliana* (de Bary) Whetzel
Asexual state: *Botrytis cinerea* Pers.:Fr

Botrytis gray mold is frequently the most common disease of beans after harvest, and losses can be serious, particularly after an inclement growing season. Overhead irrigation may result in excessive moisture and added Botrytis rot in the field. Consequently, many infected pods may be inadvertently included among harvested pods.

Infected pods develop watersoaked lesions that may spread by mycelial contact from diseased to sound pods. Disease lesions or fungus conidia on pod surfaces provide opportunities for new infections. The fungus nests by contact infections from diseased to surrounding pods (fig. 15.121).

### Snap bean white rot

Sexual state: (1) *Sclerotinia sclerotiorum* (Lib.) de Bary
(2) *Sclerotinia minor* Jagger
Asexual state: Unknown
See white rot of lettuce for added information and for synonymy of the organisms.

The two white rot pathogens are similar, differing primarily in the size of sclerotia. *Sclerotinia sclerotiorum* is much more widespread than *S. minor*.

Disease after harvest results primarily from infections that have occurred before harvest (figs. 15.122 and 15.123). Disease in the field results from infections from mycelium produced by sclerotia or from ascospores produced by apothecia (small mushrooms) that grow from sclerotia under certain climatic conditions (fig. 15.124).

### Anthracnose

Sexual state: *Glomerella lindemuthiana* Shear
Asexual state: *Colletotrichum lindemuthianum* (Sacc. & Magnus) Lams.-Scrib.

Anthracnose is common all over the world except those areas with a very dry climate, such as the western United States. Typically, diseased pods develop black, sunken lesions. The center of each lesion contains a salmon-colored ooze from the many acervuli contained in the lesion (fig. 15.125).

Control where the climate favors the disease involves field application of such fungicides as benomyl. Resistant cultivars are successful against certain strains of the fungus.

## Sweet potato

### Rhizopus rot

Pathogen:   *Rhizopus stolonifer* (Ehr.:Fr.) Lind.
            Note: See Rhizopus rot of stone fruits for synonymy.

*Rhizopus stolonifer* is a ubiquitous fungus growing on various types of organic matter as well as attacking many species of ripe fruit (berries, stone fruits, tomato, papaya) and many other commodities when not refrigerated (fig. 15.126). Little pathogenic growth of the fungus occurs at temperatures below 5°C (41°F). Germinating spores of the fungus are incapable of penetrating an uninjured host.

As the disease progresses, rotting sweet potato roots soften and the flesh is partially liquefied. White mycelium protrudes through the surface at mechanical injuries, at breaks in the skin, or at lenticels. As the fungus sporulates, the aspect changes from white to black as black sporangia form. Mycelia grow from diseased to sound roots in storage, but seldom penetrate well-cured roots.

### Black rot

Sexual state:   *Ceratocystis fimbriata* Ell. & Halst.
                = *Ceratostomella fimbriata* (Ell.& Halst.) Elliott
                = *Endoconidiophora fimbriata* (Ell.& Halst.) Davd.
                = *Ophiostoma fimbriata* (Ellis & Halst.) Nannf.
Asexual state: *Chalara* sp.

Black rot may occur in the field and after harvest. Infected roots become black and tend to remain relatively firm. Lesions are usually associated with wounds and lenticels. The fungus produces abundant asexual spores. These include endoconidia that are single celled, thin walled, cylindrical, smooth, colorless, and about $5 \times 15$ μm in size. Conidiophores are cylindrical structures often found on the surface of diseased areas.

Thick-walled, brown chlamydospores 9 to 18 × 6 to 13 μm are formed within affected tissues. Perithecia form on sprouts and roots, releasing ascospores as the cytoplasm around them dissolves.

### Java black rot

Sexual state:   *Botryosphaeria rhodina* (Cooke) Arx
Asexual state: *Lasiodiplodia theobromae* (Pat.) Griff. & Maubl.
                Note: Refer to Lasiodiplodia rot of banana for synonymy. The connection between *Botryosphaeria rhodina* and *Lasiodiplodia theobromae* is tenuous.

The Java black rot fungus *Lasiodiplodia theobromae* was at first thought to have been imported in sweet potatoes from Java. However, the fungus was determined to be present already on many hosts, and it appears that the disease is found nearly everywhere sweet potatoes are grown.

The fungus may persist in the field in debris from sweet potato or from alternate host crops or susceptible weeds. Spores are associated with the fungus growing in the soil, although transmission via air, splashing water, wind-driven mist, or insects occurs. The fungus cannot penetrate the root periderm, but must enter through wounds in the side of the root or at the ends.

Exposure of roots to chilling temperatures renders them more susceptible to the disease. Similarly, roots become more susceptible after long-term storage (5 to 8 months or more), probably as a consequence of root senescence.

### Fusarium surface rot

Sexual state:   Unknown
Asexual state: *Fusarium oxysporum* Schlect.

Fusarium surface rot affects tissues immediately below the surface, seldom extending into the root more than a centimeter or so. Rotting of tissues deep within the root may indicate that secondary organisms are present. The surface epidermal layers are not affected, but some darkening may appear.

Early symptoms are circular, slightly depressed, brown lesions. Typically, they appear slightly to moderately darkened when viewed at the surface of the root. Lesions enlarge and sometimes exhibit concentric zones (fig. 15.127).

Fusarium surface rot results from soilborne inoculum. Most infections probably occur via soil contamination of wounds caused during harvest. Exposure of roots in the field to bright sunlight for more than an hour or two enhances incidence of disease. Disease is minimized by curing to encourage wound healing.

## Tomato

### Alternaria rot

Sexual state:   Unknown
Asexual state: *Alternaria alternata* (Fr.:Fr.) Keissl.
                Note: See Alternaria rot in apple and pear for synonyms.

Lesions of Alternaria rot in tomato are flat at the fruit surface or sunken. Usually covered by the sporulating black mycelium of the fungus (fig. 15.128), the lesion extends into the flesh where it produces a firm, dry, blackened mass of tissue thoroughly ramified by the mycelium.

*Alternaria alternata* is believed to produce latent infections in the developing fruit in the field by directly penetrating the cuticle. Such infections seldom develop into disease lesions unless the fruit has been chilled. Upon chilling, however, rot lesions may develop at any point on the fruit surface. Mechanical injuries provide the fungus ready access to internal fruit tissues. A circle of lesions around the stem probably results from the tendency of fruit shoulders to be abraded during handling and transport of packed fruit.

The dramatic loss of resistance after fruit have been held for extended periods at temperatures below about 13°C (55°F) and particularly below 5°C (41°F) means that such temperatures should not be maintained for more than a few days. Mature-green fruit are most sensitive to chilling injury, followed by pink fruit, then red-ripe fruit. Resistance to Alternaria rot can be lost without the appearance of other symptoms of chilling injury.

A disease with similar symptoms is caused by a fungus having *Pleospora herbarum* (Pers.:Fr.) Rabenh. as the sexual state and *Stemphylium herbarum* Simmonds as the asexual state. Ordinarily the disease organism cannot easily be distinguished with the naked eye from the Alternaria rot organism.

## Buckeye rot

Pathogen:     *Phytophthora nicotiana* Breda de Haan var. *parasitica* (Dastur) Waterhouse
= *Phytophthora parasitica* Dastur

Buckeye rot is generally attributed to *Phytophthora nicotiana* var. *parasitica*; however, *P. capsici* Leonin and *P. drechsleri* Tucker have also been found to cause the disease. Other *Phytophthora* spp. may sometimes be involved.

*Phytophthora* spp. are soil inhabitants. During warm (18° to 22°C [64° to 72°F]), wet weather, the sporangia produced give rise to motile swimming spores that infect fruit in contact with soil. Splattering rain and wind-driven mist can deposit droplets of water containing spores on fruit.

Lesions usually are not sunken. Beneath the skin one can often see patterns resembling overlapping rose petals (fig. 15.129). Preventing contact by fruit with the soil reduces losses.

## Gray mold rot and ghost spot

Sexual state:   *Botryotinia fuckeliana* (de Bary) Whetzel
Asexual state: *Botrytis cinerea* Pers.:Fr.

The fungus causing gray mold rot or ghost spot is a common postharvest pathogen of many fruits, vegetables, and flowers in addition to tomato fruit. Its optimum growth occurs at 25°C (77°F), but it can rot produce more slowly at temperatures as low as −2° or −3°C (28° to 27°F). Consequently, disease develops readily at the lowest temperature the fruit can tolerate. The fungus can penetrate the fruit skin while still in the field. The more common loci of infection are mechanical injuries or growth cracks that often occur near the stem scar.

The most common disease symptom is a "dirty white" color of mycelium over the lesion. As the fungus sporulates the color darkens to gray or gray-brown, and the appearance is typical of *Botrytis cinerea* on many other hosts. Tissues appear watersoaked when invaded.

Ghost spots develop when conidia germinate on fruit under cool, humid conditions. Germ tubes penetrate the fruit epidermis. If the fruit is subsequently exposed to hot weather, the fungus is killed, but the "ghost spot" remains (fig. 15.130).

## Sour rot

Sexual state:   *Galactomyces geotrichum* (Butl. & Peters.) Redh. & Mall.
Asexual state: *Geotrichum candidum* Link
Note: See sour rot of citrus for synonymy.

Sour rot is present wherever tomatoes are grown. In California, it is particularly prevalent in processing tomato fields, where rotting fruit provide abundant inoculum that may be carried by vinegar flies and other insects. Fruit in contact with the ground can become infected, particularly if rain occurs or the soil is wet from irrigation.

Fresh market tomatoes (mature-green or pink) may also become diseased. On such fruit the lesions are watersoaked and bleached with the surface of the lesion dull rather than shiny (fig. 15.131). The infection often starts at the stem scar from which the lesion may extend down the side of the fruit.

Mature-green fruit that have been chilled several days at 0° to 5°C (32° to 41°F) become susceptible to sour rot (fig. 15.132). Infections occur from 5° to 38°C (41° to 100°F); optimum temperature for disease development is 30°C (86°F).

## Rhizopus rot

Pathogen:     *Rhizopus stolonifer* (Ehrenb.:F.) Vuill.
Note: See Rhizopus or "whiskers" rot of stone fruits for synonyms.

*Rhizopus stolonifer* lesions are first noted as watersoaked areas beneath the fruit skin. Often the skin ruptures after the lesion becomes large. The fungus becomes evident, first as white to gray mycelium and sporangiophores, but the aspect becomes black as the fungus sporulates (fig. 15.133). The fungus nests when a diseased fruit penetrates nearby sound fruit (fig. 15.134).

Rhizopus rot develops most rapidly at 24° to 27°C (75° to 81°F). Pathogenic growth is slow as the temperature approaches 10°C (50°F) and essentially stops at 5°C (41°F).

## Anthracnose

Sexual state:   Unknown
Asexual state: *Colletotrichum coccodes* (Wallr.) Hughes
= *Colletotrichum atramentarium* (Berk & Broome) Taubenhaus

Anthracnose caused by the above organism, and possibly by several other *Colletotrichum* spp., occurs in warm, humid growing areas. Lesions are slightly sunken and are typically about 1 to 1.5 cm (0.4 to 0.6 inch) in diameter. Acervuli in the lesions produce many conidia in a slimy mass (fig. 15.135). Conidia are capable of penetrating sound fruit. However, the fruit must be ripe for the disease to become serious.

On green fruit, a brown flecking indicates penetration but without successful lesion development.

**Other diseases.** Penicillium rot is occasionally found on tomato fruit. The organism is commonly *Penicillium expansum* but may include other *Penicillium* spp. Penicillium rot often occurs in overripe or chilled fruit.

Fusarium rot is sometimes found causing a lesion that may be mistaken for anthracnose to the naked eye. The fungus may belong to the *Fusarium roseum* group, but definite identification has not been determined. The lesions appear to occur only on ripe fruit (fig. 15.136).

# REFERENCES

1. Brown, G. E. 1975. Factors affecting postharvest development of *Colletotrichum gloeosporioides* in citrus fruits. *Phytopathology* 65:404-9.

2. Combrink, J. C., J. F. Fourie, and C. J. Grobbelaar. 1984. *Botryosphaeria* spp. on decayed deciduous fruits in South Africa. *Phytophylactica* 16:251-53.

3. Dianese, J. C., H. A. Bolkan, C. B. daSilva, and F. A. A. Couto. 1981. Pathogenicity of epiphytic *Fusarium moniliforme* var. *subglutinans* to pineapple. *Phytopathology* 71:1145-49.

4. Eckert, J. W. 1979. Pathological diseases of fresh fruits and vegetables. *J. Food Biochem.* 2:243-49.

5. Eckert, J. W., and J. M. Ogawa. 1985. The chemical control of postharvest diseases. Subtropical and tropical fruits. *Annu. Rev. Phytopathol.* 23:421-54.

6. ———. 1988. The chemical control of postharvest diseases: Deciduous fruits, berries, vegetables and root/tuber crops. *Annu. Rev. Phytopathol.* 26:433-69.

7. Eckert, J. W., and N. F. Sommer. 1967. Control of diseases of fruits and vegetables by postharvest treatment. *Annu. Rev. Phytopathol.* 5:391-432.

8. El-Goorani, M. A., and N. F. Sommer. 1981. Fungistatic effects of modified atmospheres in fruit and vegetable storage. *Hortic. Rev.* 3:412-61.

9. Farr, D. F., G. F. Bills, G. P. Chamuris, and A. Y. Rossman. 1989. *Fungi on plants and plant products in the United States.* St. Paul: APS Press. 1,252 pp.

10. Gibb, Ellen, and J. H. Walsh. 1980. Effect of nutritional factors and carbon dioxide on growth of *Fusarium moniliforme*, and other fungi in reduced oxygen concentrations. *Trans. Brit. Mycolog. Soc.* 74:111-18.

11. Harvey, J. M. 1978. Reduction of losses in fresh market fruits and vegetables. *Annu. Rev. Phytopathol.* 16:321-41.

12. Hooker, W. J., (ed.) 1983. *Compendium of potato diseases.* St. Paul: APS Press. 125 pp.

13. Ismail, M. A., and G. E. Brown. 1975. Phenolic content during healing of 'Valencia' orange peel under high humidity. *J. Am. Soc. Hortic. Sci.* 100:249-51.

14. ———. 1979. Postharvest wound healing in citrus fruit: Induction of phenylalanine ammonia-lyase in injured 'Valencia' orange flavedo. *J. Am. Soc. Hortic. Sci.* 104:126-29.

15. Ismail, M. A., R. L. Rouseff, and G. E. Brown. 1978. Wound healing in citrus: Isolation and identification of 7-hydroxycoumarin (Umbelliferone) from grapefruit flavedo and its effect on *Penicillium digitatum* Sacc. *HortScience* 13:358.

16. Jones, A. L., and H. S. Aldwinckle, eds. 1990. *Compendium of apple and pear diseases.* St. Paul: APS Press. 100 pp.

17. Kolattukudy, P. E. 1978. Chemistry and biochemistry of the aliphatic components of suberin. In *Biochemistry of wounded plant tissues*, ed. G. Kahl, 43-84. New York: Walter de Gruyter & Co.

18. Kosuge, Tsune. 1969. The role of phenolics in host response to infection. *Annu. Rev. Phytopathol.* 7:195-222.

19. Kuc, J. 1972. Phytoalexins. *Annu. Rev. Phytopathol.* 10:207-32.

20. Kuc, J., and N. Lisker. 1978. Terpenoids and their role in wounded and infected plant storage tissue. In *Biochemistry of wounded plant tissues*, ed. G. Kahl, 203-42. New York: Walter de Gruyter & Co.

21. Lakshminarayana, S., N. F. Sommer, V. Polito, and R. J. Fortlage. 1987. Development of resistance to infection by *Botrytis cinerea* and *Penicillium expansum* in wounds of mature apple fruits. *Phytopathology* 77:1674-78.

22. Maas, J. L., ed. 1984. *Compendium of strawberry diseases.* St. Paul: APS Press. 137 pp.

23. Matsumoto, T. T., P. M. Buckley, N. F. Sommer, and T. A. Shalla. 1969. Chilling-induced ultrastructural changes in *Rhizopus stolonifer* sporangiospores. *Phytopathology* 59:863-67.

24. Matsumoto, T. T., and N. F. Sommer. 1967. Sensitivity of *Rhizopus stolonifer* to chilling. *Phytopathology* 57:881-84.

25. McCracken, A. R., and J. R. Swinburne. 1980. Effect of bacteria isolated from surface of banana fruits on germination of *Colletotrichum musea* conidia. *Trans. Bact. Mycol. Soc.* 74:212.

26. Nelson, K. E. 1985. *Harvesting and handling California table grapes for market.* Univ. Calif. Div. Agric. Nat. Resour. Bull. 1913, 72 pp.

27. Sherf, A. F., and A. A. Mac Nab. 1986. *Vegetable diseases and their control*, 2d ed. New York: John Wiley & Sons. 728 pp.

28. Slabaugh, W. R., and M. D. Grove. 1982. Postharvest diseases of bananas and their control. *Plant Disease* 66:746-50.

29. Smith, J. E., and D. R. Berry. 1974. *An introduction to biochemistry of fungal development.* New York: Academic Press.

30. Snowdon, A. L. 1990. *A color atlas of postharvest diseases and disorders of fruits and vegetables. Vol. 1: General introduction and fruits.* Boca Raton, FL: CRC Press. 302 pp.

31. Sommer, N. F. 1982. Postharvest handling practices and postharvest diseases of fruit. *Plant Disease* 66:357-64.

32. ———. 1985. Role of controlled environments in suppression of postharvest diseases. *Can. J. Plant Pathol.* 7:331-34.

33. ———. 1989. Manipulating the postharvest environment to enhance or maintain resistance. *Phytopathology* 79:1377-80.

34. Sommer, N. F., R. J. Fortlage, J. R. Buchanan, and A. A. Kader. 1981. Effect of oxygen on carbon monoxide suppression of postharvest diseases. *Plant Disease* 65:347-49.

35. Sommer, N. F., R. J. Fortlage, and D. C. Edwards. 1983. Minimizing postharvest diseases of kiwifruit. *Calif. Agric.* 37(1-2):16-18.

36. Stanghellini, M. E., and M. Aragaki. 1966. Relation of periderm formation and callose deposition to anthracnose resistance in papaya fruit. *Phytopathology* 56:444-50.

37. Swinburne, T. R. 1974. The effect of store conditions on the rotting of apples, cv. Bramley's seedling, by *Nectaria galligena*. *Ann. Appl. Biol.* 78:39-48.

38. Swinburne, T. R., and A. E. Brown. 1975. The effect of carbon dioxide on the accumulation of benzoic acid in Bramley's seedling apples infected by *Nectaria galligena*. *Trans. Brit. Mycol. Soc.* 64:505-7.

39. Uritani, I., and K. Oba. 1978. The tissue slice system as a model for studies of host-parasite relationships. In *Biochemistry of wounded tissues*, ed. G. Kahl, 287-308. New York: Walter de Gruyter & Co.

40. van der Plank, J. E. 1975. *Principles of plant infection.* New York: Academic Press.

41. Whiteside, J. O., S. M. Garnesy, and L. W. Timmer, eds. 1988. *Compendium of citrus diseases.* St. Paul: APS Press. 80 pp.

42. Wilson, C. L. 1989. Managing the microflora of harvested fruits and vegetables to enhance resistance. *Phytopathology* 79:1387-90.

43. Wilson, C. L., and P. L. Pusey. 1985. Potential for biological control of postharvest plant disease. *Plant Disease* 69:375-78.

44. Wilson, C. L., and M. Wisniewski. 1989. Biological control of postharvest disease. *Annu. Rev. Phytopathol.* 27:425-42.

45. Wilson, E. E., and J. M. Ogawa. 1979. *Fungal, bacterial, and certain nonparasitic diseases of fruit and nut crops in California.* Div. Agric. Sci., Univ. Calif. 189 pp.

# GLOSSARY

**Abiotic disease.** A disease due to a nonliving cause.

**Acervulus,** pl. **Acervuli.** A fruiting body of certain imperfect (asexual) fungi. A shallow, saucer-shaped structure with a layer of conidiophores that bear conidia; found in anthracnose diseases and fungi belonging to the genus *Pezicula*

**Anamorph.** A fungus in the asexual state.

**Anthracnose.** Any disease caused by fungi that produce asexual spores in acervuli.

**Apothecium,** pl. **Apothecia.** Cuplike, ascus-containing fungus fruiting body. See figure 15.124.

**Appressorium,** pl. **Appressoria.** A bulbous or lobed swelling of a hyphal tip, often held in place by a gelatinous secretion that forms an infection peg to penetrate plant tissues. See figures 15.2 and 15.3.

**Arthrospores.** Spores formed by the simultaneous or random fragmentation of hyphae.

**Asexual state.** Production of spores without previous fusion of gametes; a form of vegetative reproduction.

**Chlamydospore.** A thick-walled spore usually resistant to adverse environmental conditions.

**Conidiophore.** Specialized portion of mycelium on which conidia are produced.

**Conidium,** pl. **Conidia.** Any asexual spore except sporangiospores and chlamydospores. Produced on specialized portion of mycelium, the conidiophore.

**Fructification.** Production of spores by fungi, fungus fruiting body, or spore-bearing structure.

**Fruiting body.** A complex fungus structure containing or bearing spores, as a mushroom, perithecium, pycnidium, etc.

**Fungicidal.** Capable of killing or inhibiting fungi.

**Fungicide.** A chemical or physical agent that kills or inhibits fungi.

**Fungistatic.** An agent that prevents development of fungi without destroying them, or term applied to this action.

**Fungi Imperfecti.** A class of fungi without a known sexual state.

**Fungus,** pl. **Fungi.** Organisms having no chlorophyll, with reproduction by sexual or asexual spores and not by fission, and usually with mycelium having well-marked nuclei.

**Gamete.** A reproductive cell with a haploid nucleus capable of fusion with that of a gamete of an opposite mating type.

**Genus,** pl. **genera.** A taxonomic group above species and below the rank of family. The first name of a binomial such as *Monilinia fructicola*.

**Germ tube.** The fungus hypha formed upon germination of a spore.

**Germination.** The swelling of a spore and the protrusion of hyphae. The beginning of growth.

**Host.** A living organism upon which a fungus or bacterium grows and obtains sustenance.

**Hyaline.** Colorless, transparent, or nearly so.

**Hypha,** pl. **Hyphae.** A single thread of fungus mycelium, a threadlike structure that increases length by growth at the tip and forms lateral branches.

**Imperfect fungus.** One lacking any sexual reproductive stage.

**In vitro.** Refers to growth occurring on nonliving substrates.

**In vivo.** Refers to growth occurring on living plants or animals.

**Incubation period.** The time between inoculation of a plant or plant tissue and the first observed disease reaction.

**Infect.** Entrance of a pathogen into a plant where it grows and obtains sustenance.

**Infection court.** The place where an infection may take place.

**Infectious.** Term applied to a disease that may be communicated from one plant, or plant part, to another.

**Inoculate.** To place inoculum in an infection court.

**Inoculum,** pl. **Inocula.** Infectious parts of a pathogen, such as a spore or bacterial cell, that can be transferred to healthy tissue and cause disease.

**Lesion.** A localized spot of diseased tissue.

**Molds.** Fungi with conspicuous mycelium or spore masses.

**Mummified.** Dried up and shriveled, as in fruit affected by the brown rot pathogen.

**Mycelium,** pl. **Mycelia.** A mass of fungus hyphae.

**Mycology.** The science dealing with fungi.

**Necrosis,** pl. **Necroses.** Death of plant tissues, as in rots, blights, and cankers.

**Parasite.** An organism that lives on or in a second organism, usually causing disease in the latter.

**Pathogen.** Any organism or factor causing disease.

**Pathogenic.** Capable of causing disease.

**Pathology.** The science of disease.

**Perfect state.** Capable of sexual reproduction.

**Perithecium,** pl. **Perithecia.** Ascospore-producing body.

**Peritrichous.** Having hairlike flagella all over the surface.

**Pycnidium,** pl. **Pycnidia.** Flasklike fruiting body containing conidia.

**Resistance.** Ability of a host plant or plant part to suppress or retard the activity of a pathogen or other injurious factor.

**Saprophyte.** An organism that feeds exclusively on lifeless organic matter.

**Sclerotium,** pl. **Sclerotia.** A resting mass of fungus tissue, often more or less spherical, usually not bearing spores.

**Sexual state.** Refers to the condition in which spores are produced following fusion of gametes.

**Sporangiophore.** Hypha bearing a sporangium.

**Sporangiospore.** A spore produced in a sporangium, as in *Rhizopus stolonifer* or *Mucor piriformis*.

**Sporangium,** pl. **Sporangia.** An organ producing nonsexual spores with a more or less spherical wall.

**Spore.** A single- to many-celled reproductive body, in fungi and/or other lower plants.

**Sporulate.** To produce spores.

**Sterilize.** To remove or destroy all living organisms on or in an object or material.

**Substrate.** The substance or object on which an organism lives and from which it obtains nourishment.

**Suscept.** Any plant susceptible to infection by a given pathogen.

**Susceptible.** Inability to oppose the operation of an injurious or pathogenic agent.

**Teliomorph.** A fungus of the sexual state.

**Toxin.** A poison formed by an organism.

**Viability.** State of being alive.

**Virulent.** Highly pathogenic; with a strong capacity for causing disease.

# 16

# Postharvest Treatments for Insect Control

F. GORDON MITCHELL AND ADEL A. KADER

Much has been learned about postharvest physiology, storage, and technology of horticultural crops, but little is known about the effect of postharvest manipulations on the large number of insects that can be carried by fruits and vegetables during postharvest handling. Many of these insect species, especially the tephritid fruit flies, can seriously disrupt produce trade among countries and even among states within the U.S. Thus, effective insect control treatments that are not harmful to the commodity, workers, or the consumer are essential for allowing unrestricted movement of fresh horticultural crops in domestic or international commerce.

## Historic Perspective

California has historically been in a position of potential rather than actual insect quarantine problems. Among fresh fruits and vegetables there is a continuing need for small-scale quarantine treatment for exports to certain destinations in Canada, Japan, and other countries. Dried fruits, nuts, and grains require large-scale control treatments against storage insects, but these are not necessarily against quarantined insects.

California horticultural industries have faced serious concerns over quarantine treatment for many years. After World War II, when the Oriental fruit fly was identified in Hawaii there was fear it might spread to California. The California legislature funded an extensive study of quarantine treatments and of product tolerance to those treatments. Information gained from that study at the University of California, Davis and Riverside, provided the base for our knowledge of product tolerance to a variety of fumigants. Ethylene dibromide (EDB) and methyl bromide (MB) emerged as the most satisfactory treatments.

In 1980 a Mediterranean fruit fly (medfly) crisis occurred in central California, and results from those early studies were useful in preparing the produce industries to use quarantine treatments. Emergency programs to expand our knowledge of insect quarantine treatment and product tolerance were initiated, primarily involving EDB and MB. During that crisis many large fumigation chambers were constructed and personnel certified as fumigation operators. Fortunately, the pest was eradicated as a result of a large-scale, expensive control program.

It now appears that the two commonly used fumigants (EDB and MB) might not be available in the future. EDB, which was on the RPAR (Rebuttable Presumption Against Registration) list in 1980, subsequently had its registration withdrawn for most quarantine use in 1984, and for all uses effective September 1987. MB continues under attack, and a lower residue tolerance or withdrawal of registration are future possibilities. Thus, alternate treatments must be developed if agriculture and the food supply are to be protected.

We must avoid complacency about the dangers of future infestations or the need for quarantine treatments. There are many "fruit flies" around the world that are potential quarantine pests. In California during the 1980's, several different fruit flies were found, some repeatedly, including the Mexican fruit fly, Caribbean fruit fly, Oriental fruit fly, Mediterranean fruit fly, peach fly, and the apple maggot (a fruit fly). There are many other insects—codling moth is an example—that are quarantined by certain countries. Should any quarantined pest be found within a production area, a quarantine zone would be established and immediate quarantine treatment would be required. Unfortunately, honest quarantine problems are often intermingled with convenient quarantines that may be imposed by some countries as trade barriers.

## Quarantine Treatment Research

Many quarantine treatment studies are being conducted today. The U.S. Department of Agriculture, Agricultural Research Service (USDA-ARS) laboratories in Hawaii, California, Washington, Florida, and Texas have worked extensively on these problems. Various state universities also became involved in some aspects of quarantine treatment research. In 1982 a regional research project, "Postharvest Biotechnology and Quarantine Treatments for Insect Control in Horticultural Crops," was begun that includes university researchers from California, Oregon, Hawaii, and Florida, along with the USDA-ARS labs, the Florida Department of Citrus, Division of Scientific and Industrial Research-New Zealand, and Agriculture Canada. Responsibilities are divided and resources and results pooled to maximize efforts on this broad project. It had been hoped that more time would be available to identify and develop viable alternatives before the commonly used fumigants were removed from use.

Many alternative treatments are under study within this regional project. Some studies are evaluating fumigants—ways of reducing residues, combination treatments to reduce concentrations, and alternate fumigants (including some natural plant volatiles) as substitutes for EDB and MB. Other researchers are working on ways of avoiding infestation, such as harvest maturity (how mature must a fruit be before the flies lay eggs?). Work is being done on heat, cold, very low oxygen levels, and very high carbon dioxide concentrations. Some work has been done on radiation including ultrasound, microwave, and the much-publicized gamma radiation. Ultimately various combination treatments must also be considered.

At the moment there are limited options available for insect quarantine treatment other than fumigation. As alternate treatments are developed their feasibility must be studied from many aspects—worker safety, product safety and tolerance, problems of commercialization, engineering and economic considerations. Certainly no one treatment will be universally applicable; each has inherent problems and limitations. Following is a brief discussion of the disinfestation methods in current use as well as those under investigation for possible future use.

Federally acceptable insect quarantine treatments are established and administered by the USDA, Animal and Plant Health Inspection Service (APHIS). In California, the Agricultural Commissioner in the county of production is the authority for currently approved treatments. The Agricultural Commissioner should, therefore, always be consulted before any treatment procedure is used.

## Insect Quarantine Treatments

### Methyl bromide fumigation

Methyl bromide remains the most widely used insect quarantine treatment for horticultural commodities. The approved concentrations, durations, and treatment temperatures depend upon the pest and the commodity. In certain cases, a lower concentration or shorter duration of fumigation combined with cold treatment is acceptable and may be less injurious to the commodity than the fumigation treatment alone.

All fumigation treatments must be done in approved and certified fumigation chambers. Trained and certified fumigation operators must be on site whenever fumigation is in progress.

Many fruit fly fumigation schedules now allow fruit pulp temperatures of 4.4°C (40°F) or above during fumigation, with the MB concentration and/or duration of fumigation decreasing with higher pulp temperatures. Injury is commonly more severe

under the low temperature-high concentration treatment. There is a potential for product injury if the product is fumigated while free water is present on its surface (as might occur from sweating). Injury may be more severe if the product has been treated with a water-soluble wax before fumigation. It may also be more difficult to purge the fumigant from many shipping containers than from field bins or lugs. These problems argue strongly in favor of fumigation before cooling or packing, especially when energy efficiency is considered.

After the commodity has received fumigation and/or cold treatment it must be protected from reinfestation. This means that any packing, storage, or loading facility handling the treated product must be isolated from the quarantined insect. The County Agricultural Commissioner should be contacted to certify compliance. Isolation requires placing an insect screen around the isolation area, using plastic strip doors for access (the type used on cold storage rooms), and positive airflow out of the openings of the isolation area that cannot be screened (cull fruit chutes, bin or box chutes, and so on). Use of traps within and in the perimeter around the packing facility may be required in certifying isolation. Depending upon destination and transport conditions, packages may also need sealing against future insect invasion or oviposition. If packages need ventilation slots for temperature protection, they may require insect-tight screening.

### Alternate fumigants

Any alternate fumigant must be carefully screened for hazards to workers and consumers. Phosphine, one fumigant receiving attention, requires 48 to 72 hours' exposure instead of the 2 hours necessary for MB. Most perishables would not tolerate that long a period out of refrigeration without unacceptable deterioration. Fumigation at low temperatures would require extensive and costly refrigerated facilities. Such a treatment would be more promising for the less perishable commodities (dried fruits, nuts, grains) where fumigation of bulk-storage facilities may be possible. There is continuing interest in some plant volatiles that may serve as fumigants against certain surface-feeding insects, although none are currently registered as alternative fumigation treatments.

### Heat treatment

The long standing, but seldom used, heat treatments in the USDA plant quarantine manual (USDA 1976) require prolonged exposure (more than 8 hours plus heating time) to moist heat at temperatures as high as 112°F (44.4°C). This treatment has been injurious enough to rule out its use on most products. A double hot-water dip treatment has been approved to control fruit flies in papayas that are less than one-fourth ripe. This process requires submersion at 42°C

(107.6°F) for 30 minutes, then 49°C (120.2°F) for 20 minutes. The short treatment time makes this alternative especially appealing. A hot-water treatment (currently 46.4°C [115.5°F] for 75 minutes) is also allowed for certain production areas and cultivars of mangoes entering the United States. These treatments would be of interest for other commodities that would tolerate them. A new dry forced-air heat treatment, which warms the seed cavity up to 41°C to 47.2°C (105.8° to 117°F) over about a 6-hour period, has been approved for Hawaiian papayas.

### Cold treatment

The quarantine manual currently allows cold treatment for control of some insects. The following cold treatments are allowed for fresh commodities from areas infested with the Mediterranean fruit fly:

10 days at 0°C (32°F) or below
11 days at 0.6°C (33°F) or below
12 days at 1.1°C (34°F) or below
14 days at 1.7°C (35°F) or below
16 days at 2.2°C (36°F) or below

Similar time-temperature combinations are approved for control of the other tropical fruit flies.

There are strict requirements for temperature monitoring in cold storage facilities to certify compliance with these treatments. Attempts are underway to develop guidelines for monitoring temperatures that would allow cold treatment to be used during marine transport.

Despite the strict monitoring requirements, this treatment would be most appropriate for commodities that are capable of extended low-temperature storage such as pear, apple, grape, kiwifruit, persimmon, and pomegranate.

Some commodities, such as avocado, mango, and papaya, are injured from extended exposure to such low temperatures. Preconditioning at warm temperatures allows some citrus fruits to tolerate the cold treatment and has been commercially used on Florida grapefruit during transport to Japan. Cold treatment is also used to control the Caribbean fruit fly in carambola shipped from Florida to California.

A period of 10 to 16 days exceeds the potential market life for many perishable commodities, such as apricot, cherry, strawberry, and cane berries. When the cold treatment must be completed before transport, there is also a logistical problem with providing enough refrigerated storage capacity in production areas during heavy shipping periods. Combination treatments, such as heat followed by cold and exposure to low-oxygen atmospheres plus cold, need study to determine if they would reduce the cold dwell time required.

### High carbon dioxide and low oxygen levels

The moderate modifications in $O_2$ and $CO_2$ that are commercially used in controlled atmosphere (CA) storage of some commodities are inadequate to control insects. Some stored-products insects are controlled by exposure for 2 to 3 days to 0.5 percent oxygen (plus 11.5 percent carbon dioxide) or to above 70 percent carbon dioxide at 27°C (81°F) and 60 percent relative humidity. At lower temperatures and/or higher relative humidity, a longer exposure is required for complete insect control. Stored dried products should tolerate such conditions and may even show slower deterioration rates. Studies by the USDA-Fresno and UC Davis suggest that CA treatment of bulk-stored almonds for insect control and quality maintenance is economical.

The possibilities for using CA as quarantine treatment of fresh commodities are much less certain. Tests by UC Davis indicate that apples, pears, peaches, nectarines, plums, cherries, valencia oranges, and strawberries have a reasonable tolerance to very low oxygen (0.25–0.50 percent) for periods ranging from 8 to 40 days depending on commodity and temperature. Tolerance to high carbon dioxide appears less promising. Studies are needed to evaluate the potential for effective control of the major quarantined insects under such atmospheres, alone and in combination with other treatments.

If these CA treatments are found effective, there is still a logistical problem in applying them to large volumes of fresh market produce, particularly if more than a day or two is required for treatment. The investment in new refrigerated CA storage facilities would be too great to make it feasible as a stand-by treatment for potential infestations. The greatest hope is that CA might be useful in combination with some other treatment to achieve more rapid insect control, and that its application during marine transport becomes feasible.

### Ultrasound and microwave

Insect control by ultrasound or microwave is in the early stages of investigation. These techniques show indications of insect control, but much more study is needed to verify effectiveness and to evaluate product tolerance and commercial potential. Some work has been done on the use of microwave heating combined with conventional heating to achieve faster and more uniform warming of the product for heat treatment.

### Radiation with gamma or X rays

By far the most publicized alternate quarantine treatment is the use of gamma or X radiation. This involves exposing the product to a radiation source (isotopic source using cobalt-60, cesium-137, or an electrically driven machine source) until it absorbs the required dose level of gamma or X rays. Successful use of this procedure is based on determining that the insect will be inactivated at a dose level that is tolerated by the host commodity.

Current interest in radiation results from the Food and Drug Administration (FDA) approval (effective April 18, 1986) of treatments up to 1 kGy (100 krad) on foods, based upon data collected over time from many sources that have satisfied the FDA of their safety.

Gamma radiation was extensively studied on fresh fruits and vegetables at UC Davis and many other locations during the 1960's. The main emphasis at that time was on postharvest disease control, which generally requires dose levels greater than the projected 1 kGy limit. These studies concluded that the dose levels required to control disease organisms were usually too phytotoxic to the host to be satisfactory. Whenever the procedure appeared marginally acceptable, less expensive alternatives were available.

During that same period certain other potentially beneficial effects of radiation were studied, including growth inhibition of certain vegetables, retardation of ripening of some tropical fruits, and insect control. These effects were found at levels generally below 1 kGy. Subsequently many studies of insect control by radiation have been conducted, primarily on potential quarantine pests. Results indicate that most insects are sterilized by doses below 0.75 kGy.

While this information makes the use of radiation for insect quarantine treatment appear promising, a number of considerations dictate extreme caution in projecting radiation as an alternative to fumigation for insect quarantine treatment. Some are listed here:

**1.** Many potential hosts (such as avocado, grape, lemon, and lime) suffer significant detrimental effects from dose levels below 1 kGy. Conflicting results reported with other potential hosts (such as orange and grapefruit) suggest that more data are needed.

**2.** The USDA Animal and Plant Health Inspection Service (APHIS) and any receiving countries must accept radiation as a quarantine treatment that sterilizes but does not kill most quarantined insects, at least to the level of control currently required by APHIS. There are indications that this will be done, using a monitoring system to ensure that treatment has occurred.

**3.** Dosimetry data must show a relatively narrow dose range between the outside and center of a mass of radiated product. If palletized fresh produce must be disassembled for treatment, the logistics become more difficult and the cost increases substantially.

**4.** Radiation imparts an injurious stress to the host commodity. Thus good handling procedures, including good temperature management, are vital to minimize deterioration. Refrigeration capability must be available throughout the handling of fresh commod-

ities, including the irradiator plant. Refrigeration before, during, and after radiation substantially increases the cost of the facility.

**5.** In an intensive horticultural area such as California, the logistics of radiation are tremendous. Many multimillion-dollar facilities would be needed to handle the volume of fruits and fruit-type vegetables that leave horticultural production areas.

**6.** Considerable engineering work is required to develop logistically sound radiation facilities. Because of the high volume of product involved, these would be of quite a different scale than is currently available. The need to incorporate refrigeration further complicates the design of radiation facilities.

**7.** Current information indicates that radiation plants must be used to capacity, essentially year-round, to be economical. Fresh fruit and vegetable production is seasonal in California. Further, because quarantine treatment is still primarily a potential rather than actual threat in California, such facilities would be too costly to build for stand-by use.

**8.** Some serious social and public policy issues must be addressed. Will local governments accept environmental impact statements and allow radiation facilities to be constructed in their areas? Will food retailers be willing to carry and display irradiated fresh produce? Will consumers be willing to purchase and consume irradiated foods?

## No Single Solution

From the studies conducted so far, it is apparent that there is no single alternative to EDB or MB fumigation for insect quarantine treatment. The choice will involve many interacting factors. Of primary concern is safe but adequate insect control. For bulk dry commodities (such as tree nuts and dried fruits), the potential for elevated carbon dioxide and/or low oxygen is promising. This quarantine procedure must be explored further with fresh produce. Some fresh commodities may tolerate brief heat treatments, which may prove to be the most rapid, least costly treatments for them. Cold treatment may be most satisfactory when storage time is not a concern and the commodity is, or can be conditioned to be, tolerant to low temperatures. For certain commodities and pests, alternate fumigants may prove safe, effective, and economical. Given the right conditions of host tolerance, year-round production, and endemic pests that are sufficiently radiation-sensitive, radiation may be useful.

Regardless of the system that is selected, California producers of horticultural crops must have available suitable postharvest insect control treatments. With changes in pesticide regulations and the increasing threat of infestations of quarantined pests, it is important not to wait for a quarantine threat before becoming prepared. Thus insect quarantine treatment capabilities will continue to be another cost in marketing many California grown horticultural crops, even though they may be available simply on a stand-by basis.

## REFERENCES

1. Aharoni, Y., J. K. Stewart, and D. G. Guadagni. 1981. Modified atmospheres to control western flower thrips on harvested strawberries. *J. Econ. Entomol.* 74:338-40.

2. Benshoter, C. A. 1987. Effects of modified atmospheres and refrigeration temperatures on survival of eggs and larvae of the Caribbean fruit fly (Diptera: Tephritidae) in laboratory diet. *J. Econ. Entomol.* 80:1223-25.

3. Burditt, A. K., Jr. 1982. Food irradiation as a quarantine treatment of fruits. *Food Technol.* 36(11):51-62.

4. Carey, J. R., and R. V. Dowell. 1989. Exotic fruit fly pests and California agriculture. *Calif. Agric.* 43(3):38-40.

5. Couey, H. M. 1983. Development of quarantine systems for host fruits of the medfly. *HortScience* 18:45-47.

6. ———. 1989. Heat treatment for control of postharvest diseases and insect pests of fruits. *HortScience* 24:198-202.

7. Gaunce, A. P., C. V. G. Morgan, and M. Meheriuk. 1982. Control of tree fruit insects with modified atmospheres. In *Controlled atmospheres for storage and transport of perishable agricultural commodities*, 383-90. Beaverton, OR: Timber Press.

8. Kader, A. A. 1986. Potential applications of ionizing radiation in postharvest handling of fresh fruits and vegetables. *Food Technol.* 40(6):117-21.

9. Kader, A. A., and F. G. Mitchell. 1981. Postharvest treatments for insect control in horticultural crops—an indexed reference list. *Perish. Handling* 47:8-28.

10. Mitchell, F. G., A. A. Kader, G. U. Crisosto, and G. Mayer. 1984. *The tolerance of stone fruits to elevated $CO_2$ and low $O_2$ levels*. Spec. Rept. to Calif. Tree Fruit Agreement. 11 pp.

11. Monro, H. A. U. 1969. *Manual of fumigation for insect control*. FAO Agric. Studies 79. 381 pp.

12. Moy, J. H., ed. 1985. *Radiation disinfestation of food and agricultural products*. Honolulu: Univ. Hawaii Press. 424 pp.

13. Soderstrom, E., and D. G. Brandl. 1990. Controlled atmospheres for the preservation of tree nuts and dried fruits. In *Food preservation by modified atmospheres*, ed. M. Calderon and R. Barkai-Golan, 83-92. Boca Raton, FL: CRC Press.

14. Sommer, N. F., and F. G. Mitchell. 1986. Gamma irradiation—a quarantine treatment for fresh fruits and vegetables? *HortScience* 21:356-60.

15. USDA. 1976 (revised). *Plant protection and quarantine programs, plant quarantine treatment manual*. U.S. Dept. Agric. Animal Plant Health Inspec. Serv.

# 17

# Transportation of Fresh Market Horticultural Crops

Robert F. Kasmire and M. Joseph Ahrens

Several modes of transportation are used to move fresh market horticultural crops from shipping points to destination markets. Airplanes, railroads, ships (marine), trucks, and combinations, such as trailer-on-flatcar (TOFC) and container-on-flatcar (COFC), are all used to transport fruits and vegetables, with trucks carrying the major portion in North America. Flowers are mostly transported by trucks and airplanes. Export shipments involve all methods, but of course only marine and airline carriers are used for transoceanic shipments. Problems, limitations, and requirements are common to all methods; however, each method has its own specific problems. Knowledge of these problems can help suppliers and users of perishables transportation equipment to select and use transportation more effectively.

## Equipment

In addition to the obvious structural features (e.g., frame, siding, doors), each mode requires certain essential equipment common to all methods except air transport.

### Components for temperature/atmosphere control

**Refrigeration source.** Mechanical refrigeration systems are generally used. Top-ice (crushed ice placed on top of the load) or ice placed in each individual box is sometimes used with or without mechanical refrigeration. Use of liquid nitrogen and carbon dioxide for refrigeration has been tried on a limited scale. Ventilation with cooler outside air is occasionally used to provide limited temperature control, but it is generally inadequate for optimum temperature management. Most mechanical refrigeration systems are also designed to heat the storage compartment when the vehicle travels during subfreezing conditions. A properly designed system can maintain a high relative humidity in the storage compartment, which is desired in most cases. Air freighters have minimal refrigeration capacity, little or no refrigerated storage space, and very low relative humidity inside at high altitudes.

**Air circulation systems.** Air circulation systems are necessary to move the refrigerated air through and around loads to absorb heat from the products (sensible and vital heat) and from external sources (mostly conducted across a vehicle's outer surfaces). In most mechanically refrigerated rail cars, trucks, and ma-

rine containers, the system circulates cold air around frozen food loads (which produce no vital heat) or circulates cold air through and around loads of fresh produce and ornamentals (where vital heat removal is important). High-density loading of trucks and rail cars often restricts air circulation and causes product deterioration and losses (see "Managing Product Transit Temperatures" later in this chapter). Air freighters have minimal air circulation capacity.

**Temperature control system.** Mechanically refrigerated systems include one or more thermostats, automatically operated (with manual override) to provide cooling or heating, and air-circulating fan speed controls. Thermostats are generally in the return-air channel, which causes some problems. Some newer units have thermostats in the discharge-air channel, or in both discharge- and return-air channels.

**Insulated product storage chamber.** Insulation restricts heat conduction across walls, floor, doors, and the roof of transportation vehicles. The storage area is tightly sealed to limit air leakage. These limit the amount of ambient heat entering the vehicle during hot weather and the amount of internal heat (mostly from the product) escaping to the outside (causing product chilling or freezing) during freezing weather. Most insulation is foamed-in-place material that deteriorates slowly over time (about 5 percent of the insulating quality per year). Rail cars have the most, trucks and container vans less, and air cargo containers very little insulation. Insulation can be damaged and its value lessened by lift truck damage during loading and unloading operations and by water that collects in the insulation.

**Air exchange system.** Ventilation of the load compartment with outside air (mostly in marine containers) is done to reduce undesirable concentrations of carbon dioxide, ethylene, or offensive odors, or to prevent depletion of oxygen to harmful levels. New exchange units used in some marine container vans can be set to within a few cubic feet per hour of air exchange. Incoming outside air is channeled directly to the evaporator coil to remove heat from it before it is circulated through the load compartment.

**Modified atmosphere accommodations.** Modified atmospheres have been used in conjunction with all modes of transport, but more commonly with rail cars, trailer-on-flatcar (TOFC) vans, and marine container vans for long haul domestic and marine export shipments. The accommodations include atmosphere-injection ports, special seals and curtains around doors, atmospheric pressure compensation apparatus, sometimes, carbon dioxide absorbers (generally lime), and specially constructed and sealed polyethylene pallet covers and bases.

## Equipment differences—problems and practices

**Truck and truck trailers.** Included are over-the-road refrigerated truck trailers, trailer-on-flatcar (TOFC or piggyback), and intermodal containers (mostly marine but also container-on-flatcar or COFC). Trucks are used to haul straight loads (one commodity) or mixed loads. Transcontinental travel time in the U.S. is 3 to 6 days for trucks; 5 to 7 days for TOFC loads. Marine shipment times range from 5 to 25 days or longer. Refrigerated truck trailer features are noted below.

Load space is intermediate in size, 70 to 100 m$^3$ (2,000 to 3,500 cubic feet) and with a net weight load capacity of about 18,000 to 20,400 kg (40,000 to 45,000 pounds). Some older marine container vans have smaller load compartments of about 40 m$^3$ (<1,400 cubic feet). Trucks are limited in gross weight by state regulations generally to 36,288 kg (80,000 pounds) maximum gross weight. Careful loading is required to distribute weight evenly to the axles. TOFC's are occasionally more heavily loaded with resulting transit temperature management problems (poor air circulation due to excessive tightness of the load).

Refrigeration units are powered by diesel engines (truck trailers and TOFC), or by diesel engines or electricity on docks and ships for marine container vans, or by diesel-electric generator sets. Some foreign ships have cooling towers from which cold air from the ship's refrigeration system is circulated to the vans via a manifold system. Most presently manufactured high-capacity units have about 44,300 to 46,400 kilojoules (42,000 to 44,000 BTU's) per hour of refrigeration capacity. They have axial-flow or centrifugal fans capable of circulating about 102 m$^3$/min (3,600 cfm) of air against zero resistance and about 57 m$^3$/min (2,000 cfm) of air against a static pressure of 2.54 cm (1.0 inch) water column, a condition comparable to a tightly loaded trailer or container.

In trailers and marine containers circulating air temperature is generally measured in the return-air stream, just before the air passes through the fan and the evaporator coil. In some newer units the air temperature is monitored in the supply airstream about 2 to 2.5 m (6 to 8 feet) into the discharge air duct. This modification reduces the possibility of freezing of the top layer (in top air units) or near the bottom of the pressure bulk-head (in bottom air units), which results when circulating air, cooled below freezing, flows directly over exposed parts of a load. A few new refrigeration units are equipped for either supply- or return-air temperature measuring. Microprocessor temperature controllers used in some modern refrigeration units incorporate thermostat control, digital thermometer, fault indication, and data recording in a self-contained controller.

Satellite tracking is used to monitor vehicle location, and is being tested to monitor and control refrigeration unit performance on marine containers aboard ship and by highway trucking companies.

*Air circulation.* The air circulation pattern in top-air delivery trailers or containers is usually lengthwise front-to-rear (fig. 17.1). Air travels from the refrigeration unit back over the top of the load, down the sides and rear of the load, back through and/or under the load, and up the front to the refrigeration unit.

Circulation capacity is designed for maintaining, not lowering product temperatures. This circulation pattern is achieved only if there is adequate air return space beneath the load and a solid air return bulkhead at the front to separate the discharge and return sides of the fan, ensuring positive air circulation around the load. Without a solid bulkhead most of the air circulates over the top of the load back to the refrigeration unit. Optimum product temperature maintenance also requires circulating air channels through the load.

Air is circulated by a fan in the refrigeration unit and aided by an air delivery chute over the top of the load. The chute delivers more air to the rear of the load and prevents or reduces chilling or freezing the front of the load, which would otherwise receive the coldest air from the refrigeration coil. Loads must be secured away from rear doors and away from flat side walls to allow air to circulate down over the load. Inner walls may be flat (as in most truck trailers) or with vertical channels (some TOFC and marine container vans). Vertical channels allow some air circulation between the side walls and the load, and provide less contact between the walls and the load, thereby reducing the amount of heat conducted through the walls.

Many newer trailers and marine container vans have vertical, bottom-to-top airflow through the load compartment, making them more effective for transporting fresh produce (fig. 17.2). They can provide better and more uniform product transit temperatures because they have a more constant and uniform airflow, greater capacity to circulate the air through the load, and shorter air channels through the load. To date, a relatively small percentage of long-haul truck trailers and TOFC trailers have bottom air delivery systems. The deep "T"-beam floors, solid-front bulkheads, and higher-capacity fans used are more costly and reduce the net payload that can be carried, but they have proved their value to carriers.

Duckboard floors, shallow and lengthwise in most truck trailers, provide inadequate air circulation for the bottom layers of nonpalletized loads. This is not a problem in trailers and marine container vans that have deeper floor racks or properly loaded palletized

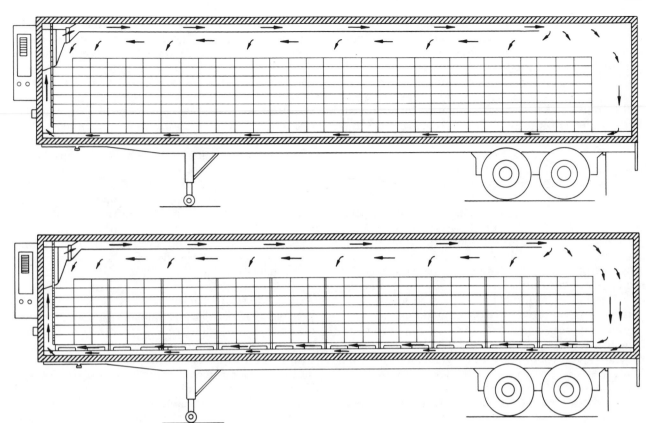

**Fig. 17.1.** *Top:* **Air circulation in and around a solid load directly on the trailer floor.** *Bottom:* **Air circulation in and around a palletized load.**

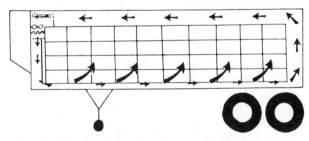

**Fig. 17.2. Air circulation in a bottom-air circulation trailer.**

loads. Marine container vans are generally not as tightly loaded and have less ambient heat conduction, especially through the floor. An increasing number of highway and TOFC trailers now have deeper floors (T-rail, duct, or ductT). Bottom layer heating can also be limited by loading products on pallets or wood racks placed on top of the trailer's floor racks. Load patterns should provide the air channels needed to maintain uniform product temperatures in the load. Tight loads restrict air circulation and thus enhance product warming. Air circulation under loads is much greater and product transit temperatures are more uniform in palletized loads or those on wood racks than in on-the-floor loads (fig. 17.1).

*Insulation.* Limited insulation in trailers and vans allows greater load widths, but allows more heat conduction across walls, roof, floor, and doors. Insulation is often damaged by lift-trucks in loading and unloading and by water accumulation. Tightness of trailers (especially doors) decreases rapidly with use and abuse. Manufacturing improvements have brought about trailers with thinner walls, greater internal load space, and improved insulating capability.

Rear and side doors are provided for loading, unloading, and inspection. Because these hinged doors are constantly used, they are easily and often damaged. Some heat leakage occurs around doors, especially those with damaged seals, hinges, or locking pins. Heat leakage is sometimes considerable around nose-mounted refrigeration units, through front floor drains, and seams.

Maintenance of modified atmospheres (MA) is not possible in most over-the-road truck trailers because they are not tight enough. However, newer marine container vans and some TOFC vans can be made tight enough for MA requirements. Injection ports must be installed in vehicles used for MA shipments. Provision for fresh air exchange is made in some truck trailers and many newer marine container vans to control buildup of $CO_2$, ethylene, and other volatiles and to prevent oxygen depletion.

*Maintenance.* Problems, such as damaged insulation, walls and doors, air-delivery chutes, and floor grooves with debris, are common. Establishing responsibility and payment for damage to trailers and refrigeration systems are continuing controversial matters. Effective maintenance programs are generally conducted by marine container services and TOFC and progressive fleet carriers (trucking companies).

**Refrigerated rail cars.** These are used primarily for long-haul (> 3,200 km) shipments to domestic and Canadian markets. Transportation times range from 6 to 10 days for transcontinental shipments in the U.S. Rail cars were originally designed for transporting frozen food, but are presently used mostly for transporting fresh produce, primarily potatoes, onions, carrots, citrus, and other less perishable commodities. Rail shipments are generally straight loads of a single commodity or limited mixed loads comprising two or three different commodities.

Mechanical refrigerated rail cars have large load compartments (> 113 m$^3$ or 4,000 ft$^3$) and can carry net loads greater than 45 metric tons (100,000 pounds). Many have heavy, movable divider doors to help partition and stabilize the load.

Diesel-powered mechanical refrigeration units are used with thermostatic control in the return air path to provide cooling (or heating) of circulating air. Vertical top-to-bottom air circulation is supplied by fans (fig. 17.3). Wall flues permit some air to flow around the load perimeter and remove heat conducted through the walls. However, the major heat sources are the sensible and vital heat of the product. Adequate air circulation and refrigeration capacity are available to provide slow product cooling, providing loads are not excessive or too tight. The floor racks have narrow spaces between slats but have 4 to 6 inches (10 to 15 cm) air space beneath them to permit adequate air circulation beneath the load. Placing loads on pallets permits more air circulation through and around loads.

Rail cars have thicker insulation and are airtight when they are new. These characteristics can sometimes allow harmful increases in carbon dioxide or decreases in oxygen levels that can cause damage to some commodities. Cars in good condition may be adapted for modified atmosphere shipments. Mechanically refrigerated rail cars do not have air exchange ports for ventilating with outside air.

As typically used, these cars haul heavy, solid loads with inadequate vertical air channels that result in freezing in top layers and product warming in the center of the load at the end of the car opposite the refrigeration unit. No new mechanical refrigerator cars have been manufactured in the United States since 1974. Much of the remaining fleet of cars is unusable because of inoperative refrigeration systems, damage, or overall deterioration. Maintenance and repair programs are minimal.

**Air transport.** Air shipment is mainly used to transport highly perishable and valuable commodities (e.g., ornamentals, berries, tropical fruits) to distant domestic and overseas markets or to supply markets with limited supplies during periods of high prices and very strong demand. Products are transported in closed (mostly unrefrigerated) container units or in net-covered pallet loads, in air freighters, or in the freight compartments of passenger airplanes. Individual containers may be loaded with packages of a single commodity or with packages of several different commodities.

Air travel time is often about 6 to 18 hours, and sometimes longer, but waiting time at origin, transfer, and/or destination terminals may be as long as 1 to 2 days, often at ambient temperatures, resulting in rapid deterioration of products. Product warming and weight loss are often serious problems in air shipments.

Most air transport containers lack refrigeration and provide only minimal air circulation. Relative humidity in jet planes at high altitudes can be as low as 4 percent. No humidity control is available. Cargo

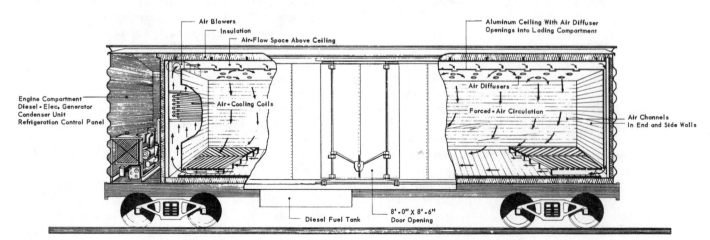

**Fig. 17.3. Mechanically refrigerated rail car. From Redit 1969.**

areas of both freight and passenger planes are refrigerated during flight by controlled inflow of outside air through the engine nacelles (to heat the air to above 0°C) and into the cargo area. Some jet freighters have small refrigerated storage compartments. A few lightly refrigerated (with dry ice or packaged eutectic substances) container units are used (e.g. Igloos or Environtainers). Some airlines use cold storage rooms at origin and destination airports, but not on a regular basis for fresh horticultural products. Products are being increasingly handled by freight forwarders with cold storage, cooling, unitizing, and shipping facilities near major airports. Freight forwarders may also be importers or exporters. Air freight charges are high, and the value of products shipped by air must justify this cost.

**Break-bulk marine transport.** Widely used for tree fruit, grapes, and bananas, this type of transportation is also used for transporting perishable commodities to and from ports lacking container-van loading facilities, or where the ships used are older or smaller vessels having common cold storage rooms. The break-bulk designation refers to the older system in which individual packages were rehandled each time the cargo was transferred from one mode of transport to another. This method can be costly because of slow loading and unloading, rough handling, and high labor costs.

In many present break-bulk systems, packed products are handled as palletized units. When a pallet load is broken apart, its inner surfaces are exposed to ambient conditions that can increase product deterioration. These problems are minimized in newer ships that can handle palletized loads and have better refrigeration systems. Packed products in palletized units are sometimes transported part of the way in break-bulk ships, transferred at a port to marine containers, and transported in container ships to final destination ports.

**Containerized marine transport.** Container vans are loaded in the holds or on decks of container vessels. Containers may be refrigerated by their own refrigerated units connected to the ship's electrical power supply or by cold air supplied via ducting from a large on-board refrigeration system. Recirculation of a common air supply around many different products could be a problem in the latter system.

## Managing Product Transit Temperatures

### Refrigeration and air circulation system

An effective system needs air channels large enough for fans to operate at near-peak performance. Anything that restricts air movement reduces the air volume output of the fans, thereby reducing the amount of circulating air for temperature maintenance. A uniform layer of top-ice 10 to 30 cm (4 to 12 inches) thick on a load blocks cold air from penetrating the ice and restricts cooling of the load. In trailers equipped with frame-type front bulkheads, this results in only the top product layer, in contact with the ice, being kept cool, while warming occurs in lower layers in tight loads. In trailers top-ice causes the thermostat to register a false, "already cold" signal. Applying top-ice in a windrow pattern, leaving the center and edges of the load uncovered, allows circulation and provides faster cooling. This causes the refrigeration unit to go to the heating cycle, which melts the top-ice and results in product warming. This problem can be corrected by placing the thermostat's sensors in the supply airstream about 2 m (7 feet) out from the evaporator coil and by setting the thermostat to 0°C (32°F).

### Transportation vehicle features

Flat side walls, shallow ribbed floors in the absence of pallets or racks, or lack of an air-return bulkhead can impede air circulation and compromise product temperature management in truck trailers. Side-wall and end-wall flues in rail cars permit some of the cooling air to bypass the load compartment.

The sharp decline in volume of air produced by the fan output as resistance to air circulation increases is also a problem. Fans must be capable of circulating the needed volume of air against the anticipated resistance in the system. Some newer model trailers have refrigeration units with fans that meet this need. The direct mechanical linkage of the refrigeration compressor and the air circulation fan results in the fan shifting to low speed whenever the compressor does. The fan should operate at high speed continuously. This feature is incorporated into some newer model refrigeration units.

### Shipping packages

Package design, construction, and use are variable. Poorly vented packages restrict air movement around products and hamper effective temperature management, especially in solid, tight loads. Partial collapse of weak cartons results in formation of a nearly solid load mass that completely prevents air circulation through the load. Also, weak packages and their enclosed contents are more easily damaged by rough handling than strong packages.

### Load patterns and load sizes

Temperature maintenance is often affected by the pattern and size of the load. Loads must have open air channels through and around them that are either vertical (for rail and bottom-air delivery systems) or lengthwise horizontal (for top air delivery systems). Loads must be assembled in ways that will maintain their integrity in transit. This is generally done by unitizing packages on pallets or slip sheets. Various types of gates, braces, and locking bars are used to maintain the integrity of loads.

Use of incentive freight rates (in which per-package-freight cost decreases as load weight increases) tends to create serious temperature maintenance problems. Larger, heavier, and tighter loads resulting from incentive rates make temperature maintenance difficult, especially with products that are not properly cooled before loading. Overloading of TOFC vans (because there is less of a problem in complying with highway load tolerances) can structurally weaken them until they cannot be used for modified atmosphere shipments. Overloading also blocks air circulation so that product temperatures increase during transit. Well-constructed palletized loads with sufficient air channels provide better assurance of proper transit temperatures in heavy loads. In hot or very cold weather, loads should be palletized or on wood racks and kept away from side walls to prevent warming or freezing in wall rows or bottom layers. Very perishable products, such as strawberries, are center-line loaded in trailers or vans: the palletized units are braced away from the side walls and the two rows of pallets contact each other along the center line of the trailer.

## Condition of transit vehicle

Maintaining product temperatures is affected by the condition of the transit vehicle. Intact side walls and insulation; clean floors and drains; refrigeration units that are properly serviced, maintained, and calibrated; intact air-delivery chutes; and tight, undamaged doors and seals are all essential. The carrier owner/operator is primarily responsible for the equipment's condition. However, users (shippers, buyers, brokers, or receivers) or their representatives are responsible for assessing the equipment's condition before loading and for damage to the vehicle caused by their workers during loading or unloading operations. Loading good products into faulty vehicles is wasteful, accelerates product quality loss, and can increase marketing losses.

## Recording thermometers

Many shippers place recording thermometers in each loaded transit vehicle; others place thermometers only in vehicles going to their most distant markets. In truck trailers the thermometer is generally secured high on a side wall at about the three-quarters length of the trailer, or on top of the load toward the rear. In rail cars they are generally placed high on a side wall, just inside the door. These thermometers measure and record only discharge air temperature at their specific locations and provide performance records of the operation of the refrigeration unit. They do not measure or record product temperatures within loads. Some newer types of recording thermometers have probes that can be inserted into the load to record product temperatures, but these are not yet in general use.

## Thermostats

Accurate temperature control is possible only when thermostats are accurately calibrated and are monitoring an airstream that is representative of the air temperature within the vehicle. Thermostats should be periodically recalibrated; operators concerned about freezing the load may set the thermostat several degrees higher to protect themselves. This avoids freezing but may result in product temperatures that are well above those desired. Self-calibrating thermostats are available but in limited use.

If the air return to the refrigeration unit is blocked (usually by the load), or if the stacking pattern does not allow air to flow through or around the packages, airflow over the evaporator coil is slowed considerably and the air is excessively cooled, often to well below 0°C (32°F). When this freezing air is discharged over the load it may result in part of the load being frozen and the rest being warmed, or it may cause excessive cycling (heating and cooling) of the refrigeration unit. Air that bypasses over the top of a load and returns directly to the refrigeration unit side or bottom ports has not absorbed enough heat to influence the thermostat. It is important not to block the air-return passage so that adequate air can circulate through or around the load without short-circuiting it.

When products are loaded warm, with pulp temperatures well above the thermostat setpoint temperature, the refrigeration unit provides maximum refrigeration until the product loses enough heat to the circulating air for the thermostat to signal for low-speed refrigeration. This may require many hours at maximum refrigeration, and parts of the load that are exposed to freezing air for prolonged periods may suffer from freezing or chilling.

## Mixed loads

Maintaining optimum temperatures in mixed loads is difficult, especially in loads containing several commodities. Commodities are generally packed in different sizes and shapes of packages that are then loaded in different load patterns in various parts of vehicles. With currently used shipping containers, these variations often result in the blockage of air circulation. When products with different optimum temperature requirements are shipped together, compromise temperature settings are used that are designed to protect the most perishable, or the most valuable, commodity in a load.

In vehicles with shallow duct floors, all products should be on pallets or racks. Vertical partitions or dividers that are used to separate wet from dry parts of the load should extend no lower than the top deckboards of the pallets or racks. Any part of a load that sits directly on the floor effectively blocks the air circulation under the entire load. When loads are assembled from multiple locations, some products may be uncooled or inadequately cooled before loading. Frequent opening of doors during loading of mixed loads causes warming of already-loaded products.

## Product Compatibility

In mixed loads certain product compatibility factors must be considered. These include the following:

**Temperature compatibility.** Differences in temperatures needed for various products in a load must be considered. For example, products that must be kept near 0°C (32°F) should not be shipped with products sensitive to chilling injury below about 12.5°C (55°F).

**Ethylene production and sensitivity compatibility.** Care must be taken not to ship commodities that produce large amounts of ethylene (e.g., apples, pears, avocados, and certain muskmelons) with commodities that are very sensitive to ethylene (broccoli, carrots, lettuce, kiwifruit, and most ornamentals). The incidence of russet spotting on lettuce (caused by exposure to ethylene) is about three times greater in mixed loads than in straight loads in truck shipments. Continuous regulated ventilation with outside air in marine containers can allow mixing of ethylene incompatible commodities in loads under certain conditions.

**Product odor compatibility.** Some commodities (e.g., onions, garlic) produce odors which can be absorbed by other products, causing the latter to have an objectionable odor and less market appeal. Some products absorb odors more readily than others.

**Moisture compatibility.** Most products benefit from a high relative humidity in the transit atmosphere. Other commodities (e.g., garlic, dry onions) benefit from intermediate humidity levels. Although humidity control at high levels is important during long transit periods, it is difficult to achieve. The use of large evaporator coils on the refrigeration unit helps to accomplish this.

## Modified Atmospheres (MA)

Some commodities benefit from MA in transit vehicles, others may even be harmed. MA may be of little or no measurable value in short-term transport but of high value in long-term transport, such as marine shipments. Benefits of MA are commodity dependent (see chapter 11 for details).

Successful MA transport depends largely upon the tightness of the transit vehicle. In general, some rail cars and newer marine container vans can be used for maintenance of MA, and older or damaged rail cars cannot. MA are applied to marine containers at ports of embarkation. Over-the-road trucks and many TOFC trailers are not tight enough to maintain MA. In these vehicles, MA can be established and maintained within pallet covers (polyethylene bags) sealed to polyethylene covers of pallet bases. In some newer trailers, desired MA are maintained by controlled purging with nitrogen from liquid nitrogen tanks.

Modified atmospheres can result when air passages (e.g., floor drains) in a tightly sealed vehicle become plugged with debris or ice, resulting in depletion of oxygen or accumulation of carbon dioxide from the product's respiration. This is a common problem in winter rail shipments of broccoli, rapini, and brussels sprouts, which are shipped with package ice and under top-ice. The problem is overcome by inserting short lengths of garden hose up through the B-end floor drains to about 5 cm (2 inches) above the floor level inside the car. The hoses must be secured to the drain pipes and the drain pipe caps should be secured in the open position.

## Conclusions

Successful transport of horticultural products to markets depends upon products being cooled to, and loaded at, their desired transit temperatures. Users and carriers must be well informed about the capabilities and limitations of each type of equipment and the condition of transportation equipment supplied. Success also depends upon the types of packages in a load, optimum load patterns, and loading methods used. The ability to maintain or prevent modified atmospheres and the compatibilities of various commodities shipped in mixed loads are also important. Failure to consider any of these factors can cause marketing losses.

## REFERENCES

1. Anon. 1985. *The transport of perishable foodstuffs*. Cambridge, UK: Shipowners Refrig. Cargo Res. Assoc. 50 pp.

2. Ashby, B. H., et al. 1987. *Protecting perishable foods during transport by truck*. U.S. Dept. Agric. Handb. 669. 94 pp.

3. ASHRAE. 1982. *ASHRAE handbook and product directory, applications volume*, sect. IV, chaps. 44-47. Atlanta, GA: Am. Soc. Heating, Refrig. Air Condit. Eng.

4. Booz, Allen, and Hamilton, Inc. 1980. *Piggyback: the efficient alternative for the 80's*. A report by Booz, Allen and Hamilton, Inc. for Transamerica Interway Inc. New York. 66 pp.

5. Hinsch, R. T., R. Rij, and R. F. Kasmire. 1981. *Transit temperatures of California Iceberg lettuce shipped by truck during the hot summer months*. U.S. Dept. Agric. Mktg. Res. Rept. 1117. 5 pp.

6. Kasmire, R. F., and R. T. Hinsch. 1987. *Maintaining optimum transit temperatures in refrigerated truck shipments of perishables*. Univ. Calif. Perish. Handling Transp. Supp. 2. 12 pp.

7. Kasmire, R. F., R. T. Hinsch, and R. Rij. 1980. *Truck inspection poster*. Univ. Calif. Coop. Ext.

8. Lipton, W. J., and J. M. Harvey. 1977. *Compatibility of fruits and vegetables during transport in mixed loads*. U.S. Dept. Agric. Mktg. Res. Rept. 1070. 7 pp.

9. McGregor, B. M. 1987. *Tropical products transport handbook*. U.S. Dept. Agric. Handb. 668. 148 pp. (In English and Spanish.)

10. Redit, W. H. 1969. *Protection of rail shipments of fruits and vegetables*. U.S. Dept. Agric. Handb. 195. 108 pp.

11. Ryall, A. L., and W. J. Lipton. 1979. *Handling, transportation, and storage of fruits and vegetables*. Vol. 1. *Vegetables and melons*, 244-93. Westport, CT: AVI Publ. Co.

12. TransFresh Corporation. 1988. *Fresh produce mixer and loading guide*. Salinas, CA: TransFresh Corp.

# 18

# Handling of Horticultural Crops at Destination Markets

Robert F. Kasmire and M. Joseph Ahrens

Handlers of fresh produce at destination markets are an integral link between shippers and consumers—they handle approximately 90 percent of the fresh market fruits and vegetables in the United States. Presently, about 60 to 65 percent of the produce is shipped directly to distribution centers of chain food stores and service wholesalers, and 5 to 10 percent is shipped directly to food service distributors, institutional receivers, and fast-food chains. The remainder is shipped to various types of terminal market handlers. Much of the growth in wholesale produce terminal markets has been in the area of food service, although some of the newer and upgraded terminal markets continue to handle a substantial part of the traditional distribution of produce.

## Roles of Wholesaling and Retailing

Wholesale operations include:

1. Buying or receiving consignment for sale and accumulating products for selling to retailers, jobbers, purveyors, and institutional outlets. This may include importing.

2. Distribution to and servicing of retailers and institutions.

3. Preparing and shipping mixed loads to other markets by terminal market operators called mixers.

4. Supplying small-volume items (specialty commodities) to chain stores and other retailers, institutions, and purveyors. This is generally done by selling consigned products on a commission basis.

5. Ripening, regrading, and prepackaging into consumer units.

6. Exporting, which generally involves purchasing, repacking (occasionally), and selling to customers in foreign markets.

Retail operations include:

1. Accumulating, preparing (trimming, sorting, and consumer packaging), and presenting products for sale to consumers. This may include preparing and selling products in salad bars or in ready-made salads.

2. Activities related to promoting various produce items.

## Wholesalers and Retailers Categorized

Wholesalers fall into several categories:

1. Chain store distribution centers that service their own stores.

2. Service wholesalers that supply produce to independent and/or chain retail stores and to food service distributors.

3. Car-lot receivers that divide and sell large quantities to retailers, brokers, jobbers, purveyors, and institutions, and may service retail stores.

4. Commission merchants who receive and sell consigned shipments on a fixed percent commission and may perform the same functions as car-lot receivers, including service of retail stores.

5. Jobbers who handle products from car-lot receivers to small, independent retailers.

6. Mixers who buy from other wholesalers, generally car-lot receivers, and make up mixed loads of various commodities for shipping on order to distant markets.

7. Importers who receive products from foreign suppliers for selling or reselling to domestic market customers. Importers may also be buyers, shippers, brokers, or exporters.

8. Exporters who buy products from domestic suppliers (shippers or terminal market operators) for shipping to customers in foreign markets. Exporters are also commonly importers, receivers, and shippers.

9. Purveyors who service restaurants, institutions, and carriers and may also be processors of prepared foods.

10. Wholesale auctions that sell certain commodities on a price-bid basis.

11. Food service wholesaler-distributors that sell a full line of products including food to restaurants, fast-food outlets, institutions (schools, prisons, hospitals, etc.), airlines, and other outlets; there is an increase in food service wholesalers purchasing produce directly from suppliers (shippers).

12. Freight forwarders that consolidate, handle necessary forms, provide temporary cold storage, and expedite transfer of products for exporters to overseas-bound carriers and from carriers to importers.

Retailers consist of:

**1.** Chain stores that may belong to corporations, individuals (families), or cooperatives. They may be large or small in size, and the number of stores per chain may vary from a few to more than a thousand.

**2.** Independent stores and other outlets that include neighborhood supermarkets, small ("Mom and Pop") retail stores, greengrocer stores that sell only produce and produce-related items, and produce carts.

**3.** Direct marketing outlets that include farmers' markets, roadside stands, "pick-your-own" operations, and "rent-a-tree" operations.

Marketing channels for fresh fruits and vegetables are illustrated in figure 18.1.

## Product Handling

Considerable variation exists in product handling practices. Good temperature management is essential to proper handling and to providing consumers with produce of the best possible quality. Unfortunately, some practices and facilities in use result in product warming or chilling, both of which cause marketing losses. Rough handling can be attributed to labor-related problems, to the use of outdated, poorly maintained, and overloaded facilities, and to poor product-handling practices. All of these factors especially affect products that are subsequently shipped in mixed loads to other markets. Improper handling is not exclusive to any one area or type of destination market; good and bad practices can be found throughout the system.

Sanitation procedures are necessary in both wholesale and retail operations. Proper disposal of decayed produce, and cleaning and sanitizing of storage facilities, preparation areas, and display bins help to maintain product quality and reduce marketing losses.

## Wholesale receivers

Most receivers handle a large number of commodities. They may have a "wet" cold room set at about 2° to 5°C (36° to 41°F) for leafy and root vegetables and a "dry" cold room set at about 0°C (32°F) for temperate fruits and other cool-season vegetables. Some have a room at a compromise temperature of 7° to 10°C (45° to 50°F), often too cold, in which chilling-sensitive commodities are stored. In other facilities, chilling-sensitive commodities are stored in a warehouse area that may be refrigerated at about 10° to 12.8°C (50° to 55°F) or not refrigerated at all. The best configuration allows for separation of commodities based on temperature and relative humidity

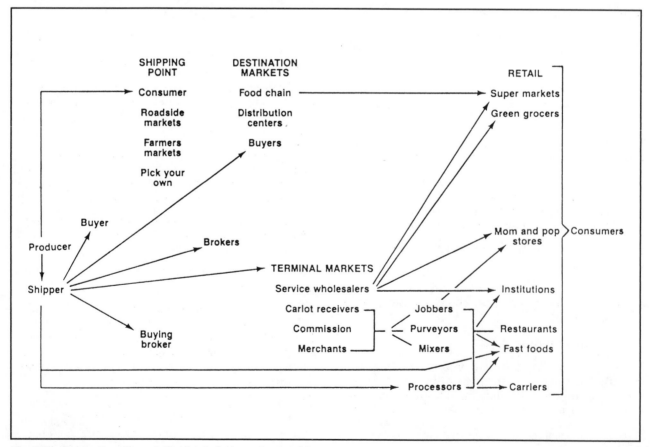

**Fig. 18.1. Marketing channels for fresh fruits and vegetables.**

requirement and by their sensitivity to or production of ethylene and other volatiles. If a wide range of produce is handled, this usually means a minimum of 4 to 5 rooms are needed to meet these conditions.

Modern chain store distribution centers and service wholesalers tend to have better designed and better maintained cold rooms than terminal market operations. This is partly because many terminal markets are old, neglected, and have inadequate space for expansion. However, large, effective cold rooms are used in new, modern terminal markets. The mixing of warm, generally local products with pre-cooled and refrigerated transported products reduces the effectiveness of these cold rooms. Fortunately, most perishable products are rarely stored for more than a few days in terminal market facilities. Products are often displayed for several hours at ambient temperatures (sometimes very cold or warm) in terminal markets, causing loss of shelf life.

### Retail handlers

Retail handling varies as much as wholesale handling. Retailers must handle a large number of produce items, and there is generally only one cold room in a store. The room is usually set at about 4.4°C (40°F), is too small, and often is not well maintained, especially in smaller stores. They basically rely on quick turnover to overcome handling problems. Product temperature management is minimal to nonexistent in many roadside markets. Many lack even adequate shade. A major problem associated with poor retail handling in the past was the lack of product handling education programs in the industry. However, extensive and aggressive training programs by individual firms and by trade associations are helping to correct this problem.

### Food service wholesalers

These tend to receive less frequent direct deliveries of fresh produce than do chain store distribution centers or terminal market operations. Produce cold room and handling facilities may be less sophisticated than in food chains' or service wholesalers' distribution centers, because fresh produce is still a relatively small part of their total volume.

### Ethylene Damage in Handling

Damaging concentrations of ethylene in ambient storage atmospheres at the wholesale and the retail levels is a serious yet largely unsolved problem. Because of limited cold storage space, ethylene-generating and ethylene-sensitive products are commonly stored in the same room for more than 24 hours: iceberg lettuce, for example, may share space in cold rooms with apples, pears, stone fruits, and muskmelons. The ethylene from the fruits and melons causes russet spotting on the lettuce. The problem is compounded by storage temperatures that are often too high. Ethylene-producing and ethylene-sensitive products are best stored in separate rooms.

Propane-powered lift trucks used in some produce warehouses are major sources of ethylene, besides causing excessive heat loads on cold room refrigeration systems. Other sources in some facilities are the release of high residual levels of ethylene during unloading of unvented ripening rooms into nearby storage areas used for produce, and the accumulation of decaying produce that generates considerable ethylene.

In some cases, periodic opening of doors for normal traffic in and out of the store room is adequate to maintain enough fresh air exchange and prevent buildup of harmful volatiles. However, many facilities need an active ventilation system. Ripening rooms may need to be exhausted to the outside of the building. Ethylene may be removed from the storehouse air by absorption using potassium permanganate pads, by UV light, or by ventilation. Two air exchanges (done during the night when the outside air is coolest) separated by an hour have shown to be the most cost effective in reducing ethylene levels in lettuce warehouses in California.

## Container Handling

The variety of shipping containers being loaded together in trucks destined for retail markets causes another major problem. The more than 500 different container sizes, shapes, and designs used in the U.S. causes stacking and unitizing problems for handlers and distributors. There have been industry attempts in the past to limit containers to a small number (e.g., 12 to 34) of modular shapes and sizes that can be handled in unitized loads during distribution. Better and less costly handling and reduced marketing losses would result from this change.

Produce shipped to distribution centers and service wholesalers on pallets other than hardwood 120 cm by 100 cm (approximately 48 inches by 40 inches) require additional, sometimes rough handling during transfer to standard-size pallets at distribution centers. The disposal of nonstandard-size pallets and inexpensive one-trip pallets is a major and costly problem for destination market handlers. Pallets made of recycled plastic (plastic lumber) and of the standard dimensions have shown promise.

Continuing education programs on postharvest handling and related subjects are conducted by trade associations and internally by many companies. Technical representatives of specific commodity shippers or associations (e.g., bananas, pineapples, and citrus) conduct programs on proper handling of their respective commodities for wholesalers, retailers, and food service distributors. Few states conduct extension programs with produce shippers, and even fewer of these programs are formally extended into distribution markets. Making such programs available to destination market handlers would help to improve the end quality of the produce.

## Problems of Wholesalers

Wholesalers' problems include:

**1.** Produce warehouse management personnel need more training in horticultural product-handling requirements.

**2.** The supply of qualified entry-level workers is diminishing for operations that are becoming increasingly technical.

**3.** Products are sometimes not of uniform quality. Quality control at shipping points and the wholesale level needs to be improved. This is being achieved through continuing industry-conducted educational programs. There is a need for better, more objective communication of product quality.

**4.** Purchased or received commodities are often immature, overripe, or of mixed maturities and require extra handling, space, and time; this causes marketing losses. Industry programs directed toward marketing of ripe fruit are gradually overcoming this problem.

**5.** Facilities in many older operations are inadequate for proper product handling, especially with respect to temperature maintenance, sanitation, and ethylene concentrations in storage atmospheres. Terminal markets that are in or near the centers of large cities have traffic and product movement problems. Even some new wholesale centers, for various reasons, are constructed with inadequate room configurations, incapable of handling products under appropriate conditions.

**6.** Transportation problems, including product temperature management and physical damage to shipping containers and products during transit, are common.

**7.** Education programs in product handling are needed for wholesale handlers. Wholesaler handling practices can either compound or alleviate previous mishandling, maturity, or temperature-management problems.

**8.** Extra handling is required for many products that are received on nonstandard-size pallets.

## Problems of Retailers

Retailers' problems include:

**1.** It is difficult to obtain uniform-quality supplies of each commodity. Quality control at both shipping points and wholesale levels is inadequate.

**2.** Nonuniform maturity and ripeness is a problem with various commodities, commonly among muskmelons, tomatoes, and some temperate fruits.

**3.** Knowledgeable personnel need to be recruited and retained. This is an increasing problem because the pool of qualified applicants is decreasing and operations are becoming more technical. Also, there is often a rapid turnover in produce handling personnel.

**4.** Physical facilities are inadequate for temporary holding and display of produce under optimum conditions.

**5.** Record-keeping systems used in most stores for tracking product loss at the retail level are inadequate and do not allow retailers to identify causes of lost sales. Currently all losses of any type, from physical injury to physiological breakdown to decay, are absorbed in the general category of "operating expense" (shrink).

**6.** Product-handling education programs for retail produce personnel are not adequately used, although they are available.

## Current Trends and Developments

Some current developments that will aid in better handling of produce in destination markets:

**1.** Better temperature management during transit from shipping point to destination markets, increased use of refrigerated trucks in distribution from wholesale warehouses or terminal markets to retail stores, increased and improved use of refrigerated storage facilities at wholesale and retail levels, and increased use of rapid cooling methods (hydrocooling, forced-air cooling, and package-icing) to cool warm produce received at distribution centers.

**2.** Improved unitization of containers in distribution from wholesale to retail. This includes unitizing containers on standard-size 100 by 120 cm (40 by 48 inches) pallets or slip sheets, using the layer concept of ordering (i.e., ordering enough packages of a given commodity to make one or more layers of containers on a pallet or slip-sheet unit), and developing a minimum number of standard-size, metric-measured, modular shipping containers that would facilitate mixing several sizes of modular containers on a pallet or slip sheet.

**3.** Increased use of computerized recording of products received, storage location in warehouses, and inventory control.

**4.** Increased use of quality control personnel at wholesale (mostly by service wholesalers, food chains, and food service wholesalers). The FOB buyer may also serve as a quality control person.

**5.** Increased use of Direct Products Profit accounting improves inventory control and profitability of individual products or product lots.

**6.** Increased use of product handling educational programs at wholesale and retail. These may be provided or conducted by the company's own staff, produce consultants, produce trade associations, commodity group representatives (e.g., California Fresh Market Tomato Advisory Board, California Iceberg Lettuce Commission, California Avocado Commission), marketing organizations (e.g., Sunkist, Blue Anchor, Dole, United Brands), or national trade associations such as the Produce Marketing Association (PMA) and United Fresh Fruit and Vegetable Association (UFF&VA).

**7.** Expanded efforts by national produce associations (PMA and UFF&VA) in promoting the nutritional value of fresh produce and in supporting studies on the effects of handling practices on maintenance of nutritional quality.

**8.** Expanded trade association efforts to assure consumers of the safety and quality of fresh produce commodities. The Center for Produce Quality is an excellent example.

**9.** A major increase in a wide range of industry educational programs by major trade associations such as PMA, UFF&VA, and the National American Wholesale Grocers Association (NAWGA).

**10.** Improved communications among handlers in the various handling steps, primarily through trade associations' programs. This includes educational activities aimed at food service produce handlers.

## REFERENCES

1. Hardenburg, R. E., A. E. Watada, and C. Y. Wang. 1986. *The commercial storage of fresh fruits, vegetables, florist and nursery stocks*. U.S. Dept. Agric. Handb. 66. 130 pp.

2. Kretchman, D. W. 1973. *Care and handling of fresh fruits and vegetables in retail markets*. Newark, DE: Prod. Mktg. Assoc. Yearb. 96, 98.

3. Lewis, W. E. 1957. *Maintaining produce quality in retail stores*. U.S. Dept. Agric. Handb. 117. 30 pp.

4. National American Wholesale Grocers' Association and PMA. 1987. *Professional produce managers manual*. 2nd ed. Newark, DE: Prod. Mktg. Assoc. 130 pp.

5. Produce Marketing Association. 1985. *The foodservice guide to fresh produce*. Newark, DE: Prod. Mktg. Assoc. 48 pp.

6. ———. 1988. *The fresh produce training manual for food service*. Newark, DE: Prod. Mktg. Assoc. (looseleaf inserts)

7. United Fresh Fruit and Vegetable Association. no date. *Basic produce department operation*. Alexandria, VA: United Fresh Fruit Veg. Assoc. 81 pp.

8. ———. various dates. *Fruit and vegetables facts and pointers*. (series of reports on each of 81 commodities). Alexandria, VA: United Fresh Fruit Veg. Assoc.

9. Volz, M. D., and J. J. Karitas. 1973. *Handling and space costs for selected food wholesalers in urban food distribution centers*. U.S. Dept. Agric. Mktg. Res. Rept. 992. 24 pp.

# 19

# Energy Use in Postharvest Technology Procedures

James F. Thompson

Production, processing, distribution, and preparation of food requires between 12 and 16 percent of the total amount of energy use annually in the U.S. About one-third of this energy (5 to 7 percent of the total U.S. energy budget) is used in postharvest procedures associated with processing and transporting food products. If energy costs rise, how should postharvest practices be modified? The answer lies in the present use of energy in postharvest handling operations and the potential there for energy conservation in these operations.

It is important to consider how a handling system affects the total energy use from production to consumption. Tables 19.1 and 19.2 summarize the direct energy use of various methods of handling potatoes and apples. Energy use for the manufacture of equipment and facilities is not included.

Total energy use varies significantly among handling systems for each commodity. Because handling fresh commodities requires no energy (or very little) for processing, one might assume this would result in less total energy use than systems that require more processing. This assumption holds true for potatoes, but fresh apples require roughly 50 percent more energy than canned apple juice or dried apples. Processing that reduces the weight of the commodity results in lower transportation costs. In the case of canned apple juice and dried apples, the reduction in transportation energy more than offsets the energy of processing. It is apparent that selecting the least energy-intensive method of handling a perishable commodity is not a simple choice.

## Energy Conservation and Costs

In the future, conservation practices should reduce energy consumption in each of the six categories of energy use: production, processing, transportation, trade, home storage, and home preparation. The categories that use large quantities of heat, processing and home preparation, hold the greatest potential for reduction. Authorities think energy use in food processing could be reduced by 50 percent. If that were done, systems that involve processing would in many cases be less energy intensive than fresh handling. Reducing energy use in home preparation would shift the advantage back toward fresh, but only a small portion of domestic energy consumption is used for cooking. The processor is more apt to make changes to reduce utility costs than the homeowner.

Transportation constitutes a significant portion of total energy costs for commodities shipped long distances. This cost can be reduced by using more efficient transportation systems or by growing commodities closer to the consumer. Table 19.3 lists the energy efficiencies of shipping methods for perishable commodities. About three-fourths of all fresh market commodity shipments originating from the West are transported by truck. A shift to rail transportation could reduce energy use by 70 percent. However, rail transit usually takes longer to reach eastern markets than truck shipment. Rail service needs to be improved before it can be used as a low-energy method of transportation for fresh perishables. New trailers-on-flatcars and special train cars designed for transporting containers in a two-high stack may offer the service improvement needed to ship perishables over the rails. Processed commodities with a long shelf life are better suited for rail.

Growing produce closer to its final market reduces transportation energy use, but it is often impractical. The major U.S. markets are east of the Mississippi River, but many of the best growing areas with long harvest seasons are in the West. Local production should be encouraged wherever possible, but consumer demand for a wide variety of fresh commodities year-round cannot be satisfied by local production. Greenhouse production of vegetables in eastern winters probably would require more energy than transporting field-grown western produce East.

Table 19.1. Energy use in potato-handling systems

| Handling system | Energy use in MJ/t | | | | | | |
| --- | --- | --- | --- | --- | --- | --- | --- |
| | Production | Processing | Transportation | Trade | Home storage | Home preparation | Total |
| Fresh | 950 | — | 2,290 | 90 | — | 3,710 | 7,040 |
| Frozen | 950 | 5,650 | 1,480 | 1,330 | 490 | 1,210 | 11,110 |
| Dehydrated | 950 | 7,950 | 420 | 170 | — | 1,060 | 10,550 |

## Table 19.2. Energy use in apple-handling systems

| Handling system | Energy use in MJ/t | | | | | | |
| | Production | Processing | Transportation | Trade | Home storage | Home preparation | Total |
|---|---|---|---|---|---|---|---|
| Fresh | 2,200 | — | 4,740 | 370 | 1,140 | * | 8,450 |
| Canned juice | 2,200 | 680 | 1,660 | 230 | 750 | — | 5,520 |
| Canned sauce | 2,200 | 3,430 | 3,640 | 600 | — | — | 9,870 |
| Dried | 2,200 | 2,090 | 710 | 140 | — | — | 5,140 |

*If fresh apples are used to make apple crisp, home preparation energy is 5,900 MJ/t

### Table 19.3. Energy use of various methods of transportation

| Transportation method | Energy use (MJ/t-km) |
|---|---|
| Water | 0.35 |
| Railroad | 0.54 |
| Motor freight | 1.78 |
| Air | 30.90 |
| Auto hauling 5 kg of fruit | 1,315.00 |

It can be argued that local production should be carried one step further by home food production. While the potential exists for home production to contribute to energy savings, correct energy-use decisions must be made to realize this potential advantage. For example, a 5-mile trip to haul 10 pounds of tomatoes to a friend makes the same per-ton energy use as a truck hauling fresh tomatoes across the country. Other poor energy-use choices are buying a rototiller for 10 hours of use per year, driving to the local chicken farm for a good deal on organic fertilizer, home canning, and so on. Consumption of homegrown produce has the potential of low energy use, but it is subject to seemingly insignificant practices that can greatly increase energy use.

## Using Energy

Tables 19.1 and 19.2 indicate that there is no energy used in processing fresh produce. This is not exactly the case. Fresh commodities are often sorted, packed, and cooled at central handling facilities and are sometimes stored. Data for a wide variety of fruits and vegetables indicate that this processing requires 120 to 240 megajoules per ton of fresh produce. That is less than 3 percent of the total energy use listed in the tables. However, energy costs are a significant part of operating costs for cooling and storage facilities. Managers have found that there are ways to economically reduce this energy use.

## Storage

Energy use in a cold storage facility is affected by the amount of heat the refrigeration equipment must remove and the efficiency of the refrigeration equipment. The main sources of heat in a facility for long-term storage are transmission through walls, evapor-

ator coil fans, lights, air leakage, and respiration of the stored commodity.

Heat entering a cold storage facility through walls can be minimized by increasing the insulation and by painting the exterior a light color. Doubling the insulation (as measured by R value) reduces transmitted heat by half. Newer facilities use insulation levels as high as R40 in walls and R60 in ceilings. In general, it is advisable to build with more insulation than utility costs may presently warrant, because energy costs are difficult to predict and it is much cheaper to install insulation during construction than after construction is completed.

Sun shining on walls and roof can dramatically increase the effective outside temperature, increasing heat flow into a storage facility. Table 19.4 shows the relationship between wall or roof orientation, color, and effective outside air temperature. (Effective outside air temperature is the normal air temperature plus a factor to account for the sun shining on a surface.) A dark, flat roof can be 42°C (75°F) warmer than the outside air temperature. Painting a south-facing wall a light color can reduce the effective wall temperature by 11°C (20°F) compared with a dark wall. Walls and roof of a cold storage facility should be painted a light color or shaded from the direct sun.

Fans are used in cold storage facilities to move air through the evaporator coils and to uniformly circulate cooled air around stored commodities. During the initial filling of a storage facility, constant air movement is needed to remove the residual field heat of the product. However, after the commodity has reached the desired temperature, air movement can be reduced. Also during the winter months, the outside air temperature drops and less heat enters

### Table 19.4. Effect of surface color and orientation on effective outside temperature*

| Color and orientation | Effective temperature | |
| | °C | °F |
|---|---|---|
| Outside air | 35 | 95 |
| Light-colored south wall | 46 | 115 |
| Dark-colored south wall | 57 | 135 |
| Dark-colored flat roof | 77 | 170 |

*Data are maximum temperatures for a clear July 21 in Fresno, CA.

through the walls compared to summer or fall conditions. Many CA apple storages have been set up to automatically cycle the fans off for at least 50 percent of the time.

Lights in the cold storage room should be turned off when not needed. Warm outside air leaking into the cold storage room increases energy use required of the refrigeration equipment. Use plastic flap doors to reduce infiltration during loading and unloading. Seal around openings for pipes and electrical conduits. Heat produced by respiration of the stored commodity can be minimized by keeping the commodity cool and using controlled atmospheric storage if appropriate.

Design of the refrigeration system significantly affects energy use. The temperature of the refrigerant fluid after it is cooled in the condenser should be as low as possible. For example, a facility maintaining 0°C (32°F) and a condensing temperature of 52°C (125°F) requires 50 percent more power than one that operates at a condensing temperature of 35°C (95°F). In warm areas, well-water cooled or evaporatively cooled condensers should be selected over air-cooled units. Maintaining highest possible suction pressures reduces compressor energy use. Use large evaporator coils and a control system that increases suction pressure as demand on the refrigeration system is reduced. Use a compressor system that operates efficiently over the required range of refrigerant flows. Screw compressors operate efficiently only at flow rates near their maximum capacity. Use several in parallel, shutting down those that are not needed, or consider using reciprocating compressors for peak loads. They operate efficiently over a large range of refrigerant flows.

Some storage facilities may reduce energy costs by using evaporative cooling, nighttime cooling, or other alternate sources of refrigeration listed in chapter 9, *Storage Systems*.

## Cooling

Energy costs for cooling facilities usually represent a small proportion of total costs. For example, in 1988, when it cost about 65 cents to have a box of lettuce vacuum cooled, the energy cost only 2 cents per box. Thus the relative energy efficiency of a particular cooling system is rarely considered in selecting a cooling system. However, the five major types of cooling systems vary significantly in their energy efficiency, and it is helpful to understand why there are differences.

Figure 19.1 compares the energy efficiency of the five systems based on a number called an energy coefficient. The coefficient equals the cooling work done divided by the energy purchased to operate the cooler. High values represent high efficiency. Vacuum cooling is the most energy efficient followed by hydrocooling, water-spray vacuum cooling, package icing, and forced air cooling. Part of the reason for the high efficiency of vacuum cooling is that it re-

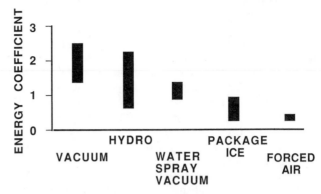

**Fig. 19.1. Energy use of various produce-cooling systems.**

**Table 19.5. Distribution of heat input to three types of fruit and vegetable coolers***

| | Total heat input | | |
|---|---|---|---|
| | Vacuum (%) | Hydro† (%) | Forced air (%) |
| Product | 100 | 54 | 47 |
| Fans or pumps | 0 | 9 | 37 |
| Infiltration, startup, conduction | 0 | 37 | 7 |
| Lift trucks | 0 | 0 | 8 |
| Lights, people, etc. | 0 | 0 | 1 |

*Based on measured or calculated heat inputs for two or three installations of each type of cooler.
†Hydrocoolers were operated at partial capacity. Average energy coefficient = 1.07 for coolers tested.

moves heat only from the product being cooled. The other types of coolers remove heat from fans, pumps, infiltration of outside air, heat conducted through exterior walls, lights, forklifts, and people working in the cooler. Table 19.5 shows a distribution of heat loads for three types of coolers. Nearly all forced air coolers are used for some short-term product storage. This contributes to their particularly low energy coefficient numbers but it is not possible to separate energy use for storage from the total.

The data in figure 19.1 show a great difference between the most efficient and least efficient coolers of a given type. For example, a well-operated hydrocooler can operate more efficiently than most vacuum coolers. However, a poorly operated hydrocooler can have nearly as low an energy efficiency as a forced-air cooler. The difference in efficiency between a given cooler and the best cooler of its type represents the potential for energy savings.

Vacuum cooler energy can be reduced by improved operation or design of the refrigeration system. Refrigeration demand varies from zero for the first 10 minutes of a cycle to maximum when product temperature drop actually begins. Many vacuum coolers use one screw compressor, which, as mentioned in chapter 9, do not operate efficiently at low refrigerant flows. Efficiency can be improved by turning off the compressor when no refrigeration is needed (although this reduces motor life) or by using reciprocating compressors. Operating the cooler with a partial load reduces energy efficiency: a half load of lettuce requires 50 percent more energy per carton to cool than a full load. Cooling products that require long cooling times greatly increases energy use. For example, a load of cauliflower, which cools in 2 to 3 hours, takes three times more energy per pound of product cooled than iceberg lettuce that requires only 20 to 30 minutes.

Hydrocooler energy use can be reduced by protecting the cooler from exterior heat gain and operating it at maximum capacity. Table 19-5 indicates that over one-third of the heat input to a hydrocooler is infiltration of warm air, outside heat conducted through walls, and cooling the water reservoir when the cooler is started up each day. Infiltration can be reduced by installing plastic flap doors and by minimizing the distance between the shower pan and the top of the product. Adding insulation and shading the cooler or painting it a light color reduces heat conduction through the walls. Using a smaller water reservoir lowers the amount of startup cooling needed. Placing the cooler in a refrigerated building reduces all sources of outside heat gain and startup losses. Energy use per unit cooled associated with pumps and with removing conduction and infiltration heat can also be reduced by operating the cooler at maximum capacity. These energy uses are dependent on the amount of time the cooler is in operation, not on the amount of product cooled. Energy use per pound of product cooled is reduced when more pounds of product are cooled per hour.

Energy use in forced-air coolers can be reduced by all of the techniques mentioned for storage facilities, but fan energy use is the most significant. In addition to their own energy consumption, fans contribute over one-third (and in some coolers, more than half) of the heat that must be removed from an average forced air cooler. Fan energy use can be minimized by turning fans off when not needed, installing evaporator coils with a minimum airflow resistance, using cartons with adequate venting area, and arranging pallets on the cooler to reduce airflow resistance.

Package ice systems lose efficiency in two ways. (1) Many operations put about 7 kg (15 pounds) more ice in a box than is needed to cool the product. This allows the product to arrive at destination with about 5 kg (11 pounds) of ice still in the box, but the extra ice is rarely needed to maintain product quality.

In a box of broccoli, this extra represents almost half of the total ice used in the cooling process. (2) Many ice systems are poorly insulated and not shaded from the sun.

Little energy testing has been done with water-spray hydrocooling. All of the recommendations for vacuum coolers should be applicable. Also, consider insulating the cooler, because water in contact with the walls during spraying and water in the reservoir allows outside heat to be transferred into the cooler.

Many cooling facilities are billed for electricity on the basis of not only the amount of energy consumed but the time of day it is consumed. Electricity is usually more expensive during the afternoon hours, which is usually when most electricity is needed for cooling operations. There are ways to shift the energy demand to hours when electricity is cheaper. Slowing the cooling rate and consequently shifting the cooling work to the night is the least expensive option for forced air coolers. This can only be used if there is excess cooling capacity available at night and if products can withstand some delay in cooling. Electric lift trucks reduce refrigeration demand because they produce less heat than propane lifts.

All coolers can use a diesel or gas-powered generator to reduce utility purchases during peak rate periods. This alternative is very cost effective when oil and gas prices are low compared to electricity prices.

Thermal energy storage is being used by several cooling operations. This equipment stores ice during the night and then uses it the following day for product cooling. Thermal energy storage is very cost effective if it is designed into the initial refrigeration system, creating a less expensive, small system that operates constantly rather than a large system that must operate only when cooling is needed.

## REFERENCES

1. Cervinka, V., et al. 1974. *Energy requirements for agriculture in California.* Joint study by Calif. Dept. Food and Agric. and Agric. Eng. Dept., Univ. Calif. Davis.

2. Thompson, J. F., and Y. L. Chen. 1988. Comparative energy use of vacuum, hydro, and forced air coolers for fruits and vegetables. *ASHRAE Trans.* 94:1427-32.

3. Thompson, J. F., and Y. L. Chen. 1989. Energy use in hydrocooling stonefruit. *Appl. Engrg. in Ag.* 5:568-572.

4. Thompson, J. F., Y. L. Chen, and T. R. Rumsey. 1987. Energy use in vacuum coolers for fresh market vegetables. *Appl. Engrg. in Ag.* 3:196-199.

5. Whittlesey, N. K., and C. Lee. 1976. *Impacts of energy price changes on food costs.* Washington State Univ. College Agric. Res. Ctr. Bull. 882. 18 pp.

# 20

# Quality and Safety Factors: Definition and Evaluation for Fresh Horticultural Crops

Adel A. Kader

Quality is defined as any of the features that make something what it is, or the degree of excellence or superiority. The word quality is used in various ways in reference to fresh fruits and vegetables such as market quality, edible quality, dessert quality, shipping quality, table quality, nutritional quality, internal quality, and appearance quality.

Quality of fresh horticultural commodities is a combination of characteristics, attributes, and properties that give the commodity value for food (fruits and vegetables) and enjoyment (ornamentals). Producers are concerned that their commodities have good appearance and few visual defects, but for them a useful cultivar must score high on yield, disease resistance, ease of harvest, and shipping quality. To receivers and market distributors, appearance quality is most important; they are also keenly interested in firmness and long storage life. Consumers consider good quality fruits and vegetables to be those that look good, are firm, and offer good flavor and nutritive value. Although consumers buy on the basis of appearance and feel, their satisfaction and repeat purchases are dependent upon good edible quality. Assurance of safety of the products sold is extremely important to the consumers.

## Components Quality

The various components of quality listed in table 20.1 are used to evaluate commodities in relation to specifications for grades and standards, selection in breeding programs, and evaluation of responses to various environmental factors and postharvest treatments. The relative importance of each quality factor depends upon the commodity and its intended use (fresh or processed). Appearance factors are the most important quality attributes of ornamental crops.

Many defects influence the appearance quality of horticultural crops. Morphological defects include sprouting of potatoes, onions, and garlic, rooting of onions, elongation of asparagus, curvature of asparagus and cut flowers, seed germination inside fruits such as tomatoes and peppers, presence of seedstems in cabbage and lettuce, doubles in cherries, floret opening in broccoli, and so on. Physical defects include shriveling and wilting; internal drying of some fruits; mechanical damage such as punctures, cuts and deep scratches, splitting and crushing, skin abrasions and scuffing, deformation (compression), and

**Table 20.1. Quality components of fresh fruits and vegetables.**

| Main factors | Components |
|---|---|
| Appearance (visual) | Size: dimensions, weight, volume |
| | Shape and form: diameter/depth ratio, smoothness, compactness, uniformity |
| | Color: uniformity, intensity |
| | Gloss: nature of surface wax |
| | Defects: external, internal |
| |     Morphological |
| |     Physical and mechanical |
| |     Physiological |
| |     Pathological |
| |     Entomological |
| Texture (feel) | Firmness, hardness, softness |
| | Crispness |
| | Succulence, juiciness |
| | Mealiness, grittiness |
| | Toughness, fibrousness |
| Flavor (taste and smell) | Sweetness |
| | Sourness (acidity) |
| | Astringency |
| | Bitterness |
| | Aroma (volatile compounds) |
| | Off-flavors and off-odors |
| Nutritive value | Carbohydrates (including dietary fiber) |
| | Proteins |
| | Lipids |
| | Vitamins |
| | Minerals |
| Safety | Naturally occurring toxicants |
| | Contaminants (chemical residues, heavy metals) |
| | Mycotoxins |
| | Microbial contamination |

bruising; growth cracks (radial, concentric); and so on. Temperature-related disorders (freezing, chilling, sunburn, sunscald), puffiness of tomatoes, blossom-end rot of tomatoes, tipburn of lettuce, internal breakdown of stone fruits, water core of apples, and black heart of potatoes are examples of physiological defects. Pathological defects include decay caused by fungi or bacteria and virus-related blemishes, irregular ripening, and other disorders. Other defects result from damage caused by insects, birds, and hail; chemical injuries; and scars, scabs, and various blemishes (e.g., russeting, rind staining).

The texture of horticultural crops is important for eating and cooking quality and is a factor in withstanding shipping stresses. Soft fruits cannot be shipped long distances without extensive losses due to physical injuries. In many cases, this necessitates harvesting fruits at less than ideal maturity for flavor quality.

Evaluating flavor quality involves perception of tastes and aromas of many compounds. Objective analytical determination of critical components must be coupled with subjective evaluations by a taste panel to yield meaningful information about flavor quality. This approach can be used to establish a minimum acceptable level. To learn consumer flavor preferences for a given commodity, large-scale testing by a representative sample of consumers is required.

Fresh fruits and vegetables play a significant role in human nutrition, especially as sources of vitamins (C, A, $B_6$, thiamine, niacin), minerals, and dietary fiber. Their contribution as a group is estimated at 91 percent of vitamin C, 48 percent of vitamin A, and 27 percent of vitamin $B_6$, 17 percent of thiamine, and 15 percent of niacin in the U.S. diet. Fruits and vegetables also supply 26 percent of magnesium, 19 percent of iron, and 9 percent of the calories. Legume vegetables, potatoes, and tree nuts contribute about 5 percent of the per capita availability of proteins in the U.S., and their proteins are of good quality as to content of essential amino acids. Other important nutrients supplied by fruits and vegetables include folacin, riboflavin, zinc, calcium, potassium, and phosphorus. Postharvest losses in nutritional quality, particularly vitamin C content, can be substantial and increase with physical damage, extended storage, high temperatures, low relative humidity, and chilling injury.

Safety factors include levels of naturally occurring toxicants in certain crops (such as glycoalkaloids in potatoes), which vary according to genotype and are routinely monitored by plant breeders so they do not exceed safe levels. Contaminants such as chemical residues and heavy metals are also monitored by various agencies to assure compliance with established maximum tolerance levels. Sanitation throughout harvesting and postharvest handling operations is essential to minimize microbial contamination; procedures that reduce the potential for growth and development of mycotoxin-producing fungi must be used.

## Interrelationships among quality components

It is important to define the interrelationships among each commodity's quality components and to correlate subjective and objective methods of quality evaluation. This information is essential for selecting new cultivars, choosing optimum production practices, defining optimum harvest maturity, and identifying optimum postharvest handling procedures. The point of all this effort is to provide high-quality fruits and vegetables for the consumer.

In most commodities, the rate of deterioration in nutritional quality (especially vitamin C content) is faster than that in flavor quality, which in turn is lost faster than textural quality and appearance quality. Thus, the postharvest life of commodity based on appearance (visual) quality is often longer than its postharvest life based on maintenance of good flavor.

Quality criteria used in the U.S. standards for grades and the California Agricultural Code (tables 21.1 to 21.5 in chapter 21) emphasize appearance quality factors in most commodities. In many cases, good appearance does not necessarily mean good flavor and nutritional quality. A fruit or vegetable that is misshapen or has external blemishes may be just as tasty and nutritious as one of perfect appearance. For this reason, it is important to include quality criteria other than appearance that more accurately reflect consumer preferences. Such quality indices must be relatively easy to evaluate, and objective methods for evaluation should be developed.

## Factors influencing quality

Many pre- and postharvest factors influence the composition and quality of fresh horticultural crops. These include genetic factors (selection of cultivars, rootstocks), preharvest environmental factors (climatic conditions and cultural practices), maturity at harvest, harvesting method, and postharvest handling procedures.

**Climatic conditions.** Climatic factors, especially temperature and light intensity, have a strong influence on the nutritional quality of fruits and vegetables. Consequently, the location and season in which plants are grown can determine their ascorbic acid, carotene, riboflavin, and thiamine content. Light is one of the most important climatic factors in determining ascorbic acid content of plant tissues. Researchers consistently find much higher ascorbic acid content in strawberries grown under high light intensity than in the same varieties grown under lower light intensity. In general, the lower the light intensity, the lower the ascorbic acid of plant tissues.

Although light does not play a direct role in the uptake and metabolism of mineral elements by plants, temperature influences the nutrient supply because transpiration increases with higher temperatures. Rainfall affects the water supply to the plant, which may influence composition of the harvested plant part.

**Cultural practices.** Soil type, the rootstock used for fruit trees, mulching, irrigation, and fertilization influence the water and nutrient supply to the plant, which can affect the nutritional composition of the harvested plant part. The effect of fertilizers on the vitamin content of plants is much less important than variety and climate, but their effect on mineral content is more significant. Increasing the nitrogen and/or phosphorus supply to citrus trees results in somewhat lower acidity and ascorbic acid content in citrus fruits, while increased potassium fertilization increases their acidity and ascorbic acid content.

Cultural practices such as pruning and thinning determine the crop load and fruit size, which can influence nutritional composition of fruit. The use of agricultural chemicals, such as pesticides and growth regulators, does not directly influence fruit composition but may indirectly affect it due to delayed or accelerated fruit maturity.

**Maturity at harvest.** This is one of the main factors determining compositional quality and storage life of fruits and vegetables. All fruits, with few exceptions, reach peak eating quality when fully ripened on the tree. However, since they cannot survive the postharvest handling system, they are usually picked mature but not ripe. Tomatoes harvested green and ripened at 20°C to table ripeness contain less ascorbic acid than those harvested at the table-ripe stage.

**Harvesting method.** The method of harvest can determine the variability in maturity and physical injuries and consequently influence nutritional composition of fruits and vegetables. Mechanical injuries such as bruising, surface abrasions, and cuts can accelerate loss of vitamin C. The incidence and severity of such injuries are influenced by the method of harvest, management of harvesting, and handling operations. Proper management to minimize physical damage to the commodity is a must whether harvesting is done by hand or by machine.

**Postharvest handling procedures.** Delays between harvesting and cooling or processing can result in direct losses (due to water loss and decay) and indirect losses (lowering of flavor and nutritional quality). The extent of such losses is related to the condition of the commodity when picked and is strongly influenced by the temperature of the commodity, which can be several degrees higher than ambient temperatures, especially when exposed to sunlight. Temperatures higher than those that are optimum for the commodity increase the loss rate of vitamin content, especially vitamin C. In general, vegetables have more loss of ascorbic acid content in response to elevated temperatures than do fruits that are more acidic (pH 4.0 or lower), such as citrus.

Chilling injury causes accelerated losses in ascorbic acid content of sweet potatoes, pineapples, and bananas, but it does not influence ascorbic acid content of tomatoes and guavas.

## Methods for Evaluating Quality

Quality evaluation methods can be destructive or nondestructive. They include objective scales based on instrument readings and subjective methods based on human judgment, using hedonic scales.

## Appearance quality (visual)

1. **Size**

   Dimensions: measured with sizing rings, calipers

   Weight: correlation is generally good between size and weight; size can also be expressed as numbers of units of commodity per unit of weight

   Volume: determined by water displacement or by calculation from measured dimensions

2. **Shape**

   Ratio of dimensions: such as diameter/depth ratio; used as index of shape in fruits

   Diagrams and models of shape: some commodity models are used as visual aids for quality inspectors

3. **Color**

   Uniformity and intensity: important appearance qualities

   Visual matching: color charts, guides, and dictionaries to match and describe colors of fruits and vegetables

   Light reflectance meter: measures color on basis of amount of light reflected from surface of the commodity; examples include Gardner and Hunter Color Difference Meters (tristimulus colorimeters) and Agtron E5W spectrophotometer

   Light transmission meter: measures the light transmitted through the commodity; may be used to determine internal color and various disorders, such as water core of apples and black heart of potatoes

   Measurement of delayed light emission: related to the amount of chlorophyll in the plant tissues; can be used to determine color-based maturity stages

   Determination of pigment content: evaluates the color of horticultural crops by pigment content, i.e., chlorophylls, carotenoids (carotene, lycopene, xanthophylls), and flavonoids (anthocyanins)

4. **Gloss (bloom, finish)**

   Wax platelets: amount, structure, and arrangement on the fruit surface affect the gloss quality; measured using a Gloss-meter or by visual evaluation

5. **Presence of defects (external and internal)**

   Incidence and severity of defects are evaluated using a five-grade scoring system (1 = no symptoms, 2 = slight, 3 = moderate, 4 = severe, and 5 = extreme) or a seven- or nine-point hedonic scale if more categories are needed. To reduce variability among evaluators, detailed descriptions and photographs may be used as guides in scoring a given defect. Objective evaluation of external defects using computer-aided vision techniques appears promising.

   Internal defects can be evaluated by nondestructive techniques, such as light transmission and absorption characteristics of the commodity, sonic and vibration techniques associated with the mass density and elasticity of the material, x-ray transmission which depends on mass density and mass absorption coefficient of the material, and nuclear magnetic resonance (NMR) imaging (also known as magnetic resonance imaging or MRI) which detects the concentration of hydrogen nuclei and is sensitive to variations in the concentration of free water and oil.

## Textural quality

1. **Yielding quality (firmness, softness)**

   Hand-held testers: Determine penetration force using testers such as the Magness-Taylor Pressure Tester and the Effegi penetrometer

   Stand-mounted testers: Determine penetration force using testers with a more consistent speed of punch such as the UC Fruit Firmness Tester and the Effegi penetrometer mounted on a drill stand

   Laboratory testing: fruit firmness can be determined by measuring penetration force using an Instron Universal Testing machine or a Texture Testing system, or by measuring fruit deformation using a Deformation Tester

2. **Fibrousness and toughness**

   Shear force—determined using an Instron or a Texture Testing system

   Resistance to cutting—determined by using a Fibrometer

   Chemical analysis—fiber content or lignin content

3. **Succulence and juiciness**

   Measurement of water content—an indicator of succulence or turgidity

   Measurement of extractable juice—an indicator of juiciness

4. **Sensory textural qualities**

   Sensory evaluation procedures—evaluate grittiness, crispness, mealiness, chewiness, and oiliness

## Flavor quality

1. **Sweetness**

   Sugar content—determined by chemical analysis procedures for total and reducing sugars or for individual sugars; indicator papers for quick measurement of glucose in certain commodities, such as potatoes

   Total soluble solids content—measured using refractometers or hydrometers; can be used as indicator of sweetness because sugars are major component of soluble solids

2. **Sourness (acidity)**

   pH (hydrogen ion concentration) of extracted juice—determined using a pH meter or pH indicator paper

   Total titratable acidity—determined by titrating a specific volume of the extracted juice; 0.1 N NaOH to pH 8.1, then calculating titratable acidity as citric, malic, or tartaric acid (depending on which organic acid predominates in the commodity)

3. **Saltiness**

   Fresh vegetables and fruits—usually not applicable

4. **Astringency**

   Determined by taste testing or by measuring tannin content, solubility, and degree of polymerization

5. **Bitterness**

   Determined by taste testing or measurement of the alkaloids or glucosides responsible for the bitter taste

6. **Aroma (odor)**

   Determined by sensory panels in combination with identification of volatile components responsible for specific aroma of a commodity (using gas chromatography)

7. **Sensory evaluation**

   Human subjects—judge and measure combined sensory characteristics (sweetness, sourness, astringency, bitterness, overall flavor intensity) of a commodity

   Laboratory panels—detect and describe differences among samples; determine which volatile compounds are organoleptically important in a commodity

   Consumer panels—indicate quality preferences

## Nutritional value

Various analytical methods are available to determine total carbohydrates, dietary fiber, proteins and individual amino acids, lipids and individual fatty acids, vitamins, and minerals in fruits and vegetables. Several companies are working to automate these analytical procedures for use in situations where nutritional labeling is required and large numbers of samples have to be analyzed routinely.

## Safety factors

Analytical procedures, using thin-layer chromatography, gas chromatography, and high-pressure liquid chromatography, are available for determining minute quantities of the following toxic substances:

1. Naturally occurring toxicants, such as cyanogenic glucosides in lima beans and cassava, nitrates and nitrites in leafy vegetables, oxalates in rhubarb and spinach, thioglucosides in cruciferous vegetables, and glycoalkaloids (solanine) in potatoes

2. Natural contaminants, such as fungal toxins (mycotoxins), bacterial toxins and heavy metals (mercury, cadmium, lead)

**3.** Synthetic toxicants, such as environmental contaminants and pollutants, and residues of agricultural chemicals. During the past few years, monitoring by the Food and Drug Administration (FDA) and the California Department of Food and Agriculture (CDFA) has confirmed the safety of fresh fruits and vegetables. No pesticide residues were detected on about 80 percent of the total samples taken, and residues within the legal tolerances occurred on about 19 percent of the samples. The remaining 1 percent had illegal residues (mostly pesticides on crops for which no tolerance had been established), and these were generally at very low levels that are unlikely to represent a health hazard. Continued and expanded monitoring of pesticide residues on fresh produce plus informing the consumers about the results are in the best interest of producers and handlers of such commodities.

**4.** Microbial contamination. Although consumers rank microbial hazards toward the bottom and pesticide residues near the top of the list of perceived hazards to public health, most scientists and food producers rank microbial hazards as most important and pesticide residues near the bottom. This major information gap on food safety must be addressed by an educational program that reaches the majority of consumers. Some fresh fruits and vegetables, such as those grown using unsterilized organic fertilizers (e.g. chicken manure) or irrigated with inadequately treated sewage water, can carry disease-causing bacteria (e.g. *Escherichia coli*, *Salmonella* spp., and *Listeria* spp.). Poor sanitation during preparation of fresh fruits and vegetables can also result in microbial contamination. Strict quality and safety assurance programs must be used by all handlers of fresh fruits and vegetables to minimize microbial hazards. A well-designed and properly-implemented hazard analysis critical control point (HACCP) program is highly recommended for all suppliers of lightly-processed produce.

## REFERENCES

1. Amerine, M. A., R. M. Pangborn, and E. B. Roessler. 1965. *Principles of sensory evaluation of food*. New York: Academic Press. 602 pp.

2. Arthey, V. D. 1975. *Quality of horticultural products*. New York: Halstead Press, John Wiley & Sons. 228 pp.

3. Bourne, M. C. 1980. Texture evaluation of horticultural crops. *HortScience* 15:51-57.

4. Carter, H. O., and C. F. Nuckton, eds. 1988. *Chemicals in the human food chain: sources, options, and public policy*. Symposium proceedings, Agricultural Issues Center, Univ. Calif. Davis.

5. Chen, P., M. J. McCarthy, and R. Kauten. 1989. NMR for internal quality evaluation of fruits and vegetables. *Trans. ASAE* 32:1747-53.

6. Doyle, M. P. 1990. Fruit and vegetable safety—microbiological considerations. *HortScience* 25:1478-82.

7. Dull, G. G., G. S. Birth, and J. B. Magee. 1980. Nondestructive evaluation of internal quality. *HortScience* 15:60-63.

8. Eskin, N. A. M. 1979. *Plant pigments, flavors and textures: the chemistry and biochemistry of selected compounds*. New York: Academic Press. 219 pp.

9. Francis, F. J. 1980. Color quality evaluation of horticultural crops. *HortScience* 15:58-59.

10. Gaffney, J. J., comp. 1976. *Quality detection in foods*. St. Joseph, MI: Am. Soc. Agric. Eng. Publ. 1-76. 240 pp.

11. Goddard, M. S., and R. H. Matthews. 1979. Contribution of fruits and vegetables to human nutrition. *HortScience* 14:245-47.

12. Gould, W. A. 1977. *Food quality assurance*. Westport, CT: AVI Publ. Co. 314 pp.

13. Gunasekaran, S. 1990. Delayed light emission as a means of quality evaluation of fruits and vegetables. *Crit. Rev. Food Sci. Nutr.* 29:19-34.

14. Heintz, C. M., and A. A. Kader. 1983. Procedures for the sensory evaluation of horticultural crops. *HortScience* 18:18-22.

15. Jen, J. J., ed. 1989. *Quality factors of fruits and vegetables--chemistry and technology*. Washington, DC: Am. Chem Soc. Ser. 405. 410 pp.

16. Kader, A. A. 1983. Postharvest quality maintenance of fruits and vegetables in developing countries. In *Postharvest physiology and crop preservation*, ed. M. Lieberman, 455-70. New York: Plenum.

17. ———. 1988. Influence of preharvest and postharvest environment on nutritional composition of fruits and vegetables. In *Horticulture and human health—contributions of fruits and vegetables*, ed. B. Quebedeaux and F. A. Bliss, 18-32. Englewood Cliffs, NJ: Prentice-Hall.

18. Liener, I. E. 1980. *Toxic constituents of plant foodstuffs*. New York: Academic Press. 502 pp.

19. Lipton, W. J. 1980. Interpretation of quality evaluations of horticultural crops. *HortScience* 15:64-66.

20. Mohsenin, N. N. 1970. *Physical properties of plant and animal materials: structure, physical characteristics, and mechanical properties*. New York: Gordon and Breach. 742 pp.

21. Pattee, H. E., ed. 1985. *Evaluation of quality of fruits and vegetables*. Westport, CT: AVI Publ. Co. 410 pp.

22. Sharples, R. O. 1985. The influence of preharvest conditions on the quality of stored fruits. *Acta Hort.* 157:93-104.

23. Spayd, S. E., ed. 1987. Naturally occurring toxins in horticultural food crops. *Acta Hort.* 207:1-70.

24. Stevens, M. A., and M. Albright. 1980. An approach to sensory evaluation of horticultural commodities. *HortScience* 15:48-50.

25. USDA. 1983. *Composition of foods: raw, processed, prepared*. In U.S. Dept. Agric. Handb. 8 (revised), sec. 9 (Fruits and fruit products) and sec. 10 (Vegetables and vegetable products).

26. Watada, A. E. 1980. Quality evaluation of horticultural crops—the problem. *HortScience* 15:47.

27. Williams, A. A. 1979. The evaluation of flavour quality in fruits and fruit products. In *Progress in flavour research*, eds. D. G. Land and H. E. Nursten, 287-305. Essex: Appl. Sci. Publ.

# 21

# Standardization and Inspection of Fresh Fruits and Vegetables

Adel A. Kader

Grade standards identify the degrees of quality in a commodity that are the basis of its usability and value. Such standards are valuable tools in fresh produce marketing because they (1) provide a common language for trade among growers, handlers, processors, and receivers at terminal markets, (2) help producers and handlers do better jobs of preparing and labeling fresh horticultural commodities for market, (3) provide a basis for incentive payments rewarding better quality, (4) serve as the basis for market reporting—prices and supplies quoted by the Federal-State Market News Service in different markets can only be meaningful if they are based on products of comparable quality—and (5) help settle damage claims and disputes between buyers and sellers.

## U.S. Grade Standards

U.S. standards for fresh fruit and vegetable grades are voluntary, except when required by state and local regulations, by industry marketing orders (federal or state), or for export marketing. They are also used by many private and government procurement agencies when purchasing fresh fruits and vegetables. The USDA Food Safety and Quality Service (FSQS) is responsible for developing, amending, and implementing grade standards.

The first U.S. grade standards were developed for potatoes in 1917. Currently there are more than 150 standards covering 80 different commodities. The quality factors used in these standards for fresh fruits, vegetables, and tree nuts are summarized in tables 21.1 to 21.5 at the end of this chapter.

The number of grades and grade names included in the U.S. standards for a given commodity vary with the number of distinct quality gradations that the industry normally recognizes and with the established usage of grade names. Currently, grades include three or more of the following: U.S. Fancy, U.S. No. 1, U.S. No. 2, U.S. No. 3, U.S. Extra No. 1, U.S. Extra Fancy, U.S. Combination, U.S. Commercial, and so on. The FSQS is gradually phasing in the first four grades as uniform grades for all fresh fruits and vegetables, to represent available levels of quality.

Steps to establish or change U.S. standards include:

1. Demonstration of need, interest, and support from the industry.

2. Study of physical characteristics, quality factors, and their normal ranges for the commodity in the main production areas.

3. Consultation among all interested parties as part of data collection.

4. Development of a proposal that is practical to use.

5. Publication of the proposal in the *Federal Register*, and publicizing of it through various means with an invitation for comments. Public hearings may be held for the same purpose.

6. Amendment of the proposal on the basis of comments received from interested parties.

7. Publication of the standards in their final form in the *Federal Register* with a specified date on which they become effective (at least 30 days after publication date).

### Applying the standards

USDA inspectors are located at most shipping points and at terminal markets. In many cases cooperative agreements between the USDA and the states are in place to allow federal-state grading by USDA-licensed state inspectors. Some inspectors are full-time employees, while others are seasonal employees hired during the peak production season in a given location.

### Methods of inspection include:

1. **Continuous inspection.** One or more inspectors are assigned to a packinghouse. They make frequent quality checks on the commodity along the packing lines and examine samples of the packed product to determine whether it meets the U.S. grade specifications for which it is being packed. The inspector gives oral and/or written reports to management so that they can correct problems.

2. **Inspection on a sample basis.** Representative samples of a prescribed number of boxes out of a given lot are randomly selected and inspected to determine the quality and condition of the commodity according to grade specifications. Automatic sampling systems are used for some commodities that are handled in bulk bins or trailers, such as tomatoes, grapes, and cling peaches destined for processing.

When inspection is completed, certificates are issued by the inspector on the basis of the applicable official standards. USDA inspectors can also inspect quality or condition based on a state grade or other specifications agreed upon by the parties involved. The cost of inspection is paid by the party requesting the service.

Each grade allows for a percentage of individual units within a lot that do not meet the standard. This reflects the practical limitations in sorting perishable products accurately into grades within a limited time. Tolerances, or the number of defects allowed, are more restrictive in U.S. No. 1 grade than in U.S. No. 2. The penalty for noncompliance with the U.S. grade specified on a given container may be rejection, resorting and repacking, or reclassification to a lower grade.

To ensure uniformity of inspection, (1) inspectors are trained to apply the standards, (2) visual aids (color charts, models, diagrams, photographs, and the like) are used whenever possible, (3) objective methods for determining quality and maturity are used whenever feasible and practical, and (4) good working environments with proper lighting are provided.

## California Standards

California is one of the few states having quality standards for horticultural crops produced within the state. The standards for fresh fruits and vegetables in the California Agricultural Code (summarized in tables 21.1 to 21.5) are mandatory minimum standards enforced by the California Department of Food and Agriculture (CDFA) Division of Inspection Services, Fruit and Vegetable Quality Control, through each county agricultural commissioner's staff. The cost of this inspection is paid by taxpayers. Noncompliance results in destruction of the commodity or its resorting and repacking to meet minimum requirements.

Steps for establishing new standards or revising existing ones are similar to those mentioned above for U.S. grade standards, except that they are carried out at the state level by the same agency resposible for inspection. Uniformity of inspection is assured by methods similar to those mentioned above for U.S. grade standards.

## Industry Standards

Some industries establish their own quality standards or specifications for a given commodity; examples include apricots, clingstone peaches, processing tomatoes, and walnuts. The standards are established by agreement between producers and processors who pay application costs. Inspection is performed by such independent agencies as the California Dried Fruit Association and the Federal-State Inspection Service.

Some companies, cooperatives, and other organizations have quality grades that are applied by their quality-control personnel. Examples include quality grades for bananas, papayas, pineapples, and lightly-processed fruits and vegetables.

## International Standards

International standards for fruits and vegetables were defined by the European Economic Commission (EEC) in 1954. Many standards have since been introduced, mainly under the Organization for Economic Cooperation and Development (OECD) scheme drawn up for this purpose. The first European International Standards were promulgated in 1961 for apples and pears, and now there are standards for 37 commodities. Each includes three quality classes with appropriate tolerances: Extra class = superior quality; Class I = good quality; and Class II = marketable quality. Class I covers the bulk of produce entering into international trade. These standards or their equivalents are mandatory in EEC countries for imported and exported fresh fruits and vegetables. Inspection and certification is done by exporting and/or importing EEC countries.

### REFERENCES

1. California Department of Food and Agriculture. 1983. *Fruit and vegetable quality control standardization*. Extracts from the Administrative Code of California. Sacramento: Dept. Food Agric. 154 pp.

2. Organization for Economic Cooperation and Development. Various dates. *International standardization of fruits and vegetables*. Paris: Org. Econ. Coop. Develop.

3. USDA. 1973. *USDA standards for food and farm products*. U.S. Dept. Agric. Handb. 341. 12 pp.

4. ———. Various dates. *U.S. standards for grades of fresh fruits and vegetables*. U.S. Dept. Agric. Food Safety and Quality Service.

**Table 21.1. Quality factors for fresh fruits in the U.S. Standards for grades (US)
and the California Food and Agricultural Code (CA)**

| Fruit | Standard (date*) | Quality factors |
|---|---|---|
| Apple | US (1976) | Maturity, color (color charts) related to grade, firmness, shape, and size, and freedom from decay, internal browning, internal breakdown, scald, scab, bitter pit, Jonathan spot, freezing injury, water core, bruises, russeting, scars, insect damage, and other defects |
| | CA (1983) | Maturity (as determined by soluble solids content [SSC] and firmness tests) |

| Cultivar | SSC (%) | Firmness (lb) |
|---|---|---|
| Red Delicious | 11.0 | 18 |
| Golden Delicious | 12.0 | 18 |
| Jonathan | 12.0 | 19 |
| Rome | 12.5 | 21 |
| Newtown Pippin | 11.0 | 23 |
| McIntosh | 11.5 | 19 |
| Gravenstein | 10.5 | — |

| Fruit | Standard (date*) | Quality factors |
|---|---|---|
| | | Size, color, flesh condition, freedom from defect (such as scald, spot, internal breakdown, water core, bruises, sunburn, russeting) and decay |
| Apricot | US (1928) | Maturity, size, and shape, and freedom from defect and decay |
| | CA (1983) | Maturity (>¾ of external surface area has a color equal to No. 3 yellowish green of the CDFA standard color chart or at least one-half has attained No. 4 yellow) and freedom from insect injury, decay, and mechanical damage |
| Avocado | US (1957) | For Florida avocados: maturity, shape, texture, skin and flesh color, and freedom from decay, anthracnose, freezing injury, bruises, russeting, scars, sunburn, mechanical damage, and other defects |
| | CA (1983) | Maturity (17 to 20.5 percent dry weight of the flesh depending on cultivar), size, and freedom from defect, insect damage, freezing injury, rancidity, and decay |
| Blueberry | US (1966) | Maturity, color, size, and freedom from defect and decay |
| Cherry, sweet | US (1971) | Maturity, color, size, shape, and freedom from cracks, hail damage, russeting, scars, insect damage, and decay |
| | CA (1983) | Maturity (entire surface with at least a solid light red color and/or 14 to 16 percent soluble solids depending on the cultivar), and freedom from bird pecks, insect injury, shriveling, growth cracks, other defects, and decay |
| Citrus Grapefruit | US (1950) | California and Arizona: maturity, color, firmness, size, shape, skin thickness, smoothness, and freedom from defect and decay |
| | US (1980) | Florida: maturity, color (color charts), firmness, size, smoothness, shape, and freedom from discoloration, defect, and decay |
| | US (1969) | Texas and other states: maturity, color, firmness, size, shape, smoothness, and freedom from discoloration, defect, and decay |
| | CA (1983) | Maturity (minimum soluble solids/acid ratio of 5.5 or 6 [desert areas] and >⅔ of fruit surface showing yellow color—0.9 GY 6.40/5.7 Munsell color) and freedom from decay, freezing damage, scars, pitting, rind staining, and insect damage |
| Lemon | US (1964) | Maturity (28 or 30 percent minimum juice content by volume depending on grade), firmness, shape, color, size, smoothness, and freedom from discoloration, defect, and decay |
| | CA (1983) | Maturity (30 percent or more juice by volume), size uniformity, and freedom from decay, freezing damage, drying, mechanical damage, rind stains, red blotch, shriveling, and other defects |
| Lime | US (1958) | Color, shape, firmness, smoothness, and freedom from stylar end breakdown, bruises, dryness, other defects, and decay |
| | CA (1983) | Maturity, and freedom from defect (freezing injury, drying, mechanical damage) and decay |
| Orange | US (1957) | California and Arizona: maturity, color, firmness, smoothness, size, and freedom from defect and decay |
| | US (1980) | Florida: maturity, color (color charts), firmness, size, shape, and freedom from discoloration, defect, and decay (used also for tangelos) |
| | US (1969) | Texas and other states: maturity, color, firmness, shape, size, and freedom from discoloration, defect, and decay |
| | CA (1983) | Maturity (soluble solids/acid ratio of 8 or higher and orange color on 25 percent of the fruit—7.5 Y 6/6 Munsell color—or soluble solids/acid ratio of 10 or higher and orange color on 25 percent of fruit—2.5 GY 5/6 Munsell color), size uniformity, and freedom from defect and decay |

*Date when standard was issued or revised.

(continued)

**Table 21.1. Quality factors for fresh fruits in the U.S. Standards for grades (US)
and the California Food and Agricultural Code (CA)** (*continued*)

| Fruit | Standard (date*) | Quality factors |
|---|---|---|
| Tangerine and mandarin | US (1948) | States other than Florida: maturity, firmness, color, size, and freedom from defect and decay |
| | US (1980) | Florida: maturity, color (color charts), firmness, size, shape, and freedom from decay and defect |
| | CA (1983) | Maturity (yellow, orange, or red color on 75 percent of fruit surface and soluble solids/acid ratio of 6.5 or higher), size uniformity, and freedom from defect and decay |
| Cranberry | US (1971) | Maturity, firmness, color, and freedom from bruises, freezing injury, scars, sunscald, insect damage, and decay |
| Date | CA (1983) | Freedom from insect damage, decay, black scald, fermentation, and other defects |
| Dewberry, blackberry | US (1928) | Maturity, color, and freedom from calyxes, decay, shriveling, mechanical damage, insect damage, and other defects |
| | CA (1983) | Maturity and freedom from decay and damage due to frost, bruising, insects, or other causes |
| Grape, table European *Vinifera* type | US (1977) | Maturity (as determined by percent soluble solids as set forth by the producing states); for states other than California and Arizona, and countries exporting to U.S.: |

| Cultivar | Minimum SSC (%) |
|---|---|
| Muscat | 17.5 |
| Cardinal, Ribier, Olivette, Blanche, Emperor, Perlette, Rish Baba, Red Malaga, and similar cultivars | 15.5 |
| All other cultivars | 16.5 |

Color, uniformity, firmness, berry size, and freedom from shriveling, shattering, sunburn, waterberry, shot berries, dried berries, other defects, and decay; bunches: fairly well filled but not excessively tight; stems: not dry and brittle, and at least yellowish-green in color

| Fruit | Standard (date*) | Quality factors |
|---|---|---|
| | CA (1983) | Maturity (minimum percent soluble solids of 14 to 17.5, depending on cultivar and production area, or soluble solids/acid ratio of 20 or higher, or a combination of a minimum soluble solids/acid ratio and percent soluble solids), and freedom from decay, freezing injury, sunburned or dried berries, and insect damage (same for Arizona) |
| American bunch type | US (1983) | Maturity (juiciness, ease of separation of skin from pulp), color, firmness, compactness, and freedom from defect and decay |
| Kiwifruit | US (1986) | Maturity (more than 6.5 percent soluble solids), firmness, cleanness, and freedom from growth cracks, insect injury, broken skin, bruises, scars, sunscald, freezing injury, internal breakdown, and decay |
| Nectarine | US (1966) | Maturity, color depending on variety, shape, and size, and freedom from growth cracks, insect damage, scars, bruises, russeting, split pits, other defects, and decay |
| | CA (1983) | Maturity (surface ground color, fruit shape), and freedom from insect injury, split pits, mechanical damage, and decay |
| Olive | CA (1983) | Freedom from insect injury, especially scale |
| Peach | US (1952) | Maturity (shape, size, ground color), and freedom from decay and defect (split pit, hail injury, insect damage, growth cracks) |
| | CA (1983) | Maturity (skin and flesh color, and fullness of shoulders and suture), and freedom from defect and decay |
| Pear Winter | US (1955) | Maturity (color, firmness), size, and freedom from internal breakdown, black end, russeting, other defects, and decay |
| Summer and fall | US (1955) | Maturity (color, firmness), shape, size, and freedom from defect and decay |
| | CA (1983) | Maturity (Bartlett: average firmness test of >23 lb, and/or soluble solids content 13 percent, and/or yellowish-green color . . . CDFA color chart), and freedom from insect damage, mechanical damage, decay, and other defects |
| Persimmon | CA (1983) | Maturity as indicated by surface color: Hachiya: blossom end's color is orange or reddish color equal to or darker than Munsell color 6.7 YR 5.93/12.7 on at least ⅓ of the fruit's length with the remaining ⅔ a green color equal to or lighter than Munsell color 2.5 GY 5/6; other cultivars: yellowish-green color equal to or lighter than Munsell color l0 Y 6/6; freedom from growth cracks, mechanical damage, decay, and other defects |
| Pineapple | US (1953) | Maturity, firmness, uniformity of size and shape, and freedom from decay, sunscald, bruising, insect damage, and cracks; tops: color, length, and straightness |
| Plum and fresh prune | US (1973) | Maturity, color, shape, size, and freedom from decay, sunscald, split pits, hail damage, mechanical damage, scars, russeting, and other defects |
| | CA (1983) | Maturity as indicated by surface color (minimum color requirements are described for 56 cultivars), and freedom from decay, insect damage, bruises, sunburn, hail damage, gum spot, growth cracks, and other defects |

*Date when standard was issued or revised.

(*continued*)

**Table 21.1. Quality factors for fresh fruits in the U.S. Standards for grades (US)
and the California Food and Agricultural Code (CA)** (*continued*)

| Fruit | Standard (date*) | Quality factors |
|---|---|---|
| Pomegranate | CA (1983) | Maturity (>1.85 percent acid content in juice and red juice color equal to or darker than Munsell color 5 R 5/12), freedom from sunburn, growth cracks, cuts or bruises, and decay |
| Quince | CA (1983) | Maturity, and freedom from insect damage, mechanical damage, and decay |
| Raspberry | US (1931) | Maturity, color, shape, and freedom from defect and decay |
| | CA (1983) | Maturity, and freedom from decay and damage due to insects, sun, frost, bruising, or other causes |
| Strawberry | US (1965) | Maturity (>½ or >¾ of surface showing red or pink color, depending on grade), firmness, attached calyx, size, and freedom from defect and decay |
| | CA (1983) | Maturity (>⅔ of fruit surface showing a pink or red color), and freedom from defect and decay |

*Date when standard was issued or revised.

**Table 21.2. Quality factors for fresh vegetables in the U.S. Standards for grades (US)
and the California food and Agricultural code (CA)**

| Vegetable | Standard (date*) | Quality factors |
|---|---|---|
| Anise, sweet | US (1973) | Firmness, tenderness, trimming, blanching, and freedom from decay and damage caused by growth cracks, pithy branches, wilting, freezing, seedstems, insects, and mechanical means |
| Artichoke | US (1969) | Stem length, shape, overmaturity, uniformity of size, compactness, and freedom from decay and defects |
| | CA (1983) | Freedom from decay, insect damage, and freezing injury |
| Asparagus | US (1966) | Freshness (turgidity), trimming, straightness, freedom from damage and decay, diameter of stalks, percent green color |
| | CA (1983) | Turgidity, straightness, percent showing white color, stalk diameter, and freedom from decay, mechanical damage, and insect injury |
| Bean, lima | US (1938) | Uniformity, maturity, freshness, shape, and freedom from damage (defect) and decay |
| Bean, snap | US (1936) | Uniformity, size, maturity, freshness (firmness), and freedom from defect and decay |
| Beet, bunched or topped | US (1955) | Root shape, trimming of rootlets, firmness (turgidity), smoothness, cleanness, minimum size (diameter), and freedom from defect |
| Beet, greens | US (1959) | Freshness, cleanness, tenderness, and freedom from decay, other kinds of leaves, discoloration, insects, mechanical injury, and freezing injury |
| Broccoli | US (1943) | Color, maturity, stalk diameter and length, compactness, base cut, and freedom from defects and decay |
| | CA (1983) | Freedom from decay and damage due to overmaturity, insects, or other causes |
| Brussels sprouts | US (1954) | Color, maturity (firmness), no seedstems, size (diameter and length), and freedom from defect and decay |
| | CA (1983) | Freedom from decay, from burst, soft, or spongy heads, and from insect damage |
| Cabbage | US (1945) | Uniformity, solidity (maturity or firmness), no seedstems, trimming, color, and freedom from defect and decay |
| | CA (1983) | Conform to U.S. commercial grade or better |
| Cantaloupe | US (1968) | Soluble solids (>9 percent), uniformity of size, shape, ground color and netting; maturity and turgidity; and freedom from "wet slip," sunscald, and other defects |
| | CA (1983) | Maturity (soluble solids >8 percent), and freedom from insect injury, bruises, sunburn, growth cracks, and decay |
| Carrot, bunched | US (1954) | Shape, color, cleanness, smoothness, freedom from defect, freshness, length of tops, and root diameter |
| | CA (1983) | Number, size, and weight per bunch, freshness, and freedom from defect and decay (tops) |
| Carrot, topped | US (1965) | Uniformity, turgidity, color, shape, size, cleanness, smoothness, and freedom from defect (growth cracks, pithiness, woodiness, internal discoloration) |
| | CA (1983) | Freedom from defect (growth cracks, doubles, mechanical injury, green discoloration, objectionable flavor or odor) and decay |

*Date when standard was issued or revised.

(*continued*)

**Table 21.2. Quality factors for fresh vegetables in the U.S. Standards for grades (US) and the California food and Agricultural code (CA)** (*continued*)

| Vegetable | Standard (date*) | Quality factors |
|---|---|---|
| Carrots with short trimmed tops | US (1954) | Roots: firmness, color, smoothness, and freedom from defect (sunburn, pithiness, woodiness, internal discoloration, and insect and mechanical injuries) and decay; leaves: (cut to <4 inches) freedom from yellowing or other discoloration, disease, insects, and seedstems |
| Cauliflower | US (1968) | Curd cleanness, compactness, white color, size (diameter), freshness and trimming of jacket leaves, and freedom from defect and decay |
| | CA (1983) | Freedom from insect injury, decay, freezing injury, and sunburn |
| Celery | US (1959) | Stalk form, compactness, color, trimming, length of stalk and midribs, width and thickness of midribs, no seedstems, and freedom from defect and decay |
| | CA (1983) | Freedom from pink rot and other decay, blackheart, seedstems, pithy condition, and insect damage |
| Collard greens and broccoli greens | US (1953) | Freshness, tenderness, cleanness, and freedom from seedstems, discoloration, freezing injury, insects, and diseases |
| Corn, green | US (1954) | Uniformity of color and size, freshness, milky kernels, cob length, freedom from defect, coverage with fresh husks |
| | CA (1983) | Milky, plump, well-developed kernels, and freedom from insect injury, mechanical damage, and decay |
| Cucumber | US (1958) | Color, shape, turgidity, maturity, size (diameter and length), and freedom from defect and decay |
| Cucumber, greenhouse | US (1985) | Freshness, shape, firmness, color, size (length of 11 inches or longer), and freedom from decay, cuts, bruises, scars, insect injury and other defects |
| Dandelion greens | US (1955) | Freshness, cleanness, tenderness, and freedom from damage caused by seed stems, discoloration, freezing, diseases, insects, and mechanical injury |
| Eggplant | US (1953) | Color, turgidity, shape, size, and freedom from defect and decay |
| Endive, escarole, or chicory | US (1964) | Freshness, trimming, color (blanching), no seed stems, and freedom from defect and decay |
| Garlic | US (1944) | Maturity, curing, compactness, well-filled cloves, bulb size, and freedom from defect |
| | CA (1983) | Size (bulb diameter) |
| Honeydew and honey ball melons | US (1967) | Maturity, firmness, shape, and freedom from decay and defect (sunburn, bruising, hail spots, and mechanical injuries) |
| | CA (1983) | Maturity, soluble solids (>10 percent), and freedom from decay, sunscald, bruises, and growth cracks; honey ball melons should be netted and should have pink flesh |
| Horseradish roots | US (1936) | Uniformity of shape and size, firmness, smoothness, and freedom from hollow heart, other defects, and decay |
| Kale | US (1934) | Uniformity of growth and color, trimming, freshness, and freedom from defect and decay |
| Lettuce, crisp-head | US (1975) | Turgidity, color, maturity (firmness), trimming (number of wrapper leaves), and freedom from tip burn, other physiological disorders, mechanical damage, seedstems, other defects, and decay |
| | CA (1983) | Freedom from insect damage, decay, seedstems, tipburn, freezing injury, broken midribs, and bursting; for sectioned, chopped, or shredded lettuce: same as intact heads plus freedom from discoloration and excessive moisture |
| Lettuce, greenhouse leaf | US (1964) | Well-developed, well-trimmed, and freedom from coarse stems, bleached or discolored leaf leaves, wilting, freezing, insects, and decay |
| Lettuce, romaine | US (1960) | Freshness, trimming, and freedom from decay and damage caused by seedstems, broken, bruised, or discolored leaves, tipburn, and wilting |
| Melon, casaba and Persian | CA (1983) | Maturity, and freedom from growth cracks, decay, mechanical injury, and sunburn |
| Mushroom | US (1966) | Maturity, shape, trimming, size, and freedom from open veils, disease, spots, insect injury, and decay |
| | CA (1983) | Freedom from insect injury |
| Mustard greens and turnip greens | US (1953) | Freshness, tenderness, cleanness, and freedom from damage caused by seedstems, discoloration, freezing, disease, insects, or mechanical means; roots (if attached): firmness and freedom from damage |
| Okra | US (1928) | Freshness, uniformity of shape and color, and freedom from defect and decay |

*Date when standard was issued or revised.

(*continued*)

**Table 21.2. Quality factors for fresh vegetables in the U.S. Standards for grades (US) and the California food and Agricultural code (CA)** (*continued*)

| Vegetable | Standard (date*) | Quality factors |
|---|---|---|
| Onion, dry<br>  Creole<br>  Bermuda-Granex-Grano<br>  Other cultivars | US (1943)<br>US (1985)<br>US (1971) | Maturity, firmness, shape, size (diameter), and freedom from decay, wet sunscald, doubles, bottlenecks, sprouting, and other defects |
| Onion, dry | CA (1983) | Freedom from insect injury, decay, sunscald, freezing injury, sprouting, and other defects |
| Onion, green | US (1947) | Turgidity, color, form, cleanness, bulb trimming, no seedstems, and freedom from defect and decay |
| Onion sets | US (1940) | Maturity, firmness, size, and freedom from decay and damage caused by tops, sprouting, freezing, mold, moisture, dirt, disease, insects, or mechanical means |
| Parsley | US (1930) | Freshness, green color, and freedom from defects, seedstems, and decay |
| Parsnip | US (1945) | Turgidity, trimming, cleanness, smoothness, shape, freedom from defects and decay, and size (diameter) |
| Pea, fresh | US (1942) | Maturity, size, shape, freshness, and freedom from defects and decay |
|  | CA (1983) | Maturity, and freedom from mechanical damage, insect damage, decay, yellowing, and shriveling |
| Pea, Southern (cowpea) | US (1956) | Maturity, pod shape, and freedom from discoloration and other defects |
| Pepper, sweet | US (1963) | Maturity, color, shape, size, and freedom from defects (sunscald, freezing injury, hail, scars, insects, mechanical damage) and decay |
|  | CA (1983) | Freedom from insect damage and decay |
| Potato | US (1972) | Uniformity, maturity, firmness, cleanness, shape, size, and freedom from sprouts, blackheart, greening, and other defects |
|  | CA (1983) | A minimum equivalent of U.S. No. 2 grade; maturity is described in terms of extent of skin missing or feathered |
| Radish (topped) | US (1968) | Tenderness, cleanness, smoothness, shape, size, and freedom from pithiness and other defects |
| Rhubarb | US (1966) | Color, freshness, straightness, trimming, cleanness, stalk diameter and length, and freedom from defect |
| Shallot, bunched | US (1946) | Firmness, form, tenderness, trimming, cleanness, and freedom from decay and damage caused by seed stems, disease, insects, mechanical and other means; tops: freshness, green color, and no mechanical damage |
| Spinach bunches | US (1987) | Freshness, cleanness, trimming, and freedom from decay and damage caused by coarse stalks or seedstems, discoloration, insects, and mechanical means |
| Spinach leaves | US (1946) | Color, turgidity, cleanness, trimming, and freedom from seedstems, coarse stalks, and other defects |
| Squash, summer | US (1984) | Immaturity, tenderness, shape, firmness, and freedom from decay, cuts, bruises, scars, and other defects |
| Squash, winter and pumpkin | US (1983) | Maturity, firmness, freedom from discoloration, cracking, dry rot, insect damage, and other defects; uniformity of size |
| Sweet potato | US (1963) | Firmness, smoothness, cleanness, shape, size, and freedom from mechanical damage, growth cracks, internal breakdown, insect damage, other defects, and decay |
|  | CA (1983) | Freedom from decay, mechanical damage, insect injury, growth cracks, and freezing injury |
| Tomato | US (1976) | Maturity and ripeness (color chart), firmness, shape, size, and freedom from defect (puffiness, freezing injury, sunscald, scars, catfaces, growth cracks, insect injury, and other defects) and decay |
|  | CA (1983) | Freedom from insect and freezing damage, sunburn, mechanical damage, blossom-end rot, catfaces, growth cracks, and other defects |
| Tomato, greenhouse | US (1966) | Maturity, firmness, shape, size, and freedom from decay, sunscald, freezing injury, bruises, cuts, shriveling, puffiness, catfaces, growth cracks, scars, disease, and insects |
| Turnip and rutabaga | US (1955) | Uniformity of root color, size, and shape, trimming, freshness, and freedom from defects (cuts, growth cracks, pithiness, woodiness, water core, dry rot) |
| Watermelon | US (1978) | Maturity and ripeness (optional internal quality criteria: soluble solids content = >10 percent very good, >8 percent good), shape, uniformity of size (weight), and freedom from anthracnose, decay, sunscald, and whiteheart |
|  | CA (1983) | Maturity (arils around the seeds have been absorbed and flesh color is >75 percent red), and freedom from decay, sunburn, flesh discoloration, and mechanical damage |

*Date when standard was issued or revised.

**Table 21.3. Quality factors for processing fruits in the U.S. Standards for grades (US)
and the California Food and Agricultural Code (CA)**

| Fruit | Standard (date*) | Quality factors |
|---|---|---|
| Apple | US (1961) | Ripeness (not overripe, mealy, or soft), and freedom from decay, worm holes, freezing injury, internal breakdown, and other defects that would cause a loss of >5 percent (U.S. No. 1) or >12 percent (U.S. No. 2) by weight |
| Berries | US (1947) | Color, and freedom from caps (calyxes), decay, and defect (dried, undeveloped and immature berries, crushing, shriveling, sunscald, insect damage, and mechanical injury) |
| Blueberry | US (1950) | Freedom from other kinds of berries, clusters, large stems, leaves and other foreign material, and freedom from damage caused by decay, shriveling, dirt, overmaturity, or other means |
| Cherry, red sour | US (1941) | Color uniformity, and freedom from decay, pulled pits, attached stems, hail marks, windwhips, scars, sunscald, shriveling, disease, and insect damage |
| Cherry, sweet for canning or freezing | US (1946) | Maturity, shape, freedom from decay, worms, pulled pits, doubles, insect and bird damage, and mechanical injury, and freedom from damage caused by freezing softness, shriveling, cracks and skin breaks, scars, and sunscald; tolerance is 7 percent (U.S. No. 1) or 12 percent (U.S. No. 2) by count |
| Cherry, sweet for sulfur brining | US (1940) | Maturity (ease of pit separation), firmness, shape, and freedom from decay and defect (bruises, bird and insect damage, skin breaks, russeting, shriveling, scars, sunscald, and limb rubs) |
| Cranberry, red sour | US (1957) | Maturity, color, firmness, size, and freedom from defect (insect damage, bruises, scars, sunscald, freezing injury, and mechanical injury) and decay |
| Currant | US (1952) | Color, stem attached, and freedom from decay and damage caused by crushing, drying, shriveling, insects, and mechanical means |
| Grape, American type for processing and freezing | US (1975) | Maturity (>15.5 % soluble solids), color, freedom from shattered, split, crushed, or wet berries, and freedom from decay and from damage caused by freezing, heat, sunburn, disease, insects, or other means |
| Grape, juice (European or *Vinifera* type) | US (1939) | Maturity (>16 to 18 % soluble solids depending on cultivar), freedom from crushed, split, wet, waterberry and redberry, and freedom from defect (insect, disease, mechanical injury, sunburn, and freezing damage) |
|  | CA (1983) | Maturity (minimum soluble solids content of 14 to 17.5 percent depending on cultivar or soluble solids/acid ratio of 20 or higher), and freedom from decay, freezing injury, waterberry, redberry, and other defects |
| Grape for processing and freezing | US (1977) | Maturity (>15.5 percent soluble solids content), and freedom from decay and defect (dried berries, discoloration, sunburn, insect damage, and immature berries) |
| Peach, freestone for canning, freezing, or pulping | US (1966) | Maturity, color (not greener than yellowish green), shape, firmness, and freedom from decay, worms and worm holes, split pits, scab, bacterial spot, insects, and bruises; grade is based on the severity of defects with 10 percent tolerance |
| Pear for processing | US (1970) | Maturity, color (less than yellowish green), shape, firmness, and freedom from scald, hard end, black end, internal breakdown, decay, worms and worm holes, scars, sunburn, bruises, and other defects; grade is based on the severity of defects with 10 percent tolerance |
| Raspberry | US (1952) | Color, and freedom from decay and defect (dried berries, crushing, shriveling, sunscald, scars, bird and insect damage, discoloration, or mechanical injury) |
| Strawberry, growers' stock for manufacture | US (1935) | Color, freedom from decay and defect (crushed, split, dried or undeveloped berries, sunscald, and bird or insect damage), size, and cap removal |
| Strawberry, washed and sorted for freezing | US (1935) | Color, cleanness, size, cap removal, and freedom from decay and defect (crushed, split, dried or undeveloped berries, bird and insect damage, mechanical injury) |

*Date when standard was issued or revised.

**Table 21.4. Quality factors for processing vegetables in the U.S. Standards for grades (US)**

| Vegetable | Standard (date*) | Quality factors |
|---|---|---|
| Asparagus, green | US (1972) | Freshness, shape, green color, size (spear length), and freedom from defect (freezing damage, dirt, disease, insect injury, and mechanical injuries) and decay |
| Bean, shelled lima | US (1953) | Tenderness, green color, and freedom from decay and from injury caused by discoloration, shriveling, sunscald, freezing, heating, disease, insects, or other means |
| Bean, snap | US (1985) | Freshness, tenderness, shape, size, and freedom from decay and from damage caused by scars, rust, disease, insects, bruises, punctures, broken ends, or other means |
| Beet | US (1945) | Firmness, tenderness, shape, size, and freedom from soft rot, cull material, growth cracks, internal discoloration, white zoning, rodent damage, disease, insects, and mechanical injury |
| Broccoli | US (1959) | Freshness, tenderness, green color, compactness, trimming, and freedom from decay and damage caused by discoloration, freezing, pithiness, scars, dirt, or mechanical means |
| Cabbage | US (1944) | Firmness, trimming, and freedom from soft rot, seedstems, and from damage caused by bursting, discoloration, freezing, disease, birds, insects, or mechanical or other means |
| Carrot | US (1984) | Firmness, color, shape, size (root length), smoothness, not woody, and freedom from soft rot, cull material, and from damage caused by growth cracks, sunburn, green core, pithy core, water core, internal discoloration, disease, or mechanical means |
| Cauliflower | US (1959) | Freshness, compactness, color, and freedom from jacket leaves, stalks, and other cull material, decay, and damage caused by discoloration, bruising, fuzziness, enlarged bracts, dirt, freezing, hail, or mechanical means |
| Corn, sweet | US (1962) | Maturity, freshness, and freedom from damage by freezing, insects, birds, disease, cross-pollination, or fermentation |
| Cucumber, pickling | US (1936) | Color, shape, freshness, firmness, maturity, and freedom from decay and from damage caused by dirt, freezing, sunburn, disease, insects, or mechanical or other means |
| Mushroom | US (1964) | Freshness, firmness, shape, and freedom from decay, disease spots, and insects, and from damage caused by insects, bruising, discoloration, or feathering |
| Okra | US (1965) | Freshness, tenderness, color, shape, and freedom from decay and insects, and from damage caused by scars, bruises, cuts, punctures, discoloration, dirt, or other means |
| Onion | US (1944) | Maturity, firmness, and freedom from decay, sprouts, bottlenecks, scallions, seedstems, sunscald, roots, insects, and mechanical injury |
| Pea, fresh shelled for canning/freezing | US (1946) | Tenderness, succulence, color, and freedom from decay, scald, rust, shriveling, heating, disease, and insects |
| Pea, Southern | US (1965) | Pods: maturity, freshness, and freedom from decay; seeds: freedom from scars, insects, decay, discoloration, splits, cracked skin, and other defects |
| Pepper, sweet | US (1948) | Firmness, color, shape, and freedom from decay, insects, and damage by any means that results in 5 to 20 percent trimming (by weight) depending on grade |
| Potato | US (1983) | Shape, smoothness, freedom from decay and defect (freezing injury, blackheart, sprouts), size, specific gravity, glucose content, and fry color |
| Potato for chipping | US (1978) | Firmness, cleanness, shape, freedom from defect (freezing, blackheart, decay, insect injury, and mechanical injury), size; optional tests for specific gravity and fry color are included |
| Spinach | US (1956) | Freshness, freedom from decay, grass weeds, and other foreign material, and freedom from damage caused by seedstems, discoloration, coarse stalks, insects, dirt, or mechanical means |
| Sweet potato for canning/freezing | US (1959) | Firmness, shape, color, size, and freedom from decay and defect (freezing injury, scald, cork, internal discoloration, bruises, cuts, growth cracks, pithiness, stringiness, and insect injury) |
| Sweet potato for dicing/pulping | US (1951) | Firmness, shape, size, and freedom from decay and defect (scald, freezing injury, cork, internal discoloration, pithiness, growth cracks, insect damage, and stringiness) |
| Tomato | US (1983) | Firmness, ripeness (color as determined by a photoelectric instrument), and freedom from insect damage, freezing, mechanical damage, decay, growth cracks, sunscald, gray wall, and blossom-end rot |
| Tomato, green | US (1950) | Firmness, color (green), and freedom from decay and defect (growth cracks, scars, catfaces, sunscald, disease, insects, or mechanical damage) |
| Tomato, Italian type for canning | US (1957) | Firmness, color uniformity, and freedom from decay and defect (growth cracks, sunscald, freezing, disease, insects, or mechanical injury) |

*Date when standard was issued or revised.

**Table 21.5. Quality factors for tree nuts in the U.S. Standards for grades (US)**
**and the California Food and Agricultural Code (CA)**

| Nut | Standard (date*) | Quality factors |
|---|---|---|
| Almond, shelled | US (1960) | Similar varietal characteristics (shape, appearance), size (count per ounce), degree of dryness, cleanness (freedom from dust, particles, and foreign materials), and freedom from decay and defect (rancidity, insect injury, doubles, split or broken kernels, shriveling, brown spot, or gumminess) |
| Almond, in-shell | US (1964) | Shell: similar varietal characteristics (shape, hardness), cleanness (freedom from loose extraneous and foreign materials), size (thickness), brightness and uniformity of color, and freedom from discoloration, insect infestation, adhering hulls, and broken shells; kernel: degree of dryness, and freedom from decay and defect (rancidity, insect damage, shriveling, brown spot, guminess, and skin discoloration) |
| Brazil nut, in-shell | US (1966) | Shell: degree of dryness, cleanness (freedom from dirt, extraneous, and adhering foreign materials), size (diameter), and freedom from damage caused by splits, breaks, punctures, oil stains, and mold; kernel: degree of development (must fill more than 50% of the shell capacity), freedom from decay and defect (rancidity, insect damage, and discoloration) |
| Filbert, in-shell | US (1970) | Shell: shape, size (diameter), cleanness, brightness, and freedom from defect (blanks, broken or split shells, stains, and adhering husk); kernel: degree of dryness (less than 10% moisture content), development (must fill more than 50% of the shell capacity), shape, and freedom from decay and defect (insect injury, shriveling, rancidity, and discoloration) |
| Mixed nuts, in-shell | US (1981) | Each species of nut must conform to a minimum size and grade (same quality criteria used for that species); grade of the mix is also determined by percent allowable for each component (almonds, brazils, filberts, pecans, walnuts) |
| Pecan, shelled | US (1969) | Degree of dryness, degree of development (amount of meat in proportion to width and length), color (plastic models for color standards are available), color uniformity, size (number of halves per pound or diameter of pieces), freedom from decay and defect (shriveling, insect damage, internal discoloration, dark spots, skin discoloration, and rancidity), and cleanness (freedom from dust, dirt, and adhering material) |
| Pecan, in-shell | US (1976) | Shell: color uniformity, size (number of nuts per pound), cleanness, and freedom from decay and defect (insect damage, dark stains, split or cracked shells, and broken shells); kernel: same as for shelled pecans (above) |
| Pistachio, shelled | US (1990) | Degree of dryness; freedom from foreign material and damage caused by mold, insects, spotting, rancidity, and other defects; and size (whole kernels, broken kernels) |
| Pistachio, in-shell | US (1990) | Freedom from foreign material, loose kernels, shell pieces, particles and dust, and blanks; freedom from non-split shells, shells not split on suture, adhering hull material, and staining; degree of kernel dryness and freedom from defects |
| Walnut, shelled | US (1968) | Color (USDA color chart), degree of dryness, cleanness (freedom from shells, dirt, dust, and foreign material), freedom from decay and defect (insect injury, rancidity, shriveling, and meat discoloration), and size (diameter of halves or pieces) |
| Walnut, in-shell | US (1976) | Shell: dryness, cleanness, brightness, freedom from decay and defect (splits, discoloration, broken shells, perforated shells, and adhering hulls), and size (diameter); kernel: same as for shelled walnuts (above) |
|  | CA (1983) | Shell: dryness, size, and freedom from blanks, decay, and defect (insect damage, adhering hulls, and perforations affecting more than ⅛ of the surface); kernel: size, and freedom from decay and defect (insect damage, shriveling, and rancidity) |

*Date when standard was issued or revised.

# 22

# Postharvest Handling Systems: Ornamental Crops

Michael S. Reid

Among horticultural crops, cut flowers and other ornamentals have perhaps the highest value and are the most perishable. Their high respiration rates, rapid deterioration, and susceptibility to damage require the utmost care during postharvest handling for quality maintenance. Ornamentals are an important part of international commerce in horticultural crops, often being produced many thousands of miles from their intended markets. Marketing ornamentals involves major changes in supply and demand, due to heavy demand for holidays such as St. Valentine's Day, and to variable supply, particularly of flowers produced out of doors (fig. 22.1). These considerations also imply the need for optimal postharvest handling of these crops. This chapter first discusses factors affecting the postharvest life of ornamentals, then outlines techniques for their postharvest handling, using the postharvest systems employed in California as an example.

## Factors Affecting Postharvest Quality of Ornamentals

Although the factors affecting the postharvest life of ornamentals are largely similar to those for fruits and vegetables, they differ in some important aspects.

### Flower maturity

Minimum harvest maturity for a cut-flower crop is the stage at which harvested buds can be subsequently opened fully and have satisfactory display life after distribution. Many flowers are best cut in the bud stage and opened after storage, transport, or distribution. Although this technique is seldom used, it has many advantages, including reduced growing time for single-harvest crops, increased product packing density, simplified temperature management, reduced susceptibility to mechanical damage, and reduced desiccation. Many flowers are presently harvested when the buds are starting to open (rose, gladiolus); although others are normally fully open or nearly so (chrysanthemum, carnation). Flowers for local markets are generally harvested much more open than those intended for storage or long-distance transport.

### Food supply

The high respiration rate and rapid development of flower buds and flowers indicate the need for a substantial carbohydrate supply to the flowers after harvest. Starch and sugar stored in the stem, leaves, and petals provide much of the food needed for cut-flower opening and maintenance. These carbohydrate levels are highest when plants are grown in high light conditions and with proper cultural management. Carbohydrate levels are generally highest in the late afternoon, after a full day of sunlight. However, flowers are preferably harvested in the early morning, because temperatures are low, plant water content is high, and a whole day is available for processing the cut flowers.

The quality and vase life of many cut flowers can be improved by pulsing them after harvest with a solution containing sugar. The cut flowers are allowed to stand in solution for a short period, usually less than 24 hours, and often at low temperature. The most dramatic example of the effect of added carbohydrate is in spikes of tuberose and gladiolus: flowers open further up the spike, are bigger, and have a longer vase life after overnight treatment with a solution containing 20 percent sucrose and a biocide to inhibit bacterial growth (fig. 22.2). Sugar is also an essential part of the solution used to open bud-cut flowers before distribution, and of the vase "preservatives" used at the retail and domestic level. Red roses opened without sugar in the vase solution turn a dark purple (blueing). Blackening of leaves of cut flowers of *Protea nerifolia* can be prevented by maintaining the flowers in high light intensity. The problem is induced by low carbohydrate status of the harvested inflorescence.

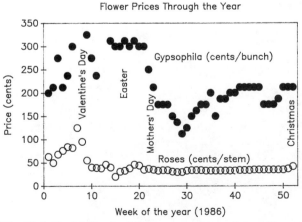

**Fig. 22.1. Variability in the flower market and the effect of holiday periods is shown by the changing wholesale value of red roses and gypsophila through the year.**

202

**Fig. 22.2. Pulsing gladiolus flowers overnight in the cold room with a 20 percent sucrose solution enhances flower size, opening, and vase life.**

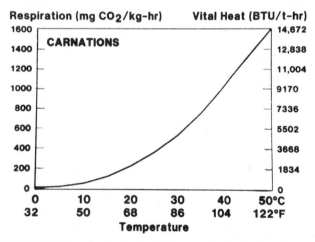

**Fig. 22.3. The respiration (and heat production) of cut carnations increases logarithmically with increasing temperature.**

## Temperature

The high respiration rate of cut flowers, as an indicator of their rate of growth and senescence, generates heat as a by-product. As in all biological systems, the respiration of cut flowers increases logarithmically with increasing temperature (fig. 22.3). For example, a flower held at 29°C (85°F) is likely to respire up to 45 times faster than a flower held at

2°C (36°F). The rate of senescence can be reduced dramatically by cooling the flowers (fig. 22.4). Rapid cooling and proper temperature maintenance are thus essential to maintain quality and satisfactory vase life of cut flowers.

The optimum temperature for storage of most common cut flowers is 0°C (32°F), just above freezing (fig. 22.4). In contrast, some tropical crops such as anthurium, bird-of-paradise, some orchids, and ginger, are injured at temperatures below 10°C (50°F). Symptoms of this chilling injury include darkening of the petals (fig. 22.5), watersoaking of the petals (which look transparent), and, in severe cases, collapse and drying of leaves and petals (fig. 22.6).

### Water supply

Cut flowers, especially those with leafy stems, have a large surface area, so they lose water and wilt rapidly. They should be stored at relative humidities above 95 percent, particularly during long-term storage. Water loss is dramatically reduced at low temperatures, another reason for prompt cooling.

Even after flowers have lost considerable water (for example during transportation or storage), they can be fully rehydrated using proper techniques. Cut flowers absorb solutions without difficulty, providing there is no obstruction to water flow in the stems. Air

**Fig. 22.4. Optimum storage temperature for most cut flowers is 0°C. Flowers kept at 10°C deteriorate three times faster than those kept at 0°C.**

**Fig. 22.5. Anthurium flowers are sensitive to low temperatures. After storage of 7 days at 0°C, the flowers show symptoms typical of chilling injury.**

**Fig. 22.6. Poinsettias, like many other tropical plants, are very sensitive to low temperature storage. Exposure for 1 week to temperatures below 7°C causes complete collapse of the plants.**

embolism, bacterial plugging, and poor water quality can reduce solution uptake.

**Air embolism.** Air embolism occurs when small bubbles of air (emboli) are drawn into the stem at the time of cutting. These bubbles cannot move far up the stem, so the upward movement of solution to the flower is restricted. Emboli can be removed by recutting the stems under water (removing about 1 inch), by ensuring that the rehydration solution is acid (pH 3 or 4), or by placing the stems in a vase solution heated to 41°C (105°F) (warm, but not hot).

**Bacterial plugging.** The quality and vase life of flowers are improved by supplying them with sugar after harvest. Sugar can, however, also act as food for the growth in the water of detrimental fungi and bacteria, whose growth can be enhanced by organic materials from the cut stem. Substances produced by the bacteria, and the bacteria themselves, can plug the water-conducting system (fig. 22.7). For this reason, buckets should be cleaned and disinfected regularly, and flower-holding solutions should contain germicides to prevent the growth of microorganisms.

Acidic vase solutions not only improve water flow by overcoming embolism, but also inhibit bacterial growth.

**Water quality.** Hard water contains minerals that make the water alkaline. Alkaline water does not move readily through stems and can substantially reduce vase life. Such water can be treated by removing the minerals using a deionizer, or by making the water acid. Commercial flower preservatives may not contain enough acid to acidify some very alkaline waters; in that case, more acid should be added.

Certain ions found in tap water are toxic to some cut flowers. Sodium ($Na^+$), present in high concentrations in soft water, is, for example, toxic to carnations and roses. Fluoride ($F^-$) is very toxic to gerbera, gladiolus, freesia, and some rose varieties; fluoridated drinking water contains enough $F^-$ (about 1–2 ppm) to damage these cut flowers.

### Ethylene

Some flowers, especially carnations and some rose cultivars, perish rapidly if exposed to minute concentrations of ethylene. In carnations and sweet peas, ethylene produced by the flower induces the normal processes of senescence. In many compound flowers, such as snapdragon and delphinium, ethylene causes flower abscission, or shattering (fig. 22.8). Ethylene is produced in large quantities by some ripening fruits and during combustion of organic materials (e.g. gasoline, firewood, tobacco). Levels of ethylene in the air above one-tenth part per million (0.1 ppm) in the vicinity of most cut flowers can cause damage. Storage and handling areas should be designed not only to minimize contamination with ethylene, but to have enough ventilation to remove any ethylene that does occur. Treatment with the anionic thiosulfate complex of silver (STS) reduces the effects of ethylene (exogenous or endogenous) in some flowers (fig. 22.9). Finally, refrigeration greatly reduces ethylene production and the sensitivity of the product to ethylene.

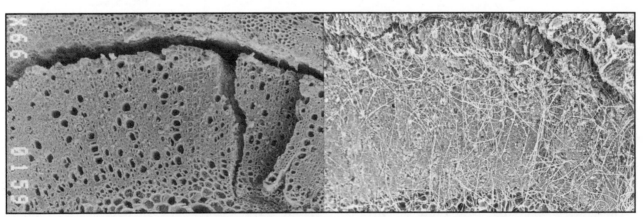

**Fig. 22.7. Scanning electron micrograph of a cross section of the stem of a rose flower before (*left*) and after (*right*) being held for 3 days in deionized water. The fine tubes conducting water to the flower have become plugged with bacterial slime and fungal hyphae.**

**Fig. 22.8. Effect of ethylene and STS on flower loss in snapdragon.**

**Fig. 22.9. Effect of STS pre-treatment on opening of cut roses (cv. Lovely Girl).**

### Growth tropisms

Certain responses of cut flowers to environmental stimuli (tropisms) can result in quality loss. Most important are geotropism (bending away from gravity) and phototropism (bending toward light). Geotropism often reduces quality in spike-flower crops like gladiolus and snapdragon, because the flowers and spike bend upward when stored horizontally (fig. 22.10). These flowers should be handled upright whenever possible. Pretreatment with an auxin transport inhibitor, naphthyl phthalamic acid (NPA), which is a common weed killer, can overcome tropisms in some cut flowers (fig. 22.10).

### Mechanical damage

Bruising and breakage of cut flowers should be avoided. Flowers with torn petals, broken stems, or other obvious injuries are undesirable for aesthetic reasons. Disease organisms can more easily infect plants through injured areas. In fact, most diseases are caused by infection through injured areas. Finally, respiration and ethylene evolution are generally higher in injured plants, further reducing storage and vase life.

### Disease

Flowers are susceptible to disease, not only because their petals are fragile, but also because the sugar solution secreted by their nectaries is an excellent nutrient supply for even mild pathogens. To make matters worse, transfer from cold storage to warmer handling areas can result in condensation of water on the harvested flowers. The common organism responsible for gray mold (*Botrytis cinerea*) can germinate wherever free moisture is present. In the humid environment of the flower head, it can even grow (albeit more slowly) at temperatures near freezing. Proper greenhouse hygiene, temperature control, and minimizing condensation on flowers all reduce losses caused by gray mold. Some fungicides, such as Ronalin, Rovral, and a copper complex, Phyton-20, have been approved for use on cut flowers and are effective against gray mold.

## Postharvest Management Techniques Used for Cut Flowers

Systems for harvesting and marketing cut flowers vary according to individual crops, growers, production areas, and marketing systems. All involve a series of steps—harvesting, grading, bunching, sleeving, storage, packing, pre-cooling, transportation, and retail marketing, not necessarily in that order. Management systems should be selected that maximize postharvest life of the flowers, a goal that usually requires prompt pre-cooling and proper temperature management throughout the marketing chain. Increasingly, producers are trying to reduce the number of steps in the postharvest handling system. For example, some field flower growers cut, grade, bunch, and pack their product in the field (fig. 22.11). The packed boxes are then taken directly to the precooler. Such systems, where appropriate, reduce damage to the flowers, and may decrease labor costs.

### Harvesting

Harvesting is normally done by hand, using shears or a sharp knife. Simple mechanical aids are used for some crops; examples are the hook-shaped comma that permits chrysanthemum harvest without stooping and rose shears that grip the flower stem after it has been cut, allowing it to be withdrawn single-

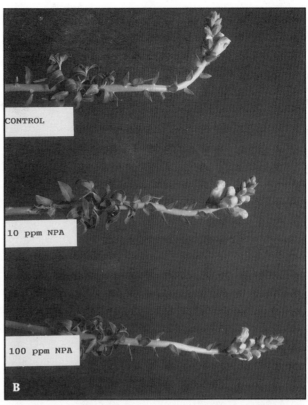

Fig. 22.10. Laying snapdragons on the bench results in unsightly curvature. Prior treatment with naphthyl phthalamic acids (NPA) prevents this. (A) Intact flowers. (B) Flowers removed to show curvature.

Fig. 22.11. Harvesting, grading, bunching, and packing are all carried out in the field, in this southern Californian production of statice.

Fig. 22.12. These specialized shears hold a rose stem after it has been cut, simplifying the task of harvesting.

handedly from the bush (fig. 22.12). At no time should harvested flowers be placed on the ground, where they may become contaminated with disease organisms. Ideally, harvesting, grading, and packing should all be done dry, without the use of chemical solutions or water. If this is not possible, clean buckets containing clean water and a biocide should be used. With hard water and for difficult-to-rehydrate flowers, clean water containing sufficient citric acid to bring the pH to 3.5 and an appropriate biocide should be used instead.

## Grading

The designation of grade standards for cut flowers is controversial. There are no U.S. federal or state standards for cut flowers or other ornamentals. Large producers, wholesalers, or retailers may have internal grade standards, but they are highly variable. The Society of American Florists has promulgated "recommended" grade standards for some major crops, but they cannot be enforced. Stem length, still the major quality standard for many flowers, may bear little relationship to flower quality, vase life, or usefulness. Straightness of stems, stem strength, flower size, vase life, freedom from defects, maturity, uniformity, and foliage quality are among the factors that should also be used in cut flower grading. Mechanical grading systems should be carefully designed to ensure efficiency and avoid damaging flowers.

## Bunching

Flowers are normally bunched, except for anthuriums, orchids, and some other specialty flowers. The number of flowers in the bunch varies according to growing area, market, and species. Groups of 10, 12, and 25 are common for single-stemmed flowers. Spray-type flowers are bunched by the number of open flowers, by weight, or by bunch size.

Bunches are held together by string, paper-covered wire, or elastic bands and are frequently sleeved soon after harvest (fig. 22.13) to separate them, protect the flower heads, prevent tangling, and identify the grower or shipper. Materials used for sleeving include paper (waxed or unwaxed) and polyethylene (perforated, unperforated, and blister). Sleeves can be preformed (although variable bunch size can be a problem), or they can be formed around each bunch, using tape, heat sealing (polyethylene), or staples.

Damage through multiple handlings can be reduced if grading, sizing, and even bunching are done in the field or greenhouse. Flowers should be graded and bunched before being treated with chemicals or being placed in storage. When the flowers are badly wilted, or when labor is not available for grading and

**Fig. 22.13. Thin plastic sleeves are pulled over the flower bunches to provide protection and keep flowers separate in the box.**

bunching, flowers should be rehydrated under refrigeration until these operations can be carried out.

## Chemical solutions

The diverse solutions in which flowers may be placed after harvest have specific purposes:

**Rehydration.** Wilted flowers should be rehydrated in a cooler, using deionized water containing a germicide. Wetting agents (0.01 to 0.1 percent) can be added, and the water should be acidified with citric acid, hydroxyquinoline citrate (HQC), or aluminum sulfate to a pH near 3.5. No sugar should be added to the solution. The solution should contain a biocide if it is not replaced daily.

**Pulsing.** The term "pulsing" means placing freshly harvested flowers for a short time in a solution formulated to extend their storage and vase life. Sucrose is the main ingredient of pulsing solutions, and the proper concentration ranges from 2 to 20 percent, depending on the crop. Some cut flowers are also pulsed with STS to reduce the adverse effects of ethylene. Flowers can be pulsed for short periods at warm temperatures (e.g., 10 minutes at 21°C [70°F]) or long periods at cool temperatures (e.g., 20 hours at 2°C [36°F]). Short pulses (10 seconds) in solutions of silver nitrate are valuable for some crops. Chinese asters and maidenhair fern respond well to solutions containing 1,000 ppm silver nitrate. Other flowers (e.g., gerberas) respond well to 100 to 200 ppm. The function of the silver nitrate is not fully understood. In some cases it seems to function strictly as a germicide (e.g., chrysanthemums). The residual silver nitrate solution should be rinsed from the stems before packing.

**Bud opening.** Bud-cut flowers are opened in bud-opening solutions before they are sold. These solutions contain a germicide and sugar. Foliage of some flowers (especially roses) can be damaged if the sugar concentration is too high. Buds should be opened at relatively warm temperatures (21 to 27°C [70 to 80°F]), moderate humidities (60 to 80 percent RH), and reasonably high light intensities (100 to 200 footcandles).

**Tinting.** Artificial coloration of flowers (tinting) is done in two ways: through the stem (for carnations) and by dipping the flower heads (for other flowers, principally daisies). In tinting carnations, proprietary dye solutions (combinations of food-type dyes with adjuvants designed to increase uptake of the solution) are mixed in a bucket and warmed to about 41°C (105°F). The carnations to be tinted (usually 'White Sim') are allowed to dry somewhat (overnight in the packing shed at 18°C [65°F]) to increase their rate of solution uptake. Dyeing is stopped before the flowers reach the desired color, because dye still in the stem is flushed into the flower by the vase solution. Tinting by dipping is done with proprietary tinting solutions containing aniline dyes dissolved in organic solvents such as rubbing alcohol (isopropanol). The heads of the flowers are dipped in a vat of the dye, shaken to remove surplus solution, and placed on a rack to dry before storage or packing.

## Packing

Most boxes for cut flowers are long and flat (fig. 22.14). This design restricts the depth of the flowers in the box, which may in turn reduce physical damage. Flower heads can be placed at both ends of the container for better use of space. Whole layers of newspaper have traditionally been used to prevent the layers of flowers from injuring each other, but using small pieces of newspaper to protect only the flower heads allows for more efficient cooling of flowers after packing.

It is critical that boxes be packed in ways that minimize transport damage. Most packers anchor the product by filling the box so that the contents are firmly held. To avoid longitudinal movement, packers use one or more "cleats," normally foam- or newspaper-covered wood pieces that are placed over the product, pushed down, and stapled into each side of the box (fig. 22.15). Padded metal straps, high-density polyethylene blocks, and cardboard tubes can also be used as cleats. The heads of the flowers are normally placed 7 to 12 cm (3 to 5 inches) from the end of the box to eliminate the danger of petal bruising should the contents of the box shift.

Gladioli, snapdragons, and some other species are often packed in vertical hampers to prevent unsightly geotropic curvature. Cubic hampers are used for upright storage of daisies and other flowers.

Specialty flowers such as anthurium, orchids, ginger, and bird-of-paradise are packed in various ways to minimize friction damage during transport (fig. 22.16). Flower heads may be individually protected by paper or polyethylene sleeves. Cushioning materials, such as shredded paper, paper wool, or wood wool may be distributed between the packed flowers

Fig. 22.14. Boxes commonly used in California for packing cut flowers are long and flat, from the *left*: a regular fiberboard box designed for a bottom air delivery trailer, a box constructed from polyurethane-sprayed fiberboard, and a polystyrene "casket."

Fig. 22.15. Cleats to hold the product (padded with newspaper and foam) are stapled to each side of the box.

Fig. 22.16. High-value flowers are often specially packed. Here, cattleya orchids with individual tubes containing floral preservative are separated and cushioned from vibration with shredded newsprint.

to further reduce damage. For some tropical flowers (anthurium, heliconia) these materials may be moistened to reduce water loss. The bases of high-value flowers, such as orchids, may be sealed in small vials containing water or a preservative solution to ensure that they do not dehydrate during transportation. This is particularly useful for chilling-sensitive flowers, because the higher handling temperatures that they require are apt to accelerate desiccation.

## Cooling

By far the most important part of maintaining the quality of harvested flowers is cooling them as soon as possible after harvest and maintaining optimum temperatures during marketing. Most flowers should be held at 0 to 2°C (32 to 36°F). Chilling-sensitive flowers (anthurium, bird-of-paradise, ginger, tropical orchids) should be held at temperatures above 10°C (50°F).

Individually, flowers cool (and warm) rather rapidly (half-cooling times of a few minutes). So, while individual flowers can be cooled quickly, it is also true that individual flowers brought out of cool storage into a warmer packing area warm quickly and develop condensation before packing. The simplest way to keep packed flowers adequately cooled and dry is, therefore, to pack them in the cool room. Although this method is not always popular with packers and may slow down packing somewhat, it ensures a cooled, dry product.

Packed boxes of cut flowers do not cool well if they are merely placed in a refrigerated room. The rapid respiration of the flowers and the insulating properties of a packed box result in heat buildup in packed flower boxes unless steps are taken to ensure temperature reduction. Forced-air cooling of boxes with end holes or closeable flaps is the most common and effective method for cooling cut flowers. Cool air is sucked or blown through the boxes (fig. 22.17). Care must be taken to pack in such a way that air can flow through the box and not be blocked by the packing material (fig. 22.18). In general, packers use less paper when packing flowers for pre-cooling. The half-cooling time for forced-air cooling ranges from 10 to 40 minutes, depending on product and packaging.

**Fig. 22.17. In forced-air pre-cooling of flowers, cool air is sucked or blown through the packed boxes.**

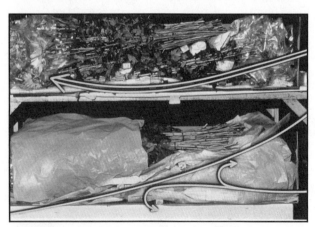

**Fig. 22.18. Proper pre-cooling of cut flowers requires that the flowers be packed so that airflow through the box is not impeded (as shown by the *arrows*). *Upper box*, correctly packed. *Lower box*, paper protection prevents satisfactory airflow.**

If the packages are to remain in a cool environment after cooling, vents may be left open to assist removal of the heat of respiration. Flowers that are to be transported at ambient temperatures can be packed in polyethylene caskets, foam-sprayed boxes, or boxes with the vents resealed. Using ice in the box is only effective if it is placed so that it intercepts heat entering the carton (i.e., it must surround the product), and care must be taken to ensure that the ice does not melt onto the flowers or cause freezing damage.

Tropical flowers shipped in a mixed load need special care. The flowers should be packed in plenty of insulating material (an insulated box packed with shredded newsprint, for example). These flowers should not be cooled. If they are to be shipped by truck, they should be placed in the middle of the load, away from direct exposure to cooling air.

## Storage

Although it would be advantageous to store cut flowers for short periods (up to a month), few cut flowers maintain their quality during cool storage. With the exception of bud-cut carnations, which can be stored for several months, flowers should not be stored for more than 2 to 3 weeks.

The factors limiting storage life of flowers are not fully understood, but development of *Botrytis* is commonly observed. Flowers intended for storage should be of premium quality and should be pre-treated appropriately (STS, sugar, growth regulators), depending on the species. Flowers to be stored for any length of time should be held dry (packed, and preferably wrapped, after cooling, in polyethylene sheet to minimize water loss). Flowers can be stored for a few days with their bases in preservative solution, but vase life is reduced if the storage period is longer. The storage room should be at 0°C (32°F) for most cut flowers, and the temperature should vary from this by no more than 0.2°C (0.5°F). To avoid larger temperature fluctuations, which are likely to cause condensation on the flowers and consequent growth of *Botrytis* and other pathogens, access to the storage room should be limited. The humidity in the room should be at least 95 percent, and ethylene-producing commodities (especially fruits) should be rigorously excluded.

In general, controlled or modified atmosphere storage has not proved to be useful with cut flowers. Carnations last well in CA conditions, but they also last well in regular storage. The vase life of daffodils falls less rapidly during storage in a nitrogen atmosphere, and some flowers (lily) can withstand high $CO_2$ concentrations in storage, which may be useful for inhibiting pathogen growth.

After storage, flowers should be rehydrated in the cool-room, using water containing a biocide, and preferably at pH 3.5.

## Retail handling

To ensure maximum quality and satisfaction to the customer, it is vital that retail florists handle flowers properly on receipt. On arrival, packed boxes should be unpacked, or placed immediately in a cool-room. After unpacking, flower bunches should be recut (preferably under water) and placed in a cool-room (in a warmed rehydration solution) for at least a few hours. If flowers are to remain in the store for several days, they should be displayed in a clean bucket or vase containing a vase preservative.

The florist's cool room should be held at 0° to 2°C (32 to 36°F). Chilling-sensitive flowers (anthurium, bird-of-paradise, ginger, tropical orchids) should be held at temperatures above 10°C (50°F). Cut flowers that are held during storage with their bases in water (wet storage) are usually kept only 1 to 3 days, while those stored dry are often kept for longer periods.

**Vase solutions.** Vase solutions normally contain a low concentration of sugar (0.5 to 2.0 percent), a chemical to keep the pH of the solution low, and a germicide. Flowers are kept in this solution indefinitely. This treatment is usually provided by the retail seller or the customer. Preservatives are supplied to the trade in bulk or may be purchased in individual sachets intended for 1 quart of solution. A typical solution contains 1 percent to 2 percent sugar, a biocide (HQC or a mixture of quaternary ammonium compounds) and 300 ppm citric acid. Various other chemicals are sometimes incorporated, including aluminum sulfate (which seems to be a good biocide, forms flocculent aluminum hydroxide to trap particles in the water, and also lowers the pH), slow-release chlorine compounds (good biocides, but sometimes toxic) and 6-benzyl adenine (a plant growth regulator that increases the vase life of some flowers). These chemicals should be used at recommended strengths.

**Flowers in the supermarket.** Flowers intended for supermarket sales require special care throughout the distribution chain. First, only flowers that have adequate vase life to remain saleable for the intended period of time should be used; thus some short-lived flowers such as iris, tulip, and narcissus should be avoided. Flowers with common postharvest problems

(roses, gypsophila) are a danger, too. It is particularly important that ethylene-sensitive flowers (notably carnations, some rose cultivars, and gypsophila) be treated with STS if they are to be marketed through produce stores.

## Drying

A number of cut flower crops are sold in both fresh and dried forms. For some, such as statice and strawflower, the drying process can be as simple as hanging bunches upside down (to keep the stems straight) in a warm, dry location. Others, such as gypsophila and silver-dollar eucalyptus, which become too brittle if dried this way, are placed in a freshly prepared solution of 30 percent glycerol. The glycerol moves through the plant in the course of a day or so. Dye can be added to the glycerol to color the dried foliage or flowers. After the glycerol treatment, the plants are hung upside down in a warm dry environment to dry. Materials dried in this way remain supple for years. Improved methods of drying and preservation are still being developed. Plastic impregnation of whole plants and freeze-drying cut flowers are two such techniques.

## REFERENCES

1. Berkholst, C. E. M., et al. 1986. *Snijbloemen: kwaliteitsbehoud in de afzetketen.* Wageningen, Neth.: Sprenger Instituut. 222 pp.

2. Carow, B. 1981. *Frischhalten von Schnittblumen.* Stuttgart: Verlag Eugen Ulmer. 144 pp.

3. Goszcynska, D. M., and R. M. Rudnicki. 1988. Storage of cut flowers. *Hortic. Rev.* 10:35-62.

4. Halevy, A. H., and S. Mayak. 1979. Senescence and postharvest physiology of cut flowers, Part 1. *Hortic. Rev.* 1:204-36.

5. ———. 1981. Senescence and postharvest physiology of cut flowers, Part 2. *Hortic. Rev.* 3:61-143.

6. Hardenburg, R. E., A. E. Watada, and C. Y. Wang. 1986. *The commercial storage of fruits, vegetables, and florist and nursery stocks.* U.S. Dept. Agric. Handb. 66. 130 pp.

7. Mayak, S., and A. H. Halevy. 1980. Flower senescence. In *Senescence in Plants*, ed. K. V. Thimann, 131-56. Boca Raton, FL: CRC Press.

8. Reid, M. S. 1990. *Handling cut flowers.* Ag. Access. In Press.

9. Rij, R. E, Thompson, J. F., and Farnham, D. S. 1979. *Handling, pre-cooling and temperature management of cut flower crops for truck transportation.* USDA AAT-W-5. 26 pp.

# 23

# Postharvest Handling Systems: Fresh Herbs

MARITA CANTWELL AND MICHAEL S. REID

Culinary herbs are leafy plant materials used in small amounts to contribute aroma and flavor to foods. Common examples are parsley, mint, basil, sage, and oregano. They have always been an important component of the human diet, providing variation and interest in flavor of staple foods and being a vital part of some food preservation techniques. Even today, the major commercial form of culinary herbs is the dried product, because it is easy to transport and market and can be stored for more than a year under proper conditions. Although freshly harvested herbs have greater and superior flavor than dried herbs, they have traditionally been available only directly from a kitchen garden or local market. The perishability of fresh culinary herbs has restricted their widespread commercialization.

Distribution of fresh herbs through the major wholesale, retail, and food-service channels has been increasing, but marketing them has not been wholly successful. Many herbs are extremely perishable; and although they are of high value, the small quantities sold mean that even a small inventory takes time to turn over. Moreover, marketing strategies have tended to use a uniform technology for all herbs. Because herbs are diverse in botanical origin and physiological state, postharvest conditions that are adequate for one fresh herb may be completely inappropriate for others. For example, herb bouquets on sale at retail markets may combine an extremely perishable herb such as chervil with long-lasting herbs such as rosemary and thyme.

Fresh herbs are complex structures consisting of stem, leaf, flower, and sometimes root tissues. Most herbs are harvested as soft or semiwoody leafy stems. Sometimes, as in the case of dill, oregano, and basil, the herb includes immature or mature flowers. Many salad herbs harvested as developing leaves (sorrel, arugula) or intact plants (mache) consist of leaves at different physiological stages. This diversity means that postharvest requirements differ for each product, although in practice they are handled much like leafy vegetables.

## Harvest and Handling

Herbs for fresh market may be grown in the field or in protective enclosures such as cloches or greenhouses. Production is typically on a small scale, and harvesting, grading, trimming, bunching, and packing are all manual operations. Harvest in the field is commonly done with scissors (fig. 23.1), and the herbs may be bunched at the time of harvest (fig. 23.2) or taken in bulk to a packing station where the herbs may be trimmed before bunching (fig. 23.3). Herbs may be packed in bulk, in bunches, or placed directly in polyethylene pouches or rigid plastic boxes designed as point-of-purchase packages (fig. 23.4).

As for other leafy greens, quality components of fresh culinary herbs are largely visual: an appearance of freshness; uniformity of size, form, and color; lack of defects such as damaged leaves, yellowing, and evidence of decay. There is no doubt that the most important quality components of culinary herbs are the expected aroma and flavor, but these usually are not considered in legal standards for fresh produce. There are presently no legal standards for quality of fresh herbs nor standard unit weights or volumes for marketing them.

**Fig. 23.1. Harvesting chives in the field with scissors.**

**Fig. 23.2. Chives trimmed and bunched in the field.**

**Fig. 23.3. Basil, brought to a packing station in bulk plastic containers (lower right), is trimmed, graded, and bunched.**

**Fig. 23.4. Herbs packed in the cold room in polyethylene bags are placed in a waxed carton.**

## Postharvest Technology

The small size of many herb operations permits attention to producing and preparing a quality product. However, the perishability of fresh herbs means that maintaining quality during marketing requires careful postharvest handling. There are six principal considerations.

### 1. Temperature

As with other perishable products, temperature is the overriding factor affecting the life of cut herbs. For most herbs, the optimal storage conditions are 0°C (32°F) and high relative humidity (95 to 98 percent). The visual quality of herbs held for 10 days at different temperatures is shown in table 23.1. Over a 10-day simulated marketing period, many herbs still have acceptable quality if held at 10°C (50°F), but holding at 20°C (68°F) greatly limits shelf life. Basil and shiso show reduced quality when stored at 0°C (32°F)—these are the two herbs presently known to be chilling sensitive. This sensitivity presents practical problems, since basil is a major component of most mixed herb shipments. Herb shipments con-

**Table 23.1. Effect of temperature and ethylene on visual quality of fresh culinary herbs after 10 days**

| | Visual Quality Score*† after 10 days at indicated temperature | | |
|---|---|---|---|
| Herb | 0°C (32°F) | 10°C (50°F) | 20°C (68°F) |
| Basil | 2 | 8 | 7 |
| Chervil | 8 | 6** | 1 |
| Chives | 9 | 6 | 3 |
| Dill | 9 | 6** | 2 |
| Epazote | 9 | 7** | 5 |
| Mache | 8 | 5 | 2 |
| Marjoram | 9 | 8** | 1 |
| Mint | 9 | 6** | 2 |
| Mitsuba | 9 | 7** | 4 |
| Rosemary | 9 | 9 | 7 |
| Sage | 9 | 8 | — |
| Shiso | 6 | 8** | 3 |
| Tarragon | 8 | 6 | — |
| Thyme | 9 | 8 | 7 |

*Quality score: 9 = excellent; 7 = good, minor defects; 5 = fair, moderate defects, limit of salability; 3 = poor, major defects; 1 = unusable.
†Two asterisks indicate herbs that showed reduced visual quality at 10°C (50°F) when exposed to 5–10 ppm ethylene.

taining basil are often held at an intermediate temperature, between 5° and 10°C (40° and 50°F), but this may still induce chilling in basil while substantially increasing the rate of deterioration of other herbs, as compared with handling them at 0°C (32°F).

Because of the small scale of herb production, facilities for temperature management are often rudimentary or lacking. If herbs are harvested cool in the early morning, the need for cooling is reduced. Some producers send their product to distributors of major volume leafy vegetables, who have the necessary facilities for better cooling and handling. Some fresh herbs (parsley, watercress, and mint) are commonly cooled with ice. Others, with tender leaves, are better forced-air or room cooled. Some growers have successfully vacuum-cooled fresh herbs.

Fresh herbs are sometimes transported in mixed loads with other leafy greens in refrigerated trailers or containers. More commonly, because of their specialty nature and the small volumes marketed, they are shipped by air to reduce transit time, and sometimes with gel-ice packs to reduce heat buildup.

### 2. Moisture loss

Many fresh herbs lose water rapidly through their leaves. The result: early wilting. Holding herbs at low temperatures reduces the rate of water loss. Some handlers place the cut stems of the herbs (e.g., mint, basil) in water, but rapid growth of microbes in the water is common, and the technique is not practical for transport and storage.

A better way to reduce water loss is to package the herbs in plastic films (fig. 23.4). For herbs packaged this way, it is necessary to maintain constant temperatures to reduce condensation inside the bag and the consequent risk of microbial growth. The bags may be partially ventilated with perforations or may be constructed of a polymer that is partially permeable to water vapor. The relative humidity of packing areas, cold rooms, and transport vehicles should be kept above 95 percent where practicable.

### 3. Physical injury

Most fresh herbs are very susceptible to damage during postharvest handling. Mishandling can result in extensive discoloration of the tender leaves of mint, basil, and coriander and may provide microbial infection sites. Careful handling and preventative packaging are means of minimizing damage. Rigid clear plastic containers such as those sometimes used for sprouts may be used for delicate herbs.

### 4. Ethylene

Like other leafy tissues, herbs are detrimentally affected by ethylene with such symptoms as yellowing (see chapter 13, fig. 13.8), leaf drop, and epinasty (petiole bending). Sensitivity to ethylene varies among herb species (table 23.1); many show epinasty, but yellowing and abscission are usually observed only in the most sensitive species. Ethylene effects can be minimized by maintaining proper storage temperatures.

### 5. Atmosphere modification (CA, MA)

Fresh herbs respond positively to reduced $O_2$ and increased $CO_2$ concentrations. Packaging in plastic films may serve not only to reduce water loss, but also to provide a beneficial storage atmosphere. Since many herbs are marketed in lighted display cabinets, photosynthesis may reduce $CO_2$ levels, making it difficult to maintain a desirable atmosphere. Pilot trials have shown that the quality of packaged herbs was reduced when they were held in the light.

### 6. Pathogens

The shelf life of fresh herbs, especially those packaged in film bags, is frequently terminated by the growth of pathogens such as gray mold and bacterial soft rot. Control measures include preharvest and handling hygiene (including use of chlorinated water) and avoidance of injury and condensation during marketing.

### REFERENCES

1. Aharoni, N., A. Reuveni, and O. Dvir. 1989. Modified atmospheres in film packages delay senescence and decay of green vegetables and herbs. *Acta Hortic.* 258:37-45.

2. Bell, L. 1987. Postharvest handling of fresh culinary herbs. In *Proc. First Annual Herb Growing and Marketing Conference*, 101-9. Lafayette, IN: Univ. Purdue Bull.

3. Cantwell, M., and M. S. Reid. 1990. Postharvest physiology and handling of fresh culinary herbs. In *Herbs, spices, and medicinal plants*, ed. L. Craker and J. E. Simon J.E., Vol. 5. Phoenix, AZ: Oryx Press (in press).

4. Hruschka, H. W., and C.-Y. Wang. 1979. *Storage and shelf life of packaged watercress, parsley, and mint.* U.S. Dept. Agric. Mktg. Res. Rept. 1101. 19 pp.

5. Joyce, D., M. Reid, and P. Katz. 1986. Postharvest handling of fresh culinary herbs. Univ. Calif., *Perishables Handling* 58:1-4.

6. Lipton, W. J. 1987. Senescence of leafy vegetables. *HortScience* 22:854-859.

7. Simon, J. E., A. F. Chadwick, and L. E. Craker. 1984. *Herbs: an indexed bibliography, 1971-1980.* Hamden, CT: Shoe String Press. 770 pp.

# 24

# Postharvest Handling Systems: Temperate Zone Tree Fruits (Pome Fruits and Stone Fruits)

F. Gordon Mitchell

California is a major producer of temperate zone tree fruits and makes large fresh shipments of most of these fruits to U.S. markets. These include pome fruits (pears and apples), stone fruits (apricots, sweet cherries, nectarines, freestone peaches and plums), and a more limited volume of several other fruits. California is also a major supplier of processed temperate zone tree fruits in the U.S. and to export markets. These include prunes, cling peaches, olives, and figs in addition to the fruits named. This chapter discusses the pome fruits and stone fruits that are major factors in fresh marketing.

## POME FRUITS

Pome fruits, especially pears and apples, have long been major crops in California, although the greatest volume has moved in processing channels. In recent years, greater emphasis has been placed on the fresh market, including increased movement of existing varieties to fresh packing and the planting of new varieties that are primarily fresh-market fruits. While small quantities of quince are shipped from California, this section focuses on pears and apples.

Pear and apple handling has become highly mechanized. Considerable investment has been made in cooling and storage facilities, including controlled-atmosphere facilities for long-term storage. Because of their capability for long-term storage, these fruits have been extensively marketed for export.

## Postharvest Diseases and Physiological Disorders

Pome fruits are subject to storage loss from Botrytis and Penicillium rots, which can develop in wounds or at the stem or calyx end. Calyx-end Botrytis rot has been a problem in some Bartlett ('Williams,' 'Bon Chretien') pears in California in recent years, possibly related to the use of water dumps. Frequent cleaning and chlorine treatment of water dumps can reduce but not eliminate the problem. In some instances rapid forced-air cooling after packing has appeared to greatly reduce the problem.

Many other fruit-rotting organisms can attack pome fruits and cause postharvest losses, but most are of minor importance in marketing. Other common rot problems, especially following long-term storage of apples and pears, include Alternaria and Stemphyllium rots in wounds and sunburned areas. To minimize losses from these problems, fungicidal treatments may precede long-term storage.

**Physiological disorders.** A number of physiological disorders can cause serious losses, especially after prolonged storage.

**Storage scald.** Apple storage scald is most severe on low-maturity fruits, but a similar disorder in pears called superficial scald is independent of fruit maturity. Harvesting apples at optimum maturity reduces the problem. These disorders, which appear as brown patches on the fruit surface, may not develop until the fruit are warmed after storage. Thorough cooling and maintaining a low storage temperature are important measures in scald control. Often diphenylamine (DPA) or ethoxyquin dips or drenches, applied according to label directions, are used before storage to reduce the incidence and severity of scald. Controlled atmosphere storage for fruit to be held beyond 3 months can delay scald development. Very low $O_2$ concentrations in CA storage (about 1 percent) can delay scald development, although care must be taken to avoid low-$O_2$ injury to the fruit.

**Senescent scald.** It can develop in pears that have been stored beyond their potential postharvest life. Scalded fruits often change their background color (from green to yellow) during storage and lose their capacity to ripen. Fruit from very early or late harvest, fruit suffering delayed cooling, and fruit held at too high a storage temperature are all more susceptible to senescent scald. Symptoms begin on the fruit surface but can progress into the flesh during ripening. Proper harvest maturity, good temperature management, and avoiding too long storage are all important in minimizing senescent scald of pears. Controlled atmosphere storage can delay fruit senescence and thus delay development of senescent scald.

**Water core.** A commonly recognized problem on apples, water core can occur on pears, especially Asian pears. The watersoaking symptoms of this disorder result from flooding of the intercellular spaces by a solution high in sorbitol. Water core is often associated with low calcium levels in the fruit and is most severe in fruit from young, vigorous, lightly cropped trees. Fruit of advanced maturity are generally more severely affected. Fruit grown at high temperatures, especially if exposed to afternoon sun, appear most prone to water core. There can be considerable disappearance of symptoms in fruit during the first few weeks in cold storage. Such fruit should not be stored for long periods, however, as affected areas from which the symptoms have cleared may develop flesh browning after long storage.

**Bitter pit.** Expressed by the development of dry, pithy spots or pits near or below the fruit surface, bitter pit is another common cause of losses in apples. Lenticel blotch pit is a similar disorder on Granny Smith apple with very shallow pits. Tree conditions that are associated with bitter pit include extreme vigor, light crop, and low calcium content. Early harvested, low-maturity fruit are most susceptible. The symptoms are sometimes visible at harvest but commonly develop during storage. Rapid cooling and high relative humidity can reduce the development of symptoms during storage. Multiple calcium sprays on fruit before harvest, following label instructions, are the most effective treatment for bitter pit. Calcium dips after harvest, used according to label instructions, may further reduce the problem. Application of pressure or vacuum, while dipping in calcium solution, has increased the penetration into the flesh and reduced bitter pit incidence in susceptible cultivars.

**Core breakdown problems.** Pears and apples can suffer losses to various core breakdown problems (sometimes called internal breakdown or brown core). Symptoms are browning and softening of the tissue in and around the fruit core. Late-harvested fruit are most susceptible to these disorders. Symptoms may develop in storage or during subsequent ripening at 15° to 25°C (59° to 77°F). A range of similar symptoms associated with specific cultivars or handling and storage treatments are described by various researchers.

Reports from cool-climate production areas describe a fruit breakdown in Bartlett pears caused by premature ripening. In Oregon, this problem can develop when daytime high temperatures do not exceed 21°C (70°F) and night temperatures are no higher than 7°C (45°F) for 2 days, 10°C (50°F) for 9 days, or 13°C (55°F) for 21 days. Affected fruit develop a pink color around the calyx (blossom) end. Night temperatures above 13°C (55°F) or day temperatures above 32°C (90°F) apparently prevent premature ripening from occurring. Bartlett pears grown in cool climates also tend to soften faster and be more susceptible to core breakdown. By contrast, Anjou pears grown in warm climates have a higher incidence of mealy breakdown of the flesh.

**Watery breakdown.** A problem associated especially with Bartlett pear, watery breakdown involves soft, watery breakdown in portions of the fruit, usually without brown discoloration during its early stages. This enzymatic softening can affect any part of the fruit and probably results from severe physiological stress on the fruit. In some seasons in California, this disorder is responsible for loss of as much as 10 percent of fruit destined for processing. Fast cooling and low storage temperatures (avoiding freezing) are effective in minimizing the problem.

All flesh breakdown develops during storage and ripening can be reduced by good postharvest temperature management. This should include rapid movement from tree to cooler, rapid cooling, and maintenance of the proper low storage temperature. If climatic, cultural, or maturity factors predispose the fruit to breakdown, then the problem may not be avoided, but storage life may be lengthened and intensity of the disorder reduced by good temperature management. In some tests with watery breakdown of Bartlett pears in California, late season, high-maturity pears developed no breakdown if fruit was cooled within 1 day of harvest, about 5 percent breakdown with 2-day cooling, and over 10 percent breakdown with 3-day cooling. In other tests with similar pears that were cooled within 1 day of harvest, stored for 5 weeks, and then ripened, breakdown incidence was 0 percent after −1°C (30°F) storage, almost 2 percent after 0.5°C (33°F) storage, and about 3.5 percent after 2°C (36°F) storage.

## Bruising

Pears and apples can be damaged by bruising. Roller bruising affects the surface of the fruit, resulting in brown areas or bands where rolling, rubbing, or vibration occurred, and damage is usually not apparent below the surface. Impact bruising affects the flesh, resulting in brown spots under the skin that may not be visible from the surface. The browning results from oxidation of phenols in the presence of an oxidizing enzyme, primarily polyphenoloxidase. Climatic differences in severity are apparently related to the level of phenols or the relative activity of the browning enzyme. Any mechanical injuries can stimulate ethylene production and thus speed fruit ripening and deterioration. Any surface damage can result in inoculation by fruit rot organisms and in more rapid water loss from the fruit.

**Roller bruising.** This can occur whenever fruit have freedom to move and rub or vibrate against some hard surface—against bins or package surfaces, packing belts, or other fruit. Pears are generally more susceptible to roller bruising damage than are apples, and Asian pears appear highly susceptible. As Bartlett and Anjou pears remain in storage they normally become more susceptible to injury, and packaging should occur within 1 to 2 weeks of harvest. Data on other cultivars are unavailable. Avoidance of roller bruising involves prevention of fruit rubbing or vibrating at all stages, from harvest to market.

**Impact bruising.** Injuries can follow any impact of the fruit, and incidence and severity of bruises increase with increasing height of drop of the fruit. Impact bruising is important because of its effect on fruit appearance, and because the injury induces higher respiratory activity of the tissue, thus reducing storage life. Some reports suggest that symptoms are less apparent in fruits that are impacted after controlled atmosphere storage.

## Maturity

Maturity changes in pears and apples include surface color, seed color, flesh firmness, soluble solids content, starch content, titratable acid content, respiratory rate, ethylene production, and production of other flavor and odor constituents. Suggested maturity indices have included all of these as well as time (days from full bloom), accumulated heat units (e.g., degree-days above 7°C [45°F]), fruit size, and various combinations of these.

Most possible maturity indices have limitations. Ethylene production and respiratory rate changes, for example, occur too late or are too variable to be useful for timing of harvest. Some other changes are too subtle to be effective. Most commonly used pome fruit maturity guides involve flesh firmness, soluble solids content, starch disappearance, surface color, seed color, or some combination of these.

In California, Bartlett pear standards utilize an index combining firmness and soluble solids content that is modified by fruit diameter. An alternate surface-color standard is available. Most legal standards define minimum maturity; California also imposes a maximum maturity standard (minimum flesh firmness level) for Bartlett pears destined for processing.

Apple maturity in California is based upon orchard surveys of flesh firmness and soluble solids content. Current apple maturity studies focus on predicting a time when changes in various maturity measures will occur. In California and elsewhere there is interest in use of a starch disappearance index for Granny Smith apples.

## Maturity versus ripening of European pears

European pears that ripen on the tree typically develop poor texture, lack juiciness, and may lack the characteristic flavor of the cultivar. Thus they are harvested when physiologically mature but still quite firm, then ripened before processing or consumption. Freshly harvested Bartlett pears, especially those harvested early in the season, ripen slowly and unevenly if placed immediately into a ripening environment (18° to 24°C [64° to 75°F]) without time in cold storage. Anjou pears can remain hard and unripe for more than a month under the same conditions. However, the same fruits ripen quickly and uniformly if first stored at low temperature (−1° to 0°C [30° to 32°F]) for 2 to 8 weeks, depending upon the cultivar and maturity. During storage, ethylene precursors form within the pear tissue so that fruit placed at ripening temperature increase their ethylene production and ripen uniformly. Bartlett pears can be ripened without cold storage by applying ethylene to the freshly harvested fruit.

A preconditioning treatment is in increasing use in California for Bartlett pears that are transported to market without cold storage. Warm, freshly harvested fruit are packed, exposed to 10 to 100 ppm ethylene at 20° to 25°C (68° to 77°F) for 24 hours, then forced-air cooled and transported to market. The fruit lose little flesh firmness during the treatment and thus do not become more subject to handling damage, but they ripen quickly and uniformly when warmed during marketing.

## Temperature management

As a general rule, pome fruits respond best to rapid cooling and storage at as low a temperature as is possible without danger of freezing or low-temperature injury. For Bartlett pears the recommended temperature is 0° to 0.5°C (32° to 33°F) for up to 1 month of storage and −1°C (30°F) if fruit is to be stored more than one month. The lowest safe storage temperature is related to the extent of temperature fluctuation in the room and the soluble solids content of the fruit. Freezing injury in pears and apples usually starts in the core area, which has the lowest soluble solids content. Therefore, knowledge of the soluble solids content of the core tissue is necessary to predict the lowest safe storage temperature.

Some apple cultivars are subject to internal browning from low-temperature injury if held near 0°C (32°F). Specific storage temperature recommendations are available for some cultivars, but they may need confirmation under localized growing conditions. In California, at present, Granny Smith storage is at 0.5°C (33°F) and Yellow Newtown storage is at 3.5° to 4.5°C (about 38° to 40°F).

Cooling delays are associated with shortened storage life, flesh softening, increased physiological disorders, and postharvest diseases. Thus fruit should be cooled as soon as possible. Bartlett pears in California should be cooled to near storage temperature within 24 hours of harvest. The importance of cooling speed varies greatly with cultivar. Evidence that faster cooling benefits apples is increasing, especially for long-term storage. When antioxidants are used to delay storage scald development, a cooling delay of up to one day may be needed. Rapid cooling of the fruit after this delay then becomes more important.

Controlled atmospheres are often used when apples and pears are stored longer than 3 months. Most fruits are CA stored in field bins, often with unsorted, field-run fruit. However, some apples are dumped, sorted, sized, and refilled into their bins before moving to CA storage.

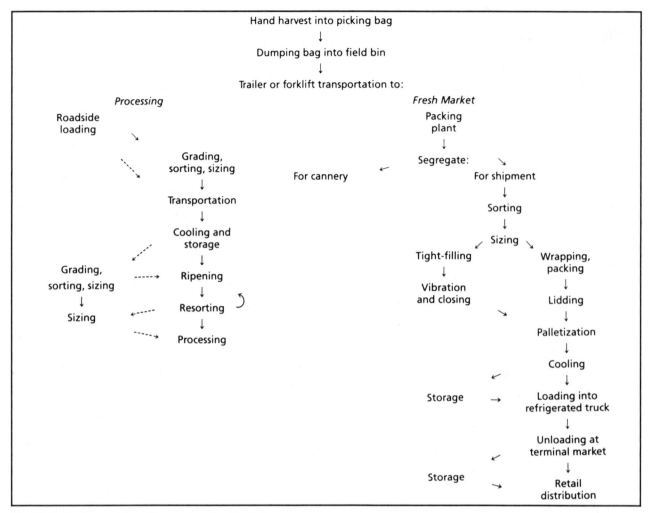

**Fig. 24.1. Postharvest handling system for Bartlett pears.**

## Relative humidity

A relative humidity of about 95 percent, considered ideal for pears and apples, is difficult to maintain using large-surface cooling coils alone. Supplemental humidification, especially with fog spray nozzles, is now widely used. Unfortunately the water added in such a system increases the coil defrost problem. Using perforated polyethylene bags or liners to maintain humidity inside the shipping container, once a common practice, is now seldom done in California because it interferes with temperature management. Wax or plastic coatings on corrugated containers are extensively used to provide a moisture barrier for California Bartlett pears.

## Handling

Fruit are hand picked into bags, gently transferred into field bins, and transported to the packing facility. Apples, which are often stored in field bins to await later packing, may be drenched with a scald inhibitor and fungicide and sometimes treated with calcium chloride for bitter pit control before storage. Use of

these treatments for short-term storage in California is limited, as costs often exceed the benefits.

Fruit to be packed for fresh market are dumped (water submersion dumps are common), washed, presized to eliminate undersize fruits, and sorted. Fruit may be segregated for fresh market, processing, and by-product use. Frequently over 50 percent of California Bartlett pears are diverted from a fresh market sorting line to processing.

Apples in the U.S. are mostly tray-packed, often with fruit in top trays individually wrapped. Smaller apples may be bagged. Many California apples are volume-filled into corrugated containers for fresh marketing.

Pears may be wrap-packed or tight-fill packed. Many California Bartlett pears are tight-fill packed into corrugated containers. Most other pears are wrap-packed into corrugated containers, although some wood containers are still in use. Many Asian pears are wrapped, using paper or plastic materials, and some are packed into plastic trays.

Packed containers are segregated by fruit size and stacked onto unitized pallets, which are designed and manufactured to be moved directly to the receiving markets. Final cooling of these pallet loads is done after packing, commonly by forced-air cooling in California.

A flow diagram for the handling of California Bartlett pears (fig. 24.1) shows the steps involved in fresh market handling and processing of this commodity.

## STONE FRUITS

Stone fruits, as a group, constitute one of the largest fresh market crops in California. Although a large volume of stone fruits also moves into processing (prunes, cling peaches, apricots, and cherries), fresh market production and handling of stone fruits is a distinct industry. Fresh nectarines, peaches, and plums constitute the bulk of this industry, and current shipments approach 50 million packages of more than 450 cultivars of these three fruits. Much smaller volumes, and only a few cultivars, of sweet cherries and apricots are also shipped fresh. In the major shipping districts of California, harvest of early cultivars of all of these fruits starts in May, and harvest of late cultivars of nectarines, peaches, and plums is completed in September.

### Deterioration problems

Stone fruits are characteristically soft fleshed and highly perishable, and have a limited market life potential. Because of the large number of cultivars spanning a 5-month harvest season, long-term storage has historically not been a concern in California. Potential opportunities for export marketing, combined with the desire to store some late-season cultivars to extend the marketing season, has increased interest in procedures to extend postharvest life.

**Fruit decay organisms.** Postharvest loss of stone fruits to rotting organisms is considered the greatest deterioration problem. Some market studies, however, show that much fruit rotting is secondary to another problem, especially mechanical injury of the fruit. Worldwide, the most important pathogen of fresh stone fruits is Botrytis rot, caused by the fungus *Botrytis cinerea*. In California an even greater cause of loss is brown rot, caused by the fungus *Monilinia fructicola*. Good orchard control is vital to minimizing the problem after harvest. It is common to use a postharvest fungicidal treatment against these diseases. The fungicide is often incorporated into a fruit wax for uniformity of application. Careful handling to minimize fruit injury, sanitation of packing equipment, and rapid, thorough cooling as soon after harvest as possible are also important for effective disease suppression.

**Internal breakdown.** Another major cause of deterioration is a low-temperature or chilling injury problem generically called internal breakdown. The disorder can manifest itself as dry, mealy, woolly, or hard textured fruit, flesh or pit cavity browning, flesh translucency usually radiating through the flesh from the pit, intense red color development of the flesh ("bleeding"), also usually radiating from the pit, and complete loss of characteristic fruity flavor. Only sweet cherries appear unaffected, although there are large intercultivar differences in susceptibility among the other stone fruits.

Most susceptible cultivars show the greatest problems at temperatures between about 2° and 7°C (36° and 45°F), and symptom development is slowed at 0°C (32°F) or below. Fruit therefore should be cooled and held near or below 0°C (32°F) if possible. Storage or transport at around 5°C (41°F) can significantly reduce the postharvest life. Several treatments to delay development of the disorder, including heating fruit before storage (pretreatment) or during storage (intermittent warming), and various controlled-atmosphere treatments have been under study, but none are currently recommended.

**Mechanical injuries.** Stone fruits are susceptible to mechanical injuries including impact, compression, and abrasion (or vibration) bruising. Careful handling and packing equipment that minimizes such injuries are important because the injuries result in reduced appearance, accelerated physiological activity (thus shorter potential market life), potentially more inoculation by fruit decay organisms, and greater water loss.

Protection against abrasion bruising involves procedures to reduce vibrations during transport and handling and immobilize the fruit so they are not affected by vibrations. Steps can include installing air suspension systems on axles of field and highway trucks and plastic side liners inside field bins, installing special bin top pads before transport, avoiding abrasion on the packing line, and using packing procedures that immobilize the fruit within the shipping container before they are transported to market.

**Water loss.** Loss of 4 to 5 percent of water can cause visual shrivel in stone fruits. While there are large intercultivar differences in susceptibility to water loss, all stone fruits must be protected for best postharvest life. Rapid cooling after harvest followed by storage under constant low temperature and high relative humidity are the main means of limiting water loss. Fruit waxes that are commonly used as carriers for postharvest fungicides can reduce the rate of water loss by up to 35 percent. Because fruit shrivel results from cumulative water loss throughout handling, it is important to maintain low temperature and high relative humidity throughout harvesting, packing, storage, transport, and distribution.

### Harvest maturity

The maturity at which stone fruits are harvested greatly influences their ultimate quality. Harvest maturity controls the fruit's flavor components, physiological deterioration problems, susceptibility to mechanical injuries, resistance to moisture loss, susceptibility to invasion by rot organisms, and ability to ripen.

Stone fruits that are harvested too soon may fail to ripen properly or may ripen abnormally. Immature fruit typically soften slowly and irregularly, never reaching the desired melting texture of fully matured fruit. Green color may never fully disappear. Because immature fruit lack a fully developed surface cuticle, they are more susceptible to water loss than properly matured fruit. Immature fruit have lower soluble solids content, higher acids, and higher starch content than properly matured fruit, all of which contribute to inadequate flavor development. Low-maturity fruit are more susceptible to both mechanical bruising and the development of internal breakdown symptoms than properly matured fruit.

Overmature fruit have a shortened postharvest life, primarily because they are already approaching a senescent stage at harvest. Such fruit have partially ripened, and the resulting flesh softening renders them highly susceptible to mechanical injury and microbial invasion. By the time such fruit reach the consumer they may have become overripe, with poor eating quality including off-flavors and mealy texture.

The optimum maturity for stone fruit harvest must be defined for each cultivar. The highest maturity at which a cultivar can be successfully harvested is influenced by postharvest handling and temperature management procedures. Maturity selection is more critical for distant markets than for local markets, but this does not necessarily mean lower maturity. When stone fruits are harvested at low maturity to reduce senescent breakdown problems during long-distance marketing, they become more susceptible to losses from internal breakdown.

### Field handling

Physical wounding of stone fruits can occur at any time from harvest until consumption. Good worker supervision assures adequate protection against impact bruising during picking, handling, and transport of fruit. Protection against roller bruising may require modifications of transport equipment and procedures, as described in chapter 6, *Preparation for Fresh Market*. If severe injuries are encountered, consider using a top bin pad that maintains a slight tension across the top fruit. It is also helpful to grade farm roads to reduce roughness, avoid rough roads during transport, and establish strict speed limits for trucks operating between orchards and packinghouses.

### Packing and handling

Stone fruits are transported from orchard to packinghouse or cooler as soon as possible after harvest. Fruit should be shaded during any delay between harvest and transport. Stone fruits are often cooled as soon as they arrive from the orchard, then packed cool the next day. Some fruit are packed upon arrival from the orchard and cooled immediately after packing.

At the packinghouse the fruit are dumped (mostly using dry bin dumps) and cleaned. Here trash is removed and fruit may be detergent washed. Peaches are normally wet-brushed to remove the trichomes (fuzz), which are single cell extensions of epidermal cells. Cherries pass through a cluster cutter, which cuts stems to separate the fruit. Waxing and fungicide treatment may follow. Water-emulsifiable waxes are normally used, and fungicides may be incorporated into the wax. Waxes are applied cold and no heated drying is used.

Sorting is done to eliminate fruit with visual defects and sometimes to divert fruit of high surface color to a high-maturity pack. Attention to details of sorting line efficiency, as described in chapter 6, *Preparation for Market*, is especially important with stone fruits where a range of colors, sizes, and shapes of fruit can be encountered. Sizing segregates fruit by either weight or dimension. Sorting and sizing equipment must be flexible to efficiently handle large volumes of small fruit or smaller volumes of larger fruit.

Most California stone fruits are now volume-fill packed, with the fruit automatically filled by weight into shipping containers. Some fruits are packed into trays (especially peaches and some nectarines). Mechanical place-packing units use hand-assisted fillers where the operator can control the belt speed to match the flow of fruit into plastic trays.

Limited volumes of stone fruits are "ranch packed" at point of production. In a typical operation, fruit are picked into buckets, which are carried by trailer to the packing area. There packers work directly from the buckets to select, grade, size, and pack fruit into plastic trays. Ranch pack operations are typically nonmechanized, and the fruit are not washed, brushed, waxed, or fungicide treated. Because of the lesser handling of the fruit, a higher maturity standard can be used, and growers can benefit from increased fruit size and greater yield.

### Temperature management

Cooling requirements depend in part upon the scheduling of the packing operation. Fruit can be cooled in field bins using forced-air cooling or hydrocooling. Forced-air cooling in side-vented bins can be by either the tunnel or the serpentine method (see chapter 8, *Cooling Horticultural Commodities*. Hydrocooling is normally done by a conveyor type hydrocooler.

Cooling of packed fruit is normally by forced air, using either the tunnel or cold wall method. Cherries are sometimes hydrocooled in bulk after they have been prepared for packing but before they are sized or filled into the container. This "in-line" hydrocooling method is only feasible with cherries, which because of their small size can be completely hydrocooled in a few minutes.

Fruit in field bins can be cooled to intermediate temperatures (5° to 10°C [41° to 50°F]) provided packing will occur the next day. If packing is to be delayed beyond the next day, then fruit should be thoroughly cooled in the bins to near 0°C (32°F).

Fruit in packed containers should be cooled to near 0°C (32°F). Even fruit that were thoroughly cooled in the bins will warm substantially during packing and should be thoroughly recooled after packing. Forced-air cooling is normally indicated after packing (sweet cherries may be hydrocooled with the in-line method described above). An exception to the need for cooling after packing would be a system that handles completely cold fruit and provides protection against rewarming during packing.

Stone fruit storage and transport should be at or below 0°C (32°F). Maintaining these low temperatures requires knowledge of the freezing point of the fruit and of the temperature fluctuations in the storage system. Holding stone fruits at these low temperatures minimizes both the losses associated with rotting organisms and the deterioration resulting from internal breakdown.

A flow diagram for the handling of California stone fruits (fig. 24.2) shows the steps involved in mechanized and ranch packing.

## REFERENCES

### Pome fruits

1. Bartsch, J. A., and G. D. Blanpied. 1984. *Refrigeration and controlled atmosphere storage of horticultural crops.* Ithaca, NY: Northeast Reg. Agric. Eng. Serv. Publ. 22. 42 pp.

2. Blanpied, G. D., E. D. Markwordt, and C. D. Londington. 1962. *Harvesting, handling and packing apples.* Cornell Univ. Ext. Bull. 750. 32 pp.

3. Blanpied, G. D., and R. M. Smock. 1982. *Storage of fresh market apples.* Cornell Univ. Info. Bull. 191. 19 pp.

4. Cappellini, R. A., M. J. Ceponis, and G. W. Lightner. 1987. Disorders in apple and pear shipments to the New York market, 1972-1984. *Plant Dis.* 71:852-56.

5. Porritt, S. W., M. Meheriuk, and P. D. Lidster. 1982. *Postharvest disorders of apples and pears.* Agr. Can. Publ. 1737/E. 66 pp.

6. Ryall, A. L., and W. T. Pentzer. 1982. *Handling, transportation, and storage of fruits and vegetables. Vol. 2. Fruits and tree nuts.* 2nd ed. Westport, CT: AVI Publ. Co. 610 pp.

7. Tyler, R. H., W. C. Micke, D. S. Brown, and F. G. Mitchell. 1983. *Commercial apple growing in California.* Univ. Calif. Coop. Ext. Leaflet 2456. 20 pp.

### Stone fruits

1. Ceponis, M. J., R. A. Cappellini, J. M. Wells, and G. W. Lightner. 1987. Disorders in plum, peach and nectarine shipments to the New York market, 1972-1985. *Plant Dis.* 71:947-52.

2. LaRue, J. H., and R. S. Johnson, eds. 1989. *Peaches, plums and nectarines—growing and handling for fresh market.* Univ. Calif. Davis Publ. 3331. 246 pp.

3. Lill, R. E., E. M. O'Donoghue, and G. A. King. 1989. Postharvest physiology of peaches and nectarines. *Hortic. Rev.* 11:413-52.

4. Mitchell, F. G. 1987a. Influence of cooling and temperature maintenance on the quality of California grown stone fruit. *Internat. J. Refrig.* 10:77-81.

5. ———. 1987b. Preparing peaches and nectarines for export markets. *Orchardist of N. Zeal.* 60(5):150-52.

6. Ryall, A. L., and W. T. Pentzer. 1982. *Handling, transportation and storage of fruits and vegetables. Vol. 2. Fruits and tree nuts.* 2nd ed. Westport, CT: AVI Publ. Co. 610 pp.

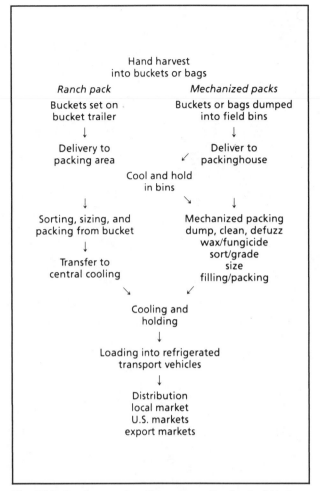

**Fig. 24.2. Postharvest handling system for fresh shipping stone fruits.**

# 25

# Postharvest Handling Systems: Small Fruits (Table Grapes, Strawberries, Kiwifruit)

F. Gordon Mitchell

California has a long history of producing and marketing small fruits; among these are table grapes, strawberries, bushberries, and kiwifruit. These fruits are typically tender and subject to severe losses without careful handling and protection. Handling problems and procedures for table grapes, strawberries, and kiwifruit are described in this chapter.

## TABLE GRAPES

The table grape, a nonclimacteric fruit with a relatively low rate of physiological activity, is subject to serious water loss following harvest, which can result in stem drying and browning, berry shatter, and even wilting and shriveling of berries. Botrytis rot infection (gray mold), caused by the fungus *Botrytis cinerea*, requires constant attention and treatment during storage and handling. The bloom (natural wax) on the grape berry's surface is a primary appearance quality factor. Rough handling and rubbing destroys this bloom, giving the skin a shine rather than the more desirable luster effect.

### Cultivars

Of the many table grape cultivars, only a few constitute most of the commercial shipments. In California the major cultivar is Thompson Seedless (Sultanina), marketed mostly during the summer months, usually soon after harvest. This cultivar is capable of relatively short storage. Longer storage is achieved with some fall cultivars, including Emperor, Ribier, Almeria, and Calmeria. Present interest centers around other recently introduced seedless cultivars. The red Flame Seedless has made the greatest impact, but many others are in development or trial. Little is known about the specific postharvest requirements of these new cultivars.

### Maturity

Grapes are harvested when mature, based upon the soluble solids content (or ° Brix or Balling) of the cluster. Titratable acidity and sugar-to-acid ratio are also used as maturity indices. The minimum requirements vary with cultivar and growing area. Colored cultivars also have minimum color requirements based on the percentage of berries on the cluster that show a certain minimum color intensity and coverage.

### Water Loss

Because stems and fruit are susceptible to deterioration from water loss, grapes are normally cooled as soon as possible after harvest. Even a few hours' delay at field temperatures can cause severe drying and browning of cluster stems, especially in the hottest growing regions. Rapid cooling, mostly by forced air, has become widely used. Grapes do not tolerate the wetting associated with hydrocooling.

### Fruit Rots

Botrytis rot of grapes is not sufficiently avoided by fast cooling alone. It is standard practice in California to fumigate with sulfur dioxide ($SO_2$) immediately after packing followed by lower dose treatments at periodic intervals during storage. Formulas for calculating the initial and subsequent $SO_2$ fumigation dosages are available.

The usual method of fumigation in California is periodic introduction of $SO_2$ into the storage room or fumigation chamber. In some other growing areas, a continuous low level of $SO_2$ is maintained in the storage. There is some use of $SO_2$-generating pads, especially during export marketing where grapes are in ocean transport for extended periods. These pads have sodium metabisulfite incorporated into them to allow slow release of $SO_2$ during transit and marketing.

One problem associated with $SO_2$ fumigation of grapes is the constant potential for injury to the fruit and stems. Injured tissue first shows bleaching of color, followed by sunken areas where accelerated water loss has occurred. These injuries first appear on the berry where some other injury has occurred, such as a harvest wound, transit injury, or breakage at the cap stem attachment. Symptoms may also be seen around the cap stem, and slowly spreading over the berry. Careful attention to $SO_2$ treatment procedures is necessary to minimize this damage.

Another problem with $SO_2$ fumigation of grapes is the level of sulfite residue remaining at time of final sale. Sulfur dioxide was once included on the "Generally Recognized As Safe" (GRAS) list of chemicals, for which no registration is required. Heavy usage of sulfites in some other foods has caused a change in regulation, because some people are dangerously allergic to sulfites. While much work is underway on the residues associated with gaseous fumigation and sodium metabisulfite pads, a temporary tolerance of 10 ppm sulfite residue has been established, and registration for use at this level is pending. This is subject to change on short notice as further information is accumulated.

## Packing Systems

The various methods of handling table grapes in California are summarized in figure 25.1. Most California table grapes are now packed in the field. A few grapes are "vine packed;" that is, the picker does all of the quality selection, trimming, and sorting and packs the fruit directly into shipping containers. This system minimizes rehandling of the fruit but makes supervision and consistent quality control difficult.

**Field packing.** The most common field-packing system is the "avenue pack." The fruit is picked and placed into picking lugs. Usually, the picker also trims the cluster. The picking lug is then transferred a short distance to the packer, who works at a small portable stand in the avenue between vineyard blocks. It is common for the packer and several pickers to work as a crew. Packing materials are located at the packing stand, which also shades the packer. With many packing stands around the vineyard, supervision is more difficult than in a packing shed, but easier than for vine packing (described above). Lidding is done in the field for both packing systems. Substandard fruit can be accumulated in field lugs for transport to wineries.

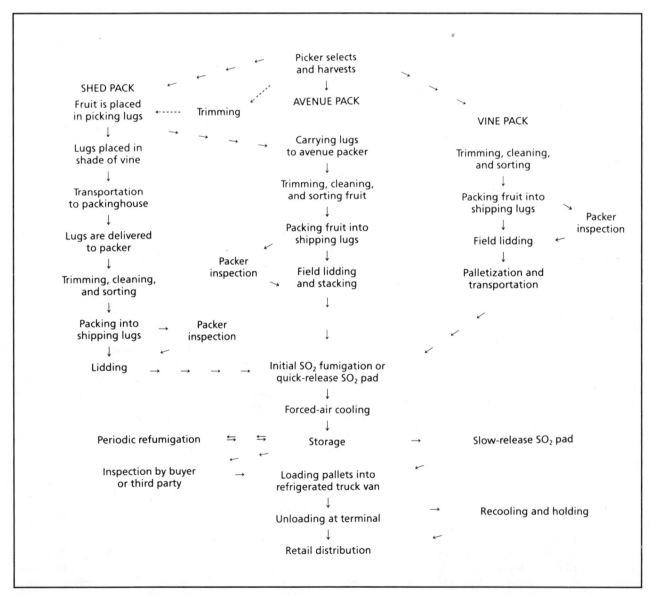

**Fig. 25.1. Handling system for table grapes in California.**

**Shed packing.** Shed-packed fruit is harvested by pickers and placed in field lugs without trimming, then placed in the shade of the vines to await transport to the shed. At the packing shed the field lugs are distributed to packers who select, trim, and pack the fruit. Often two different grades are packed simultaneously by each packer to facilitate quality selection.

Grapes are nearly always packed on a scale to facilitate packing to a precise net weight, whether field or shed packed. Packed lugs are subject to quality inspection and check weighing before lidding.

Some grapes for export and some high-quality grapes for domestic markets are wrap packed. Here, individual clusters are wrapped in tissue before placement in the shipping container. Some grapes are also filled into small plastic bags before packing into containers. The plastic bags often have some slitting to facilitate airflow.

**Packages.** The California grape industry has been slow to adopt corrugated containers. One reason is the ease of handling wooden containers in the field, where much packing is done; another is concern over container strength during long-term storage at high relative humidity. Corrugated container use by the California grape industry is increasing, especially for grapes that are not stored for long periods. Some polystyrene containers are used for California table grapes.

**Palletization.** After packing and lidding, grapes are palletized, usually on disposable pallets. Often loaded pallets coming from the field pass through a "pallet squeeze," a device that straightens and tightens the stacks of containers. These pallet loads are unitized, usually by strapping or netting. Some use of palletizing glue has begun in shed-packing operations. This glue bonds the containers vertically on the pallet so that only horizontal strapping is required.

## Cooling and Storage

After palletization is complete, the pallets are moved either to a fumigation chamber for immediate $SO_2$ treatment or to a forced-air cooler where fumigation is done at the end of the day's packing. In either case, cooling starts as soon as possible. Many grape forced-air coolers in California are designed to achieve $7/8$ cooling in 6 hours or less.

After cooling is completed, the pallets are moved to a storage room to await transport. Ideally the storage room operates at $-1°$ to $0°C$ ($30°$ to $32°F$) and 90 to 95 percent RH, with a moderate air flow. Fumigation is most commonly repeated every 7 to 10 days during the storage period using a lower concentration of $SO_2$. Grapes should be regularly monitored during storage for physiological deterioration, fruit rot, $SO_2$ injury, and stem drying.

When grapes are loaded for transport they often receive an additional $SO_2$ fumigation. Their market life is limited, however, because seldom is repeat fumigation available in receiving markets. Unless $SO_2$ fumigation is available, the receiver must order grapes for immediate needs, and must complete distribution and marketing within a reasonable time after arrival. An exception would be when $SO_2$-generating pads are placed in the container before shipment.

**Sulfur dioxide fumigation.** The proper use of sulfur dioxide is necessary to control growth and spread of Botrytis rot during storage and distribution. Pending U.S. registration limits residues in grapes to 10 ppm. There are also limits on the number of repeat fumigations allowed, depending upon cultivar.

Various recommendations have been made for application of the initial fumigation and repeat lower dose fumigation during storage. Formulas are available for calculating the proper amount of $SO_2$ to apply. Typically the initial dose is 5,000 ppm or lower, and subsequent fumigations are at 2,500 ppm or lower, with the fumigant vented or scrubbed after 20 minutes. Wall materials, type of container, and fill density of the room may all affect the amount applied to achieve the desired concentration.

Some studies have been done with more frequent fumigations at lower concentration, and some commercial use of the procedure has begun. Typically 200 to 400 ppm is applied three times a week, so total $SO_2$ usage is less than with the older system.

To maintain grape quality and limit losses to Botrytis rot with any of these fumigation procedures, good management and supervision of table grape production and handling are vital. Fruit coming from vineyards should be as free of infections and injuries as possible. Thorough and rapid cooling should be done as soon as possible after harvest. Fruit should be stored at $-0.5°$ to $0°C$ ($31°$ to $32°F$) pulp temperature throughout its postharvest life.

## STRAWBERRIES

Strawberries are among the most perishable fresh fruits, yet worldwide they are being marketed successfully in increasing volume. Much of the marketing is at a great distance from the point of production, thus effective handling procedures are required to prevent excessive deterioration. In California since about 1950, a large fresh strawberry shipping industry has become established based largely on fruit delivery to markets 3,000 to 5,000 kilometers (2,000 to 3,000 miles) distant.

California strawberries are produced primarily in the coastal valleys and plains of central and southern California. Production starts in late winter in the south and continues through late fall in the north. The fresh market is the primary outlet, followed by the freezing and jam manufacturing market. The temperature in production areas is generally cool, although warm summer days with highs of 30° to 35°C (86° to 95°F) sometimes occur.

The primary market for fresh strawberries extends across the U.S. and into southern Canada. Some fruit are exported overseas by air. At one time, about one-third of the volume was shipped by air, but this has declined to less than 10 percent, the balance moving by surface transport (highway trucks). Surface-transported strawberries must have a 5- to 7-day market life when shipped to Eastern cities.

## Problems in Postharvest Handling

Strawberries have very tender skins and are easily injured. They are subject to invasion by fruit-rotting organisms and have one of the highest respiratory activity rates of all fresh fruits. Each of these factors poses a severe loss potential to the fruit.

**Fruit rots.** The greatest single cause of loss to fresh strawberries is "gray mold" or Botrytis rot, caused by the fungus *Botrytis cinerea*. This organism can invade the berry blossom and remain latent until ripening commences, or it can enter wounds that occur during harvesting or handling, or mycelium of the fungus growing in the soil may penetrate berries in contact with the soil. Surface mycelia on infected berries directly penetrate nearby berries to produce an ever-enlarging "nest" of rotting berries. The organism continues growth at 0°C (32°F) but at a very slow rate.

Many infections have probably occurred by the time the fruit are picked. Often small lesions are overlooked by pickers; sometimes fairly large lesions are overlooked. Supervision of picking and grading operations to try to avoid all visible lesions in the picked fruit is critical to minimizing postharvest losses.

Another fruit-rotting organism, *Rhizopus* sp., can cause severe losses at warm holding temperatures, but the advent of good temperature management has made it a relatively minor problem because it does not grow at recommended low holding temperatures.

**Fruit shrivel.** Strawberries are subject to rapid water loss, which can cause the fruit to shrivel and appear old and deteriorated, and cause the fruit calyx (the green cap or hull) to wilt and dry. These symptoms affect the fruit's sales appeal before they affect actual eating quality.

**Overripeness.** Because of their high rate of physiological activity, strawberries quickly pass from ripeness to an overripe or senescent state if held at warm temperatures.

**Bruising.** Strawberries are subject to serious injury during harvest and postharvest handling. Studies show that much of the injury occurs in the field during picking and packing, and variation in the magnitude of mechanical injuries caused by human pickers is so great as to mask any other causes of deterioration.

Vibration (or roller) bruising is not usually considered serious for strawberries. Because of its elasticity, the fruit apparently absorbs much of the motion rather than transmitting it and thus limits its exposure to damage in transport.

Often, berries are cut by basket rims as a result of being packed over the tops of the open-mesh plastic baskets that are commonly used for packaging in California. Other overpacked fruit are injured by abrasion against the corrugated crate. Injuries from this type of packing cause both a direct loss and subsequent Botrytis rot losses. A redesign of the pack to avoid overfilling and good picker supervision would help reduce berry injuries and the associated fruit rot problems.

## Handling Methods

In California, strawberries are harvested, graded, and packed in the field by the picker, who is also responsible for removing rotted or overripe fruit from the plants (fig. 25.2). This combined operation precludes rehandling of the fruit at packing facilities

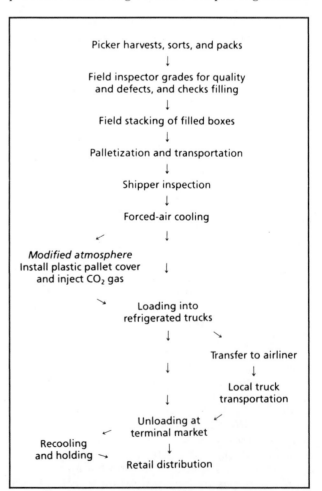

**Fig. 25.2. Postharvest handling system for strawberries.**

and reduces the time delay between harvest and cooling. Major attention must, however, be directed to picker supervision if good quality control is to be achieved.

**Maturity.** Strawberries are usually at least three-fourths colored when harvested for the fresh market. Riper fruit can sometimes be harvested for nearby markets where handling delays can be avoided. Fruit that become too ripe for fresh market handling are diverted to the freezer. Depending on weather conditions, harvesting may be necessary several times per week for uniform fruit maturity.

**Containers.** An open corrugated crate (outer container) that holds 12 1-pint (about ½-liter) baskets is most commonly used for strawberries. When filled, crates are bundled into pairs, using special wire ribs. The ribs serve as carrying handles and stacking tabs for palletization. Most of the baskets are open-mesh, ribbed plastic. Side ventilation of the crates facilitates air movement for initial cooling and for temperature maintenance.

**Palletization.** Most strawberries are shipped on unitized pallets. The "one-way disposable" wooden pallets are placed on flatbed trucks in the field and stacked with berry crates immediately after harvest, and they remain intact until final distribution in the market area. The tight stacking required for stability and transport-load density must be compensated for in cooling and handling operations.

**Cooling delays.** Delays of more than one hour between harvest and cooling can accelerate fruit deterioration. Growers should schedule frequent deliveries of small or partial loads of fruit to the cooler to minimize delays.

**Cooling method.** Forced-air cooling is virtually universal for California strawberries. This is the only method that meets the need for rapid cooling and avoids fruit wetting, which strawberries do not tolerate. Most strawberry coolers are capable of seven-eighths cooling in 2 to 3 hours. Rapid cooling slows the rate of fruit deterioration and prepares the berries for immediate shipment.

**Modified atmospheres.** Many loads are shipped under modified atmosphere transit. The most common procedure is to cover the entire loaded pallet with a plastic bag that is sealed to a plastic pallet cover beneath the berry crates. After sealing, the bag is injected with carbon dioxide to create an approximate 15 percent $CO_2$ atmosphere. If sealing is done correctly, this atmosphere can be maintained throughout transit, with the $CO_2$ produced by fruit respiration about matching $CO_2$ leakage from the bag. Pallet bags are installed after cooling, usually just before loading for transport.

Some shippers use the carbon dioxide treatment extensively, others sparingly. One shipper, evaluating a series of arrivals, found significant benefit (primarily fruit-rot reduction) only when the treatment was used after a foggy or rainy period. That shipper has subsequently used the treatment only for shipments judged to have a significant rot potential. This procedure is apparently successful.

In a laboratory study of Botrytis rot on strawberries, elevated-$CO_2$ atmospheres yielded no measurable benefit at or below 2°C (36°F). Thus, when these transit temperatures can be maintained, the addition of carbon dioxide appears to be unnecessary. Realistically, however, it is common for strawberries to encounter an average 5°C (41°F) temperature during transport, and at that temperature the high $CO_2$ atmosphere does suppress Botrytis rot spread.

**Loading.** All surface transport is now by refrigerated truck vans, and virtually all consists of unitized pallet loads. Because of heat leakage through the van walls, pallets should be center-loaded; that is, an air space should be left along the two sides of the van, between the walls and the pallets rather than down the center. Stabilizing blocks prevent the load from shifting during transit. When pallets are placed against the side walls, berry temperatures sometimes are 5° to 10°C (9° to 18°F) warmer near the wall than in the midst of the load. The open pallets under the fruit also facilitate air movement and help avoid warming of the load bottom.

Strawberries are often shipped in mixed loads with other fruits and vegetables. It is vital that all other products in the load have storage requirements similar to those of strawberries. Other products' temperatures should always be monitored before loading so that the strawberries are not warmed by contact with an improperly cooled product.

Many strawberry loading docks are refrigerated and designed so the refrigerated truck van can be entered directly from the refrigerated dock. The truck van should be thoroughly refrigerated before loading begins. These procedures ensure low fruit temperatures during the loading operation.

## Transportation

### Surface transport

Some refrigerated truck vans that transport strawberries have air-suspension systems that reduce transit vibrations more than 50 percent from spring suspension systems. This greatly reduces the potential for injury to the fruit.

Many truck van refrigeration systems cannot maintain fruit temperatures below about 5°C (41°F), which is too warm for good strawberry protection. Often, temperature-regulating equipment lacks the accuracy to achieve a lower temperature without danger of fruit freezing. While transport conditions need improvement, berries should still be at 0°C (32°F) pulp temperature when loaded to achieve the lowest possible average fruit temperature during transport.

Air circulation capacity is also limited in many transport vehicles. If the fruit is not thoroughly cooled before loading, or was allowed to warm during the loading operation, proper cooling may never be achieved.

Transport time by highway truck can range up to 4 days for the most distant markets. An additional 1 to 3 days are needed for marketing. Strawberry carriers that regularly service those more distant markets are often more particular in their transport equipment requirements because they recognize by experience the danger of heavy losses when handling this commodity.

### Air transport

The 10 to 12 percent of fresh strawberry volume that moves by air is mostly destined for export or distant domestic markets (East Coast cities). Strawberries for air transport are thoroughly cooled before loading and often protected with modified atmospheres, using pallet bags and carbon dioxide injection. The fruit is normally transported from the production areas to air freight terminals in small refrigerated trucks. After unloading at the terminal it is sometimes placed in cold storage when long delays (several hours) are expected. Unfortunately, however, it often remains out of refrigeration until it reaches the terminal market receiver.

Studies of the air shipment of strawberries indicate that unrefrigerated transit time is considerable. During the time in flight, about one-third of the total transit time, the temperature of the cargo hold is warmed to protect other freight. On the ground at intermediate terminals the temperature may be quite high, as much of the transporting occurs during summer. The fruit is normally warm upon arrival, often much warmer than at harvest, and considerable deterioration can occur. It is possible, therefore, for losses to be greater in air shipment than in truck transport despite the shorter transit time. Tests show that even though strawberries warm during transport, it is better to provide cooling whenever possible than not to cool at all.

### Terminal market handling

Market handlers of strawberries have historically sustained heavy fruit losses (25 percent in some cases) because of fruit deterioration. Improvements in postharvest cooling, combined with better surface transport conditions, have greatly reduced these losses. This has changed the attitude of many receivers toward strawberry handling and prompted them to seek further improvements.

Some market handlers provide special handling for strawberries to minimize losses. This occasionally includes establishing "strawberry rooms" at the distribution warehouse, where refrigeration and airflow capacities allow reasonably fast recooling of the berries on arrival. Warm berries arriving via air transport benefit greatly, as do many berries arriving by highway truck because transport temperatures are often higher than desired.

Terminal market handlers commonly load produce trucks for store distribution several hours before the start of delivery. Strawberries are often listed among the few items that receive special handling, and are held under refrigeration until just before the truck departs. This preserves temperature protection for the fruit, improves the receiver's chances of making a profit, and increases the likelihood of consumer satisfaction.

## KIWIFRUIT

The kiwifruit is a fairly recent introduction into commercial fruit production and marketing. Although California acreage is limited, this fruit has received considerable attention as a new fruit because it complements other fruits in the marketing system, and because much needs to be learned about its culture and handling.

Most production in California is located in the hot Central Valley. Because of the hot dry climate, kiwifruit must be well irrigated, but it develops very high soluble solids and matures earlier than in cooler climates. Most of the production area is relatively free of heavy winds so wind scarring of fruit is minimal, but care must be taken to avoid sunburn. Most of the production is of the Hayward cultivar.

### Physiology

Both carbon dioxide and ethylene production show a slight peak during early storage, then level out and continue a fairly constant rate through a long period of storage. This small early peak in respiratory activity may be a response to handling injuries (even the wound caused by picking the fruit), rather than a real acceleration in physiological activity related to maturation. Kiwifruit have an initial high starch content throughout the flesh before ripening, which disappears with time in storage. Soluble solids content (SSC) increases sharply in kiwifruit after harvest and may more than double during the first one to two months of storage. This SSC increase coincides with a decrease in starch content and with as much as a 50 percent decrease in titratable acid content.

Most dramatic of all changes during early storage is a reduction in flesh firmness. Flesh firmness typically declines by one-third to half during each month of air storage at 0°C (32°F), until the fruit is fully ripe (near 0.9 kilogram-force or 2 pounds-force using an 8-mm tip). This flesh softening parallels softening in other kiwifruit that remain on the vine. In maturity tests, fruit from four different harvests, taken over a six-week period, show essentially the same softening curve when the data are plotted by harvest date. The only variation from this curve occurs when fruit are harvested at an earlier, apparently immature stage and placed in storage. In those instances, softening is delayed for several weeks, and fruit remain firmer than later harvested fruit. Such fruit, however, have low SSC levels and ripen to unacceptable quality. Thus, reduced fruit quality nullifies any possible advantage in early harvesting to achieve delayed fruit softening during storage.

## Postharvest Disease Management

Most kiwifruit rotting problems are a result of infection by *Botrytis cinerea*. This organism grows and spreads slowly at 0°C (32°F) storage, but because of the long storage duration for kiwifruit it is a major cause of fruit loss. Botrytis rot can invade the fruit directly, but it also enters through wounds, invades dead floral parts or other organic matter on the fruit, and spreads from infected fruit to healthy surrounding fruit (nesting). It is important to maintain cleanliness in the vineyard, to avoid fruit injuries during handling, to brush the fruit to remove dead floral parts and other material on the fruit surface, to avoid contamination (such as juice from soft fruit), to cool the fruit rapidly, and to maintain a constant 0°C (32°F) storage temperature. Because Botrytis rot is associated with soft fruit, any practices that maintain flesh firmness during storage decrease the fruit-rotting problem.

## Maturity

Minimum maturity for kiwifruit in California is set at 6.5 percent SSC. Taste evaluations indicate that satisfactory flavor is achieved when ripe fruit reaches at least 14 percent SSC. In most vineyards this occurs by the time the SSC of freshly harvested fruit reaches 6.5 percent. This may thus be a reasonable minimum maturity index for kiwifruit, provided only freshly harvested fruit are sampled. Taste evaluations also show that flavor improves as the SSC of the ripened fruit increases to at least 16 percent; thus, delaying the harvest could improve eating quality of the fruit provided there is a continuing increase in the SSC level of the ripe fruit samples. By the time the ripe fruit SSC levels off, any further increase in the SSC level of freshly harvested fruit is a result of starch-to-sugar conversion, and not a real increase in the total carbohydrate content of the fruit. Further harvest delays beyond this time would not result in improved fruit quality.

Measurement of SSC of ripe fruit requires about 5 to 6 days to complete. While this can be useful as a management tool it is too slow to use for inspections. Because total solids of the fruit parallels SSC of ripened fruit it could be useful to estimate the SSC level. A microwave drying procedure can supply a total solids measurement in about one hour. A total solids level of 16.2 percent would compare with a 14 percent SSC level in ripened fruit.

Flesh softening also continues as fruit mature, and flesh firmness measurements may help growers determine the start of harvest. One feature of the flesh firmness changes is that the variability among fruit from a given location increases as softening on the vine continues. Variability among fruit becomes excessively large when flesh firmness drops below about 6.5 kilogram-force or kgf (14 lbf). Thus, there would be more soft fruit if harvest occurs when flesh firmness drops below this level. To avoid possible increased fruit handling injuries, harvest should occur by the time the average flesh firmness reaches 6.5 kgf (14 lbf).

The accumulating evidence indicates that kiwifruit should be harvested with SSC of freshly harvested fruit at or above 6.5 percent, when ripened fruit attain 14 percent SSC or greater (at least 16.2 percent total solids), while flesh firmness remains at least 6.5 kgf (14 lbf), determined as penetration force with an 8-mm tip.

Late harvested kiwifruit will retain their flesh firmness during storage better than early harvested fruit. Even though these fruit are least firm at harvest, they will emerge after 4 to 6 months storage more firm than earlier harvested fruit. Thus growers should not rush to harvest fruit destined for long storage. A good rule for kiwifruit storage management appears to be "last harvested-last marketed."

## Injury

The influence of fruit flesh firmness level on injury susceptibility has been studied, along with an evaluation of fruit response to that injury. Both impact bruising (bruising caused by dropping) and abrasion or vibration bruising (injury caused by fruit movement during transport) have been evaluated. When firm fruit (≥6 kgf [13 lbf]) are impacted, a light, whitish bruise results. The white color results from failure to convert starch to sugar in the injured cells, and this can be demonstrated by staining the injured areas with an iodine/potassium iodide solution, which stains the starch black. When the fruit softens to about 3 kgf (6 lbf), then a translucent bruise results. This injured flesh no longer contains starch. At intermediate firmnesses between about 6 and 3 kgf (13 and 6 lbf), no visual bruising symptoms appear.

Kiwifruit injured at above 6 kgf (13 lbf) flesh firmness do not show a physiological response in either elevated carbon dioxide or ethylene production. However, below that firmness, there is a sharp increase in ethylene production which persists for more than 2 weeks after injury. This provides another reason for completing fruit harvest before flesh firmness drops to this level.

Vibration bruising of kiwifruit usually results in only minor signs of surface injury, but could cause severe internal flesh injury. Such injury occurs when fruit soften to about 2.3 kgf (5 lbf). Concurrent with the injury is a sharp increase in ethylene production that persists for at least a week. Opportunity for vibration bruising can be expected during transport from storage to distribution market. This provides a compelling reason for attempting to market kiwifruit at firmnesses above 2.3 kgf (5 lbf).

The results of these injury studies suggest the following steps:

**1.** Pick fruit at or above 6.5 kgf (14 lbf) flesh firmness penetration force with an 8-mm tip.

**2.** Avoid drops and abrasion to the fruit throughout handling.

**3.** Transport the fruit to market at ≥2.3 kgf (5 lbf) flesh firmness penetration force with an 8-mm tip.

Other studies show that kiwifruit should be brushed either before cooling or after several days in storage.

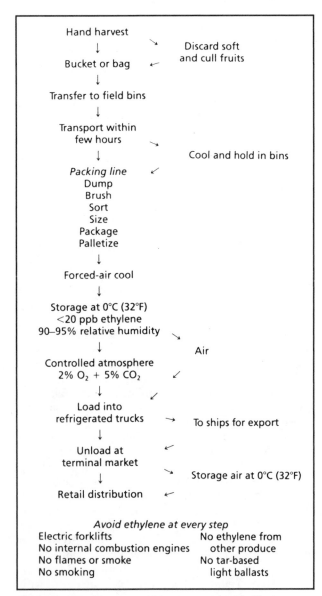

**Fig. 25.3. Postharvest handling system for kiwifruit.**

## Storage and Cooling

### Cooling

Kiwifruit flesh softening is rapid during the first few weeks of storage, even at 0°C (32°F). Exposure to ethylene during any cooling delay can substantially accelerate flesh softening during subsequent 0°C (32°F) air storage. The greater the ethylene concentration during this cooling delay period, the greater the effect on fruit softening. The softening effect increases as the cooling delay period lengthens. Advancing fruit maturity also increases the sensitivity of freshly harvested fruit to ethylene exposure.

The best protection of kiwifruit after harvest thus requires cooling to near storage temperature within 6 hours of harvest, avoiding any ethylene exposure, and storing at 0°C (32°F). Because other studies show increased fruit rot problems associated with fruit wetting, this fast cooling can best be achieved with kiwifruit using forced-air cooling.

### Air storage

Kiwifruit should be stored at a flesh temperature of 0°C (32°F). Flesh softening is substantially faster at even 2.5°C (36.5°F) storage temperature. Ethylene exposure accelerates flesh softening at either temperature, and the softening rate at 2.5°C (36.5°F) without ethylene will approximately equal the softening rate at 0°C (32°F) with ethylene. Thus the temperature and ethylene composition of the storage environment influence the rate of flesh softening of kiwifruit during storage.

During 0°C (32°F) storage, flesh softening is substantially slowed when the ethylene level is held below 10 ppb. Thus, while kiwifruit storage at no more than 20 ppb has been recommended, an even lower ethylene concentration would result in slower flesh softening.

During long-term storage, some individual kiwifruit become rotted, particularly from *Botrytis cinerea*. Fruit infected with Botrytis rot produce ethylene at a higher rate and this can affect flesh softening of healthy fruit. Even a single soft, decayed kiwifruit in the center of a flat can speed the softening of surrounding fruit. Even fruit furtherest from the rot can soften more rapidly than fruit in trays that are rot-free.

Thus, for air storage of kiwifruit, it is important to cool the fruit thoroughly within 6 hours of harvest, store at a constant 0°C (32°F), and maintain below 10 ppb ethylene if possible. Control of water loss and shrivel of kiwifruit requires a relative humidity of the storage room at 90 to 95 percent, and airflow during storage should be no higher than needed to maintain fruit temperature.

## Controlled-atmosphere storage

Extensive studies have been conducted on the potential benefits of controlled-atmosphere (CA) storage for kiwifruit. The major benefits are a delay in flesh softening during storage and reduction in Botrytis rot problems. Best results have been obtained in a CA atmosphere of about 5 percent $CO_2$ and 2 percent $O_2$. The fruit must be promptly cooled and placed under the CA conditions as soon after harvest as possible. There is little benefit if delays in establishing CA conditions exceed one week. Upon removal from CA conditions the fruit are firmer and retain a greater market life.

Ethylene accumulation must be avoided in CA storage as in air storage. Ethylene at low levels causes rapid flesh softening, and a fruit injury problem involving an ethylene-$CO_2$ interaction has been identified. Avoiding ethylene contamination is therefore important, and only CA generating equipment that is free of ethylene should be used. It is possible that ethylene scrubbing may be required during CA storage, but no reliable commercial CA storage system for kiwifruit has been critically monitored.

## Kiwifruit ripening

Because there is increasing interest in early season marketing, there may be a need to ripen kiwifruit to achieve satisfactory market acceptance. Freshly harvested kiwifruit ripen poorly unless exposed to ethylene.

A limited study was conducted in which kiwifruit were harvested and air stored at 0°C (32°F) for 6 weeks. At harvest and at weekly intervals thereafter, samples were transferred to 20°C (68°F) for ripening. During the first 24 hours they were exposed to ethylene at various levels from 0 to 10 parts per million (ppm), and flesh firmness was monitored after 7 days of ripening. The benefits from ethylene exposure were greatest during the first few weeks of storage and decreased over the 6-week period of the test. In all cases, 10 ppm ethylene exposure aided subsequent fruit ripening. Although more study is needed, it appears that freshly harvested kiwifruit can be effectively ripened for early season marketing by exposure to 10 ppm ethylene for 24 hours at 20°C (68°F) followed by maintenance at 20°C (68°F) until soft.

Studies are needed to evaluate the possibility of partially softening the fruit (to a flesh firmness of 3.5 to 4 kgf [about 8 to 9 lbf]) followed by rapid cooling and transport. If this partial softening could be achieved, such fruit could be capable of rapid ripening after transport is completed.

## REFERENCES

### Grapes

1. Cappellini, R. A., M. J. Ceponis, and G. W. Lightner. 1986. Disorders in table grape shipments to the New York market, 1972-1984. *Plant Dis.* 70:1075-79.

2. Harvey, J. M., and W. T. Pentzer. 1960. *Market diseases of grapes and other small fruits.* U.S. Dept. Agric. Handb. 189. 37 pp.

3. Harvey, J. M., and M. Uota. 1978. Table grapes and refrigeration: fumigation with sulfur dioxide. *Internat. J. Refrig.* 1:167-71.

4. Nelson, K. E. 1978. Pre-cooling—its significance to the market quality of table grapes. *Internat. J. Refrig.* 1:207-15.

5. ———. 1980. Improved harvesting and handling benefit table grape markets. *Calif. Agric.* 34(7):34-36.

6. ———. 1985. *Harvesting and handling California table grapes for market.* Univ. Calif. Div. Agric. Nat. Resour. Bull. 1913. 72 pp.

7. Reynaud, E., and P. Ribereau-Gayon. 1971. The grape. In *The biochemistry of fruits and their products,* ed. A. C. Hulme, vol. 2, 172-206. New York: Academic Press.

8. Ryall, A. L., and W. T. Pentzer. 1982. *Handling, transportation and storage of fruits and vegetables. 2. Fruits and tree nuts.* 2nd ed., 257-62, 529-42. Westport, CT: AVI Publ. Co.

9. Winkler, A. J., J. A. Cook, W. M. Kliewer, and L. A. Lider. 1974. *General viticulture.* Berkeley, CA: Univ. Calif. Press. 710 pp.

### Strawberries

1. Ceponis, M. J., R. A. Cappellini, and G. W. Lightner. 1987. Disorders in sweet cherry and strawberry shipments to the New York market, 1972-1984. *Plant Dis.* 71:472-75.

2. Green, A. 1971. Soft fruits. In *The biochemistry of fruits and their products,* ed. A. C. Hulme, vol. 2, 375-410. New York: Academic Press.

3. Harvey, J. M., and C. M. Harris. 1973. *Strawberries—market quality in relation to postharvest handling and shipping practices.* ASHRAE Symp. LO-73-7, 5-9. Washington, DC: Am. Soc. Heating, Refrig. Air Condit. Eng.

4. Harvey, J. M., C. M. Harris, W. J. Tietjen, and T. Serio. 1980. *Quality maintenance in truck shipments of California strawberries.* U.S. Dept. Agric. SEA Adv. Agric. Tech. AAT-W-12. 13 pp.

5. Mitchell, F. G., E. C. Maxie, and A. S. Greathead. 1964. *Handling strawberries for fresh market.* Univ. Calif. Agric. Exp. Sta. Circ. 527. 16 pp.

6. Mitchell, F. G., and G. Mayer. 1978. Effects of basket design on cooling and holding strawberries. *Calif. Agric.* 32(5):17-18.

7. Ryall, A. L., and W. T. Pentzer. 1982. *Handling, transportation, and storage of fruits and vegetables. 2. Fruits and tree nuts,* 2nd ed., 255-57, 525-29. Westport, CT: AVI Publ. Co.

### Kiwifruit

1. Arpaia, M. L., F. G. Mitchell, A. A. Kader, and G. Mayer. 1982. The ethylene problem in modified atmosphere storage of kiwifruit. In *Proc. 3rd Natl. Controlled Atmos. Res. Conf.,* eds. D. G. Richardson and M. Meheriuk, 331-35. Beaverton, OR: Timber Press.

2. Mitchell, F. G. 1990. Postharvest physiology and technology of kiwifruit. *Acta Hortic.* 282:291-307.

3. Mitchell, F. G., M. L. Arpaia, and G. Mayer. 1982. Modified atmosphere storage of kiwifruit (*Actinidia chinensis*). In *Proc. 3rd Natl. Controlled Atmos. Res. Conf.,* eds. D. G. Richardson and M. Meheriuk, 235-38. Beaverton, OR: Timber Press.

4. Sale, P. R. 1985. *Kiwifruit culture.* Wellington, NZ: V. R. Ward, Govt. Printer. 96 pp.

5. Scott, K. J., S. A. Spraggon, and R. L. McBride. 1986. Two new maturity tests for kiwifruit. *CSIRO Food Res. Q.* 46:25-31.

6. Sommer, N. F., R. J. Fortlage, and D. G. Edwards. 1983. Minimizing postharvest diseases of kiwifruit. *Calif. Agric.* 37(1-2):16-18.

# 26

# Postharvest Handling Systems: Subtropical Fruits

Adel A. Kader and Mary Lu Arpaia

Subtropical fruits include avocado, carob, cherimoya, citrus fruits (orange, grapefruit, lemon, lime, pummelo, tangerine and mandarin, kumquat), date, fig, jujube, kiwifruit, loquat, lychee (litchi), olive, persimmon, and pomegranate. Some of these fruits are also grown in tropical and temperate zones. Subtropical fruits are diverse in morphological and compositional characteristics and in postharvest requirements. Subtropical fruits can be grouped according to their relative perishability as follows:

**Highly perishable**—fresh fig, loquat, lychee

**Moderately perishable**—avocado, cherimoya, olive, persimmon

**Less perishable**—citrus fruits, carob (dry), dried fig, date, jujube (Chinese date), kiwifruit, pomegranate

This chapter relates the general characteristics of subtropical fruits to their postharvest biology and handling requirements. The emphasis is on avocado and citrus fruits, the fruits that are most important commercially. Commercial production of citrus fruits in this country is limited to Arizona, California, Florida, and Texas. The United States produces about 30 percent of the world's production of lemons and 40 percent of its oranges. Florida is the leading U.S. producer of citrus fruits, most of which (>80 percent) is processed. Most California citrus fruits are marketed fresh. California accounts for almost all U.S. production of dates, figs, kiwifruit, olives, persimmons, and pomegranates.

Three strains of cultivated avocados are: Mexican (e.g., Bacon), Guatemalan (e.g., Hass, Reed), West Indian (e.g., Pallock, Waldin), and their hybrids (e.g., Fuerte). Fresh avocados from Florida (West Indian cultivars) are available from July through February; California avocados (Mexican, Guatemalan, and Mexican-Guatemalan hybrid cultivars) are available year-round. California produces more than 80 percent of U.S. avocados.

## Morphological and Compositional Characteristics

Avocados are one-seeded berries. Cultivars vary in size. Usually pear shaped, avocados can also be round or oval. The flesh has more energy value than meat of equal weight, and is a good source of niacin and thiamine. Of all tree fruits, avocados and olives are the highest in protein and fat content. In California,

minimum maturity of the avocado is defined in terms of dry weight, which is highly correlated with oil content of the flesh. The minimum dry weight required before harvesting can begin varies with cultivar.

Citrus fruits rank first in their contribution of vitamin C to human nutrition in the United States. Botanically, citrus fruit are classified as a hesperidium, a specialized berry. The rind has two components: the pigmented part, called the flavedo (epidermis and several subepidermal layers), and the whitish part, called the albedo. The juicy part consists of segments filled with juice sacs. Minimum maturity requirements of citrus fruits are based on juice content (lemon and lime) or soluble solids content, titratable acidity, and the ratio of the two (orange, grapefruit, and tangerine).

## Postharvest Physiology

Avocados have a relatively high respiration and ethylene production rate, and both rates exhibit a climacteric pattern. Citrus fruits are nonclimacteric and have low respiration and ethylene production rates. Postharvest compositional changes in citrus fruits are minimal, whereas avocados undergo many changes in composition, texture, and flavor associated with ripening.

Avocados do not ripen on the tree. The exact nature of the ripening inhibitor is not known, but it continues to exert its effect for about 24 hours after harvest. Avocado ripening can be hastened by exposure to 10 ppm ethylene at 15°C to 17°C (59° to 62.6°F) and 85 to 90 percent relative humidity; temperatures up to 25°C (77°F) can be used if faster ripening is desired. On the other hand, removal or exclusion of ethylene from the storage environment helps extend the storage life of avocados by delaying softening, onset of chilling injury, and decay incidence.

Cold nights, followed by warm days, are necessary for loss of green color and appearance of yellow or orange color in citrus fruits. This is why citrus fruits remain green after attaining full maturity and good eating quality in tropical areas. Occasionally, regreening of Valencia oranges occurs in certain production areas after they have reached full orange color.

Degreening of citrus fruits essentially results in removal of chlorophyll from the flavedo but does not influence composition of the fruits' edible portion. The need for and duration of degreening treatments depend on the cultivar and the fruit's condition at harvest—the amount of chlorophyll to be removed. Lemons are usually degreened at 16°C (60.8°F) with or without added ethylene; higher temperatures may be used for faster degreening. Recommended conditions for degreening California oranges and grapefruits are:

Temperature: 20°C to 25°C (68° to 77°F)

Relative humidty: 90 percent

Ethylene concentration: 5 to 10 ppm

Air circulation: One room volume per minute

Ventilation: One to two air changes per hour, or sufficient changes to keep $CO_2$ below 0.1 percent

In Florida, a temperature of 27°C to 29°C (80.6° to 84.2°F) and an ethylene concentration of 1 to 5 ppm are recommended.

## Physiological Disorders

### Chilling injury

Susceptibility to chilling injury varies in subtropical fruits according to species and cultivar. For example, grapefruit, lemon, and lime are much more susceptible to chilling injury than orange and mandarin. Orange cultivars grown in Florida are reportedly less sensitive to chilling injury than those grown in California and Arizona. Date, fig, kiwifruit, and Hachiya persimmon are not sensitive to chilling injury. Fuyu persimmon, pomegranate, olive, and other subtropical fruits are chilling sensitive. Ripe avocado fruit tolerate lower temperatures than unripe fruit, without danger of chilling injury. Symptoms of chilling injury on selected subtropical fruits are summarized in table 26.1 and illustrated in figures 26.1 and 26.2 for orange and grapefruit, respectively.

Table 26.1. Chilling injury symptoms
on avocados and some citrus fruits

| Fruit | Minimum* safe temperature | | Symptoms |
| | Celsius (°C) | Fahrenheit (°F) | |
|---|---|---|---|
| Avocado | 5–10 | 41–50 | Grayish-brown discoloration of flesh, discoloration of the vascular tissue, softening, pitting, development of off-flavors |
| Grapefruit | 10–13 | 50–55 | Pitting, scald, watery breakdown |
| Lemon | 10–13 | 50–55 | Pitting, membranous stain, red blotch |
| Lime | 10–13 | 50–55 | Pitting, accelerated decay |
| Orange | 3–5 | 37–41 | Pitting, brown stain |

*Varies with cultivar, maturity stage, and duration of storage.

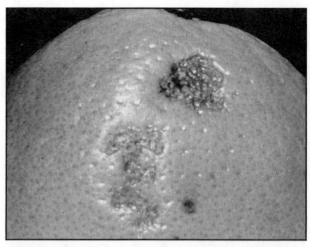

Fig. 26.1. Chilling injury symptoms on oranges. Darker areas are brown.

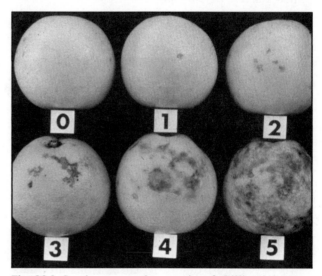

Fig. 26.2. Scoring system for severity of chilling injury on grapefruit. Darker areas are brown.

### Other disorders

Fruit exposed to temperatures below their freezing point before or after harvest can suffer serious injuries; for example, injured citrus fruits become dry and useless, and have to be separated in the packinghouse by flotation or x-ray techniques. High-temperature disorders resulting from preharvest exposure to sun can result in sunburned avocado and citrus fruits. Exposure of avocado to temperatures above 25°C (77°F) may cause uneven softening, skin discoloration, flesh darkening, and off-flavors.

Citrus fruit peel disorders, other than chilling injury, include (1) oil spotting or oleocellosis (breaking of oil cells, causing the oil to extrude and damage surrounding tissue) as shown in figure 26.3, (2) rind staining of navel orange (an indication of peel overmaturity that can be controlled by preharvest application of gibberellin), (3) stem-end rind breakdown of orange, (4) stylar-end breakdown of lime, and (5)

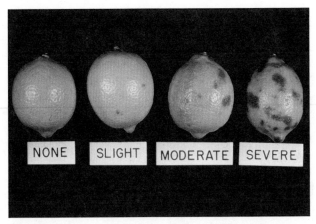

**Fig. 26.3. Scoring system for severity of oil spotting on lemons. Darker areas are green.**

shriveling and peel injury around the stem end indicating aging.

## Pathological Breakdown

### Avocado

Avocado fruit can be affected by one or more pathogens. *Dothiorella gregaria* (probably the asexual state of *Botryosphaeria ribis*) is a postharvest rot of California avocados. Anthracnose occurs particularly in humid areas such as Florida. Neither organism is usually a serious problem in California unless the weather has been unusually wet at or near harvest time. Stem-end rots (*Diplodia natalensis*, *Phomopsis citri*) can also be serious in Florida and other humid growing areas.

### Citrus fruits

Postharvest diseases also limit the postharvest life of citrus fruits. Blue mold (*Penicillium italicum*) and green mold (*Penicillium digitatum*) occur in citrus fruits in all production areas. In humid areas, stem-end rots (*Diplodia* spp. and *Phomopsis* spp.) and anthracnose (*Colletotrichum gloeosporioides*) are common. Sour rot (*Geotrichum candidum*) affects lemons during long-term storage, especially during wet seasons. Phytophthora brown rot occurs in California following cool, wet weather, and can be controlled by heat treatment. Alternaria stem-end rot (*Alternaria citri*) usually follows senescence of the calyx of the fruit.

Citrus diseases can be controlled by the following procedures:

**Reduce the pathogen population in the environment.** Use an effective preharvest disease control program to reduce postharvest incidence of stem-end rots and anthracnose. Use chlorine (e.g., sodium hypochlorite) in wash water. Regularly disinfest field containers, packinghouse equipment, and storage facilities using a fog of 1 percent formaldehyde solution or quaternary ammonium products. Circulate *Penicillium* spore-laden air through filters in a special box-dumping room for stored lemons.

**Maintain fruit resistance to infection.** Minimize mechanical injuries during harvesting and postharvest handling. Use proper temperature and relative humidity management throughout postharvest handling. Use 2,4-D treatment (200 ppm) on lemon to maintain vitality of button tissue and reduce development of stem-end rots; the use of gibberellic acid (50 ppm in the storage wax), as an alternative to 2,4-D, is pending registration; this treatment decreases the incidence of *Geotrichum* during storage. Use postharvest fungicides, such as sodium orthophenylphenol (SOPP), thiabendazole (TBZ), sec. butylamine, and imazalil. New fungicides are continually being evaluated. Choosing a fungicide depends upon whether it has been approved for use and whether it has been accepted by importing countries. Judicious use of fungicides is a valuable component of a disease control program, as resistance to fungicides can develop quickly.

## Alternatives to Postharvest Fungicides

Without the use of available postharvest fungicides or replacements, storage life of citrus fruits would be significantly reduced. Consequently, exports of fresh citrus fruits, other than air shipped, would be curtailed and postharvest losses would increase in both domestic and export marketing. Short-term and long-term options and alternatives to currently used postharvest fungicides are listed below.

### Short-term alternatives

**1.** More careful handling during harvest and postharvest operations to reduce mechanical injuries will reduce fungal infection and losses due to decay. This, coupled with providing the optimum temperature and relative humidity and expedited handling during all marketing steps, can extend the postharvest life of citrus fruits about 2 to 3 weeks, depending on the cultivar.

**2.** Controlled or modified atmospheres (including carbon monoxide) can be used during transport and temporary storage. Carbon monoxide at 5 to 10 percent, added to 5 percent oxygen, provides adequate fungistatic control of many fungi causing citrus postharvest diseases. Use of carbon monoxide requires strict safety precautions to protect transport and storage facility workers. Also, the cost of maintaining such atmosphere is greater than the cost of treatments with postharvest fungicides. More research is needed, however, to evaluate the fruit's tolerance to controlled or modified atmosphere conditions.

### Long-term alternatives

**1.** Treatment with ionizing radiation for decay control has been suggested. However, the dose needed to effectively control decay is between 1.5 and 2.0 kGy (150 and 200 krad). Such doses can result in rind injuries and increase fruit softening. Furthermore, the currently approved upper limit for irradiating fresh fruit is 1 kGy (100 krad). Combining irradiation with heat treatments may reduce the dose required and consequently the resulting detrimental effects.

**2.** Heat treatments, such as dipping citrus fruits in 44°C (111°F) water for 2 to 4 minutes, have been tested as possible means to kill fungal spores on fruit surfaces and reduce decay. The limiting factor for heat treatments is the narrow margin between the time-temperature combinations that reduce decay and those that cause fruit injury.

**3.** Biological control (use of antagonistic microorganisms) has been investigated, but no treatment has yet provided as good decay control as postharvest fungicides. If a successful procedure is found, it will require thorough testing for safety considerations before receiving EPA approval, and will also need to be economically feasible.

**4.** Breeding new citrus cultivars with resistance of their fruit to decay-causing fungi is a long-term option.

## Postharvest Handling Procedures

### Harvesting

Research into mechanical harvesting of citrus fruits (especially in Florida for processing fruit) has been extensive, but no satisfactory system is available. Chemicals that promote abscission will probably be part of any mechanical harvesting system. Several harvest aids, such as mobile ladders and picker platforms, have been tested, but few are in commercial use. California avocados (figure 26.4) and citrus fruits are harvested with hand clippers. Some Florida citrus fruits are snap-picked (twist and pull method), but

**Fig. 26.4. Harvesting avocados.**

this may increase their susceptibility to decay. Some Florida processing oranges and grapefruit are picked and dropped on the ground. This practice is detrimental to the fruit even though they are processed within a day or two after harvest.

### Packinghouse operations

A flow diagram of the postharvest handling system used for avocados in California is shown in figure 26.5. The dumping, sizing, and packing operations

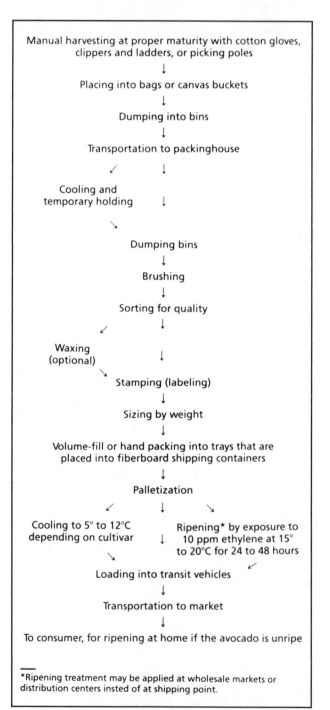

Manual harvesting at proper maturity with cotton gloves, clippers and ladders, or picking poles
↓
Placing into bags or canvas buckets
↓
Dumping into bins
↓
Transportation to packinghouse
↓
Cooling and temporary holding
↓
Dumping bins
↓
Brushing
↓
Sorting for quality
↓
Waxing (optional)
↓
Stamping (labeling)
↓
Sizing by weight
↓
Volume-fill or hand packing into trays that are placed into fiberboard shipping containers
↓
Palletization
↓
Cooling to 5° to 12°C depending on cultivar          Ripening* by exposure to 10 ppm ethylene at 15° to 20°C for 24 to 48 hours
↓
Loading into transit vehicles
↓
Transportation to market
↓
To consumer, for ripening at home if the avocado is unripe

*Ripening treatment may be applied at wholesale markets or distribution centers insted of at shipping point.

**Fig. 26.5. Postharvest handling system for avocados.**

**Fig. 26.6. Dry-dumping of avocados at the packinghouse.**

**Fig. 26.8. Packing avocados from an accumulation bin.**

**Fig. 26.7. Sizing avocados by weight.**

are illustrated in figures 26.6, 26.7, and 26.8, respectively.

The packinghouse operations for citrus fruits are summarized in figure 26.9. Figures 26.10 to 26.14 illustrate surface drying, quality sorting, hand packing, pattern packing, and packing in bags. Lemons are usually sorted into four color classes (dark green, light green, silver, and yellow) by electronic sorters based on their light reflectance (figure 26.15). Some orange and tangelo cultivars are colored with a certified food dye in Florida, but this treatment is not allowed in California. Cooling methods include hydrocooling, forced-air cooling, and room cooling. Attention to proper and fast cooling for citrus fruits is badly needed in most citrus-handling facilties.

Seal packaging (wrapping with various types of plastic film) of individual citrus fruit has been extensively tested and is currently used by a few shippers. The treatment reduces water loss and maintains the vitality of the peel because of the high relative humidity maintained around the fruit. It also prevents the spread of decay from fruit to fruit. For decay contol, fruit must be treated with fungicides before wrapping. While seal packaging of individual fruit may allow short-term holding of citrus fruits without refrigeration, it must be combined with refrigeration for long-term storage to maintain good quality and reduce losses.

Citrus fruits produced in certain areas must be treated for insect control before shipment to some markets. The main disinfestation method in use was once fumigation with ethylene dibromide (EDB) against fruit flies. Since EDB was completely withdrawn from the Environmental Protection Agency's list of approved chemicals in 1987, cold treatments or fumigation with methyl bromide or phosphine have been used. These treatments result in some phytotoxicity. Losses due to phytotoxicity as a result of cold treatment can be mitigated by conditioning the fruit for one week at 16°C (61°F). Other alternatives to EDB fumigation being evaluated include heat treatments, irradiation, and controlled atmospheres. The citrus cultivar and the stage of maturity at harvest influence the fruit's response to the quarantine treatment.

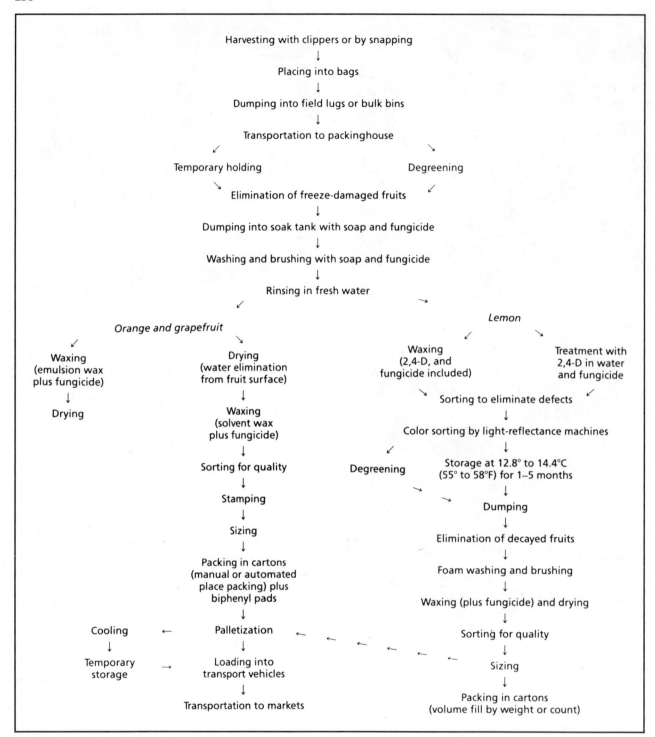

**Fig. 26.9. Postharvest handling systems for citrus fruits.**

## Quality and Storage Life of Citrus Fruits

The composition and quality of citrus fruits at harvest and the fruits' potential for storage are influenced by many pre- and postharvest factors. Preharvest factors include rootstock and cultivar, fruit maturity at harvest, harvesting season, tree condition (vigor), weather conditions (temperature, relative humidity, rain), and cultural practices (fertilization, irrigation, pest control, growth regulators). Harvesting methods influence the uniformity among fruit and the extent of mechanical injuries due to rough handling.

Postharvest factors that influence the postharvest lifespan of citrus fruits include delays between harvest, packing, and cooling; degreening conditions;

Fig. 26.10. Drying citrus fruits with warm air to remove surface moisture.

Fig. 26.11. Quality sorting of citrus fruits.

Fig. 26.12. Rapid packing of citrus fruits.

fungicidal treatments; waxing; seal-packaging; growth-regulator treatments; temperature and relative humidity management; and presence of ethylene and other volatiles in storage.

Fig. 26.13. Automated pattern packing system for oranges.

Fig. 26.14. Packing citrus fruits into consumer packages.

## Storage

Some citrus cultivars may be left on the tree for up to 5 months after attaining legal maturity. Depending on the cultivar, avocados will remain attached to the tree for 3 to 12 weeks after maturity before excessive abscission begins. The duration of "on-tree storage" depends on the cultivar. Citrus fruit and avocado quality may, however, deteriorate during on-tree storage. For successful postharvest storage, maintain the conditions summarized in table 26.2. These recommendations also apply for optimum transport and temporary storage conditions.

**Fig. 26.15. Light reflectance machine used for sorting lemons by color.**

**Table 26.2 Optimum storage conditions for avocado and citrus fruits**

| Commodity | Temperature | | Approximate storage life* (weeks) | Modified atmospheres if used† | |
|---|---|---|---|---|---|
| | Celsius (°C) | Fahrenheit (°F) | | O₂ (%) | CO₂ (%) |
| Avocado, unripe§ | 5–12 | 41–54 | 2–4 | 2–5 | 3–10 |
| Avocado, ripe§ | 5–8 | 41–46 | 1–2 | — | — |
| Grapefruit | 12–14 | 54–57 | 4–8 | 5 | 5–10 |
| Kumquat | 4–8 | 39–46 | 2–4 | 5 | 0–5 |
| Lemon | 12–14 | 54–57 | 16–24‡ | 5 | 0–5 |
| Lime | 10–12 | 50–54 | 6–8 | 5 | 0–10 |
| Orange | 4–8 | 39–46 | 4–8 | 5 | 0–5 |
| Pummelo | 8–10 | 46–50 | 8–12 | 5 | 5–10 |
| Tangerine | 5–8 | 41–46 | 2–4 | 5 | 0–5 |

*Under optimum temperature and 85 percent to 90 percent relative humidity.
†MA use on citrus is limited; 5 percent to 10 percent CO may be added to MA for decay control during transport to export markets.
‡Storage life for dark-green lemons; for other stages: light-green, 8 to 16; silver, 4 to 8; yellow, 3 to 4.
§Response to temperature and modified atmospheres is dependent upon cultivar.

## REFERENCES

1. Biale, J. B., and R. E. Young. 1971. The avocado pear. In *The biochemistry of fruits and their products*, ed. A. C. Hulme, Vol. 2, 2-64. New York: Academic Press.

2. Bower, J. P., and J. G. Cutting. 1988. Avocado fruit development and ripening physiology. *Hort. Rev.* 10:229-71.

3. Ceponis, M. J., R. A. Cappellini, and G. W. Lightner. 1986. Disorders in citrus shipments to the New York market, 1972-1984. *Plant Disease* 70:1162-65.

4. Dezman, D. J., S. Nagy, and G. E. Brown. 1986. Postharvest fungal decay control chemicals: Treatments and residues in citrus fruits. *Residue Rev.* 97:37-92.

5. Eaks, I. L. 1977. Physiology of degreening: Summary and discussion of related topics. *Proc. Int. Soc. Citriculture* 1:223-26.

6. Eckert, J. W., and I. L. Eaks. 1989. Postharvest disorders and diseases of citrus fruits. In *The Citrus Industry*, eds. W. Reuther et al., vol. 5, 179-260. Oakland: Univ. Calif. Div. Agric. Nat. Resources Publ.

7. Grierson, W., and T. T. Hatton. 1977. Factors involved in storage of citrus fruits: A new evaluation. *Proc. Int. Soc. Citriculture* 1:227-31.

8. Grierson, W., W. M. Miller, and W. F. Wardowski. 1978. *Packingline machinery for Florida citrus packinghouses*. Inst. Food Agric. Sci., Univ. Fla. Coop. Ext. Serv. Bull. 803. 30 pp.

9. Hatton Jr., T. T., and D. H. Spalding. 1974. Maintenance of market quality in Florida avocados. *ASHRAE Trans.* 80:335-40.

10. Lee, S. K., R. E. Young, P. M. Shiffman, and C. W. Coggins, Jr. 1983. Maturity studies of avocado fruit based on picking dates and dry weight. *J. Am. Soc. Hort. Sci.* 108:390-94.

11. Lindsey, P. J., S. S. Briggs, K. Moulton, and A. A. Kader. 1989. Postharvest fungicides on citrus: Issues and alternatives. In *Chemicals use in food processing and postharvest handling: Issues and alternatives*, 23-38. Davis: Univ. Calif. Agric. Issues Center.

12. Nagy, S., and J. A. Attaway, eds. 1980. *Citrus nutrition and quality*. Symposium Series 143. Washington: American Chemical Society. 456 pp.

13. Nagy, S., and P. E. Shaw. 1980. *Tropical and subtropical fruits: Composition, properties, and uses*. Westport, CT: AVI Publ. Co. 570 pp.

14. Ohr, H. D., and J. W. Eckert. 1985. *Postharvest diseases of citrus fruits in California*. Univ. Calif. Coop. Ext. Leaflet 21407, 6 pp.

15. Smoot, J. J., L. G. Houck, and H. B. Johnson. 1971. *Market diseases of citrus and other subtropical fruits*. U.S. Dept. Agric. Handb. 398. 115 pp.

16. Ting, S. V., and J. A. Attaway. 1971. Citrus fruits. In *The biochemistry of fruits and their products*, ed. A. C. Hulme, vol. 2, 107-71. New York: Academic Press.

17. Ting, S. V., and R. L. Rousett. 1986. *Citrus fruits and their products—Analysis and technology*. New York: Marcel Dekker. 312 pp.

18. Wardowski, W. F., S. Nagy, and W. Grierson, eds. 1986. *Fresh citrus fruits*. Westport, CT: AVI Publ. Co. 571 pp.

# 27

# Postharvest Handling Systems: Tropical Fruits

Noel F. Sommer and Mary Lu Arpaia

The most important tropical fruit by far in temperate-zone markets is the banana, having a per capita consumption in the U.S. higher than any other fruit. Its popularity among consumers is enhanced by a reasonable price, high fruit quality, and availability throughout the year.

The other fresh tropical fruits most commonly found in the temperate-zone markets—mango, papaya, and pineapple—are more recent arrivals. They are categorized as gourmet fruits and are purchased only occasionally by most consumers.

The regular appearance of papayas in mainland U.S. markets was stimulated by the development of small-fruited cultivars, such as 'Solo' and 'Sunrise', weighing only about ½ kg (1 pound), and the establishment of regular air cargo service. The small size permitted the fruit to be marketed profitably at prices comparing favorably to those of other gourmet foods. To develop markets for papaya, advertisement and other market-development activities were concentrated in certain chosen cities. Market development was further aided by increased tourism to Hawaii. High air transport costs and expanding markets have stimulated increased shipments via refrigerated marine containers.

The introduction of fresh pineapples to the mainland market was aided by consumers' familiarity with the canned product. Tourism in producing areas has been an important factor in increasing consumers' awareness of tropical fruits. In addition, the influx of many people from tropical countries into the U.S. has increased the potential demand for mangoes and other tropical fruits.

## BANANAS

The banana is a large herbaceous plant. The underground tuberous stem or "bulb" gives rise to leaves and the fruit bunch. The aboveground "trunk" of the banana tree is in reality a pseudostem consisting of tightly oppressed leaf bases. The pseudostem dies after a fruit bunch has been produced and is replaced by one of the young pseudostems that have emerged. The banana inflorescence contains three kinds of flowers. The fruits originate from female flowers that are produced first, followed by hermaphrodites, and then male blossoms.

The postharvest operations required for bananas include transportation to packinghouse, dehanding, washing to remove dirt and latex, disease control, packaging, transportation to market, ripening, and retail sale (fig. 27.1).

## Harvesting

Bunches are examined about 3 months before harvest. Those that have completed their female (fruit-producing) stage have their buds removed to prevent further floral development. One or two apical hands are also removed at this time to promote development of the remainder. Removed buds may be consumed or discarded. A single finger is retained, apical to the position of the hands that were removed. This terminal finger continues to grow, preventing the stalk from rotting.

Each bunch is covered with a perforated polyethylene bag. The top of the bag may be secured to the stalk with a colored ribbon. With different colors used each week a ready record of age is maintained. The plastic film protects the bananas from leaf scarring and keeps dust off, and an insecticide may be placed within the bag to reduce insect damage.

During fruit development, the bunches may require props to support their weight (fig. 27.2). Sometimes a pseudostem is provided with guys and twine from its crown to the bases of nearby pseudostems.

Bananas are harvested green and are ripened in market areas. Fruits that are allowed to ripen on the tree often split, and tend to be mealy. The maturity at which bananas are harvested depends on the time required to get them to market. Fruits shipped from Central America to Europe are usually harvested less mature than those shipped to North America. A penalty in lost yield is, of course, paid when bunches are harvested before fingers are fully developed. Sizes and shapes of finger sections at various stages of maturity are illustrated in figure 27.3.

At harvest, crews pass through the plantation, usually at 3 or 4 day intervals, selecting bunches for harvest. Colored ribbons provide information regarding age. The diameter (caliper) of fruits is monitored.

Harvesting and field handling vary with location. However, harvesting is usually a two-person operation. The cutter makes a cut with a machete, partially severing the pseudostem at about its midpoint. A backer, positioned under the bunch, catches and braces the bunch firmly. The cutter then severs the bunch from the pseudostem, just below the basal hands.

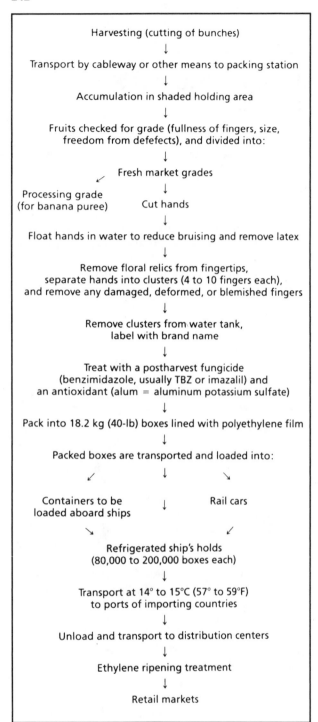

Harvesting (cutting of bunches)
↓
Transport by cableway or other means to packing station
↓
Accumulation in shaded holding area
↓
Fruits checked for grade (fullness of fingers, size, freedom from defefects), and divided into:

Fresh market grades

Processing grade
(for banana puree)          Cut hands
↓
Float hands in water to reduce bruising and remove latex
↓
Remove floral relics from fingertips,
separate hands into clusters (4 to 10 fingers each),
and remove any damaged, deformed, or blemished fingers
↓
Remove clusters from water tank,
label with brand name
↓
Treat with a postharvest fungicide
(benzimidazole, usually TBZ or imazalil) and
an antioxidant (alum = aluminum potassium sulfate)
↓
Pack into 18.2 kg (40-lb) boxes lined with polyethylene film
↓
Packed boxes are transported and loaded into:

Containers to be          Rail cars
loaded aboard ships

Refrigerated ship's holds
(80,000 to 200,000 boxes each)
↓
Transport at 14° to 15°C (57° to 59°F)
to ports of importing countries
↓
Unload and transport to distribution centers
↓
Ethylene ripening treatment
↓
Retail markets

Fig. 27.1. Postharvest handling system for bananas.

## Transportation to packing station

Many banana plantations have been equipped with a system of cableways. A backer carries a cut bunch to the nearest cableway, where the bunch is attached by its base to a roller on the cable. The bunches are separated by spacer bars to prevent contact. A train of up to 75 or 150 bunches of 30 to 60 kg (66 to 132 lb) each forms and is pulled along the cableway to

Fig. 27.2. Banana fruit bunches are enclosed in plastic bags, and guys are run to prevent pseudostems from falling (near Puerto Limon, Costa Rica).

Maturity Stages of Banana Fruits

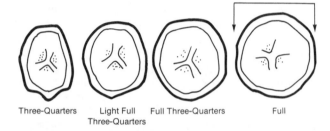

Three-Quarters    Light Full    Full Three-Quarters    Full
                  Three-Quarters

Fig. 27.3. Changes in size and shape of banana fruit sections at various stages of maturity.

Fig. 27.4. A train of banana stems moves to the packing-house by aerial tramway (La Lima, Honduras).

the packing area by a small tractor . Bananas waiting to be packed are shaded (fig. 27.4).

A tractor that hangs from the cable has the advantage of not requiring a roadway or bridges to cross drainage ditches. Such tractors have been developed and are used in some areas.

In Queensland and New South Wales, Australia, winter temperatures limit banana production to sunny, northern exposures, often on very steep hills. Some cableway systems have been installed that use gravity as the means of locomotion. Otherwise, bunches accumulate at roadways that have been bulldozed across the slopes. Bunches are placed on small trucks or on trailers, usually two or three bunches deep. Padding is placed on the vehicle bed and between bunches to limit damage. Low tire pressure and low speed during transit are important injury avoidance measures.

## Packinghouse operations

Upon entering the packinghouse, bananas are checked for finger fullness and length, and for blemishes from leaf rub, insect activity, pathogens, and handling bruises (fig. 27.5). Those not meeting a fresh fruit grade are processed as puree or discarded.

Most bananas from the American tropics are shipped as hands in fiberboard cartons. Hands are removed from the stalk by the dehander, using a sharp curved knife (fig. 27.6). When the hand is cut away, latex flows from the wound for a time. If allowed to coat the surface of the fingers, latex deposits and the resulting stains would seriously detract from appearance. Consequently, the hands are immediately placed in water to coagulate the exuded latex and reduce staining (fig. 27.7). Dust and dirt are removed at the same time.

Selectors at the dehanding tank remove dead floral parts still adhering to the fruit and sort out undersized, damaged, deformed, and blemished fingers. Large hands are divided into smaller clusters to facilitate packing and provide a convenient unit for the consumer. The bananas may be floated in a second water tank for an additional 10 or 15 minutes to permit further exudation of latex.

The water tank is a potential source of serious disease problems. Fungus spores on dead floral parts may accumulate in the water and contaminate the cut surfaces of freshly cut hands. In particular, crown

Fig. 27.6. A worker cuts the hands from banana stems. Hands are placed in water to clean the fruit and avoid latex stains (La Lima, Honduras).

Fig. 27-7. Bananas are floated in water to coagulate latex exuding from cut surfaces. Withered floral parts are removed from the fingers (La Lima, Honduras).

rot results from inoculation by mixed spores of *Colletotrichum musae, Fusarium roseum, Nigrospora sphaerica, Thielaviopsis paradoxa, Botryodiplodia theobromae,* and other fungi.

Sodium hypochlorite (75 to 125 ppm) in the water will kill spores, reducing the likelihood of inoculation. Concentrations of the chemical must be carefully monitored and maintained. Indicator papers provide a simple and inexpensive means of checking chlorine concentrations.

As an alternative, a fungicide, such as TBZ or imazalil, is added to the wash water instead of as a separate treatment. Relatively large amounts of the fungicide are usually required because the wash water becomes dirty and must be changed from time to time. The fungicide must be included with fresh water.

Fig. 27.5.Inspecting banana stems (La Lima, Honduras).

A fungicide may be applied as a separate dip or spray treatment after washing (fig. 27.8). Alum, which may be applied at the same time, serves as an antioxidant to prevent subsequent latex exudations from darkening and staining the fruit surface. This is the last step before packing.

## Packing and shipping

In the American tropics, clusters of fruit are packed in corrugated fiberboard cartons lined with polyethylene (fig. 27.9), which are then moved to conveyors for loading onto trucks or trains for transportation to the wharf. Cartons move by conveyer to an elevated conveyer, or gantry, that lowers them into the hold of a ship. The ship's refrigeration system cools the fruit and holds it at 13° to 14°C (55° to 57°F).

Marine shipment of bananas in refrigerated containers on container ships (fig. 27.10) continues to increase. Faster loading and unloading is an economic advantage for container ships. Containers allow the use of pallets and forklift trucks, and are adaptable to intermodal transportation. Fruits can be loaded into containers at the packing stations and remain untouched until arrival in the market area.

Fig. 27.8. Banana hands emerge from a plastic enclosure where the fruit were sprayed with TBZ for disease control and alum as an added precaution against latex staining (La Lima, Honduras).

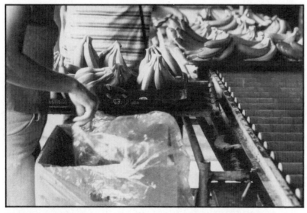

Fig. 27.9. Bananas are hand packed into polyethylene-lined corrugated paper containers (La Lima, Honduras).

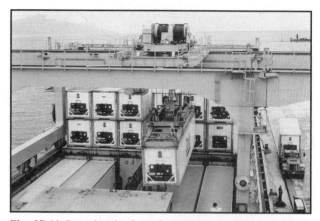

Fig. 27.10. Containerized marine transport of bananas.

Fig. 27.11. Ripening rooms for bananas (San Francisco, California).

Presumably, that reduced handling minimizes fruit bruising.

## Handling in market areas

Banana fingers should be green on arrival in market areas to provide time for ripening and distribution to retail markets. Facilities for controlled ripening are often provided at produce distribution centers (figs. 27.11 and 27.12). Ripening is triggered by exposure to ethylene. By scheduling the time at which fruit lots are exposed to ethylene and adjusting subsequent temperatures, distributors have a measure of control over the stage of ripeness of fruits delivered to retail stores.

## Ripening facilities

Ripening rooms require close temperature control. The rooms must be well-insulated and provided with heating and refrigeration. Vigorous air circulation is required to thoroughly disperse the ethylene and remove respiration heat from the fruit. Rooms are as nearly airtight as possible to contain the ethylene. High relative humidity (90 to 95 percent) is essential in ripening rooms to avoid fruit dehydration. Moisture may be introduced automatically in the form of steam or a mist or spray of water at ambient temper-

**Fig. 27.12. Cartons of bananas are spaced to ensure air circulation and uniform exposure to ethylene (San Francisco, California).**

ature. In the absence of a special system, walls and floors may be wetted before closing the room to initiate ripening.

The requirements for ripening rooms closely resemble those for refrigeration rooms. Commonly, rooms are prefabricated, have polystyrene foam insulation between sheets of steel or aluminum and are installed on a concrete slab. Insulated wood or steel frames and concrete block construction are common. Any wood must be treated to prevent rot in the high-humidity.

The size of ripening rooms varies with the volume of bananas handled. In modern facilities, bananas are normally handled in palletized boxes. Often, all pallets are placed on the floor. Valuable floor space is better used if rooms are designed for pallets stacked two or three high. Supporting framework is required to support the upper pallets.

To provide ripening bananas continuously to retail stores, a minimum of three ripening rooms is required; bananas can be delivered from each room on two successive days. Fruits leaving on the second day are somewhat riper than the first day's output. With six or more ripening rooms, all fruits from a single room can be delivered the same day.

### Controlled ripening

The objective of controlled ripening is to provide retail stores with bananas at a stage of ripeness desired by consumers. The state of ripeness is judged primarily by color, using a 1 to 7 scale common in the industry. At color number 1, the finger is hard and completely green; number 2 is green but with some traces of yellow color; number 3 is about half green and half yellow; number 4 is more yellow than green; number 5 is yellow with green tips; number

6 is fully yellow; and number 7 is yellow with brown spots.

Fruits should be ripened at least to color number 3 before delivery to retail stores, or ripening may not continue normally. Generally, the fruits are not riper than number 4 when shipped from the distribution center to the retail store, because fruits may suffer handling injury if they are too ripe.

Ripening is initiated by releasing ethylene into the ripening room for 24 hours with fruit pulp temperatures at 15.5° to 16.5°C (60° to 62°F). Enough gas is used to provide a concentration, by volume, of 100 to 1,000 ppm in air with vigorous air circulation, such as forced-air systems to ensure uniform distribution of ethylene.

The ripening rate varies to some extent between lots. Cloudy conditions or low temperature during growth may slow the rate of ripening. Temperature conditions during handling and transit may also affect the ripening rate. In particular, if elevated temperatures have caused some bananas to ripen en route, as shown by slight color changes or slight softening of pulp, the rate of ripening will be much faster than average. The maturity of fruit affects ripening time—hard-green fruits at the three-quarter stage take noticeably longer to ripen than fruits at full three-quarters.

Steady shipments of ripening bananas over a period of 2 weeks or more require considerable judgment based on experience. Bananas can be held at 13° to 14°C (56° to 58°F) for a number of days, depending on condition, before being placed in ripening rooms. Once begun, ripening can be slowed by lowering the temperature to 13°C (56°F) or speeded by raising the temperature to 18.5°C (65°F).

### Special considerations

Ethylene in concentrations greater than 30,000 ppm in air (3.02 to 34 percent by volume) is explosive. Take care not to reach dangerous concentrations.

Fruits exposed for extended periods to temperatures below about 13°C (56°F) exhibit symptoms of chilling injury, including a muddy yellow instead of bright yellow color of ripened bananas. The color difference is due to necrosis of vascular tissues in the peel. Symptoms are most severe if exposure occurs before ripening has started. Length of exposure affects the severity of symptoms.

## PAPAYAS

Intensive culture of papayas for long-distance marketing has been developed mainly in Hawaii. Production is mostly of small-fruited Solo cultivars. Fruits generally average about 0.5 kg (1.1 pounds). This size facilitates all handling operations and also fosters consumer acceptance as the size is appropriate for small North American families.

Handling and market preparation requirements are influenced greatly by the susceptibility of papayas to certain diseases. The most important of these is anthracnose, caused by the fungus *Colletotrichum gloeosporioides* (see chapter 15, *Postharvest Diseases of Selected Commodities*). Infections may occur by direct penetration during fruit development on the tree, but disease development does not proceed at this time because of the almost-complete immunity of the fruit flesh. Infections remain latent until the start of the climacteric rise in respiration rate. The disease becomes evident in ripe or ripening fruits. Control requires the use of postharvest treatments. Fungicidal sprays in the orchard during fruit development do not eliminate the need for postharvest heat treatment, but they may reduce the disease pressure considerably.

Other diseases include stem-end rots caused by *Ascochyta caricae-papayae* (see chapter 15), *Phomopsis caricae-papayae* and *Phytophthora nicotiana* var. *parasitica*. Although characteristically colonizing the fruit stem, these fungi often invade wounds caused by handling. Rhizopus rot (*Rhizopus stolonifer* and, possibly, related species) is an important cause of rot among fruit marketed in Hawaii. It is not common among Hawaiian fruit marketed in North America.

The presence of the Mediterranean, oriental, and melon fruit flies in Hawaii necessitates postharvest treatment to eliminate flies before shipment to non-infested areas. Fruit handling after disinfestation must be in screened areas. Packages must be sealed to avoid reinfestation, unless they can be loaded into marine containers with screen protection.

## Harvesting

Papaya trees are normally trained to a single trunk. Buds form progressively higher as the tree grows. Consequently, the lowermost fruits are the oldest. The fruit's position on the tree, therefore, provides an indication of relative maturity (fig. 27.13).

Fruits are generally picked according to the change in color from mature green (deep green) to color break (light green), exhibiting a slight overall loss of green color, with some hint of yellow at the blossom

Fig. 27.13. Developing papayas are produced successively on the stem. The lowest fruit are the oldest. Size is variable despite position (Kahului, Hawaii).

end. More mature fruits are categorized as one-quarter, one-half, and three-quarters yellow.

For long-distance shipment, fruits are generally harvested at color break or between color break and one-quarter color. To obtain maximum fruit life for long-distance surface shipments to mainland North America, the fruits are better harvested at the mature-green stage. It is difficult to distinguish mature-green from immature-green fruits. Immature-green fruits will not ripen after long-distance refrigerated transport.

Fruits can be harvested by pickers standing on the ground while trees are small. As the fruit-bearing area progresses higher, a long-handled suction cup (plumber's friend) is positioned by the picker over the stylar end of the fruit. A twist of the handle breaks the fruit's pedicel, and the picker catches the fruit if it falls from the suction cup. Fruits are placed in a pail and transferred to field boxes or bins.

In recent years various picking aid machines have been used to elevate the pickers to fruit level. Bins or other containers on the picking aid machines permit accumulation of the fruit.

## Transportation from orchard to packinghouse

Bulk bins have largely replaced 18.2 kg (40 pound) field boxes. Extreme care is essential when filling bins or boxes and transporting them to the packinghouse, because of the fragile nature of the fruit. Compression or impact bruises result from careless handling or transportation over rough roads. Sometimes equally serious is abrasion damage to the fruit's tender skin. Damaged skin generally does not degreen when the fruit ripens.

## Packinghouse operations

Some operations include preliminary washing and sorting to separate cull fruits and ripe fruits (which are diverted to processing) from those suitable for fresh marketing. This reduces the volume of fruit that must be heat-treated and handled within the insect-proof packinghouse.

Papayas are sorted for off-size or defective fruits as they move on conveyor belts. Fruits can be sized by eye, although automatic weight sizers are increasingly replacing eye sizing (fig. 27.14). Uniformly sized fruits are hand packed into shipping cartons (fig. 27.15).

When ethylene dibromide fumigation was first utilized for fruit fly disinfestation, a serious increase in disease resulted. Once handlers realized that the vapor-heat treatment was fungicidal as well as insecticidal, substitute heat treatments were explored. The currently approved disinfestation methods for fruit flies may substitute for a fungicidal vapor heat treatment.

Bins or pallet loads of fruit may be submerged for 20 minutes in 46° to 50°C (114.8° to 122°F) water

**Fig. 27.14. A weight sizer is often used with papayas (Hilo, Hawaii).**

**Fig. 27.15. Workers hand pack papayas (Hilo, Hawaii).**

with vigorous circulation. As the fruit warms, the water may cool several degrees.

The heat treatment operation has been integrated into the packinghouse line. At the speed at which fruit moves through such lines, the dwell period in the hot water (without excessively long tanks) would necessarily be very brief, perhaps only 20 to 30 seconds. To obtain enough heat penetration for fungicidal effectiveness, the temperature of the water must be very high—60.6°C (141°F) for 20 seconds. Positive movement of fruits through such a bath to insure that each fruit is exposed to an exact dwell time is essential. A fruit with a short dwell period would not receive an effective treatment, while those with long periods would sustain heat injury. In papayas, heat injury results in failure to degreen. A rot caused by *Dothiorella* sp. often develops, usually from the blossom end, in heat-injured fruit.

## Disinfestation procedures for fruit flies

**Fumigation.** Until recently ethylene dibromide (EDB) was used in concert with a hot water treatment as a disinfestation procedure. The use of ethylene dibromide (EDB) within the United States as a fumigant on fruits and vegetables for consumption in the United States was prohibited effective September 1, 1984. Papayas can no longer be treated with EDB if destined for U.S. markets, but the fumigant can be used for fruit exported to certain countries.

**Heat treatments.** There are currently three approved heat treatments for papayas: vapor heat (from Hawaii only), hot water, and forced hot air. The vapor heat is administered in a room with accurate temperature control and adequate air circulation. Fruit temperature is raised by saturated water vapor at 44.4°C (112°F) until the center of the fruit reaches that temperature, and then held for 8¾ hours.

An alternative treatment allows for a short-term two-stage hot water treatment. The treatment only allows the use of papayas less than one-fourth ripe and must be completed within 18 hours of harvest. Papayas must be kept at 18°C (65°F) or above until treated. The fruit are first submerged for 30 minutes at 42°C (107.6°F). This is followed within 3 minutes by a 49°C (120.2°F) dip for 20 minutes. Some phytotoxicity problems have been reported with this treatment in Hawaiian papaya, with the occurrence of hard spots in the fruit that will fail to ripen. In addition, this treatment has failed to control fruit flies in some strains of 'solo' papaya, because of blossom end deformity. This may lead to modification or cancellation of the hot water treatment.

More recently (1989) a "high temperature forced-air heat treatment" has been approved. The schedule for multi-stage treatment is:

**Multi-stage treatment schedule**

| Air temp °C(°F) | | Seed cavity temp °C(°F) | | Time (hrs) |
|---|---|---|---|---|
| 43 | (109.4) | 41 | (105.8) | 2 |
| 45 | (113.0) | 44 | (111.2) | 2 |
| 46.5 | (115.7) | 46 | (114.8) | 2 |
| 49 | (120.2) | 47.2 | (117.0) | Final |

After reaching the final stage the fruit is then immediately cooled in 20 to 25°C (68 to 77°F) water. Commercial use of this procedure is currently limited, but use in the near future will probably increase, especially if the two-stage hot water treatment is cancelled.

## Transportation and storage temperatures

Papayas are sensitive to chilling injury. Symptoms include sensitivity to *Alternaria* sp., failure to ripen normally, and, sometimes, water-soaked tissues. Immature and mature green papayas are most sensitive to chilling temperatures. Lowering the temperature below a critical point and lengthening the exposure period increases the injury. The critical temperature above which injury does not occur can be as low as 6°C (43°F), but Alternaria rot has occurred at that temperature in fruits harvested at the color-break stage and shipped distances requiring up to 2 weeks in transit. Under these conditions, temperatures above about 13°C (55.4°F) appear to be required.

# MANGOES

The essential operations that prepare mangoes for market include harvesting, transport to packing area, washing, sorting, sizing, packaging, and transport to market. Frequently mangoes must be treated for quarantine purposes before fruits can enter certain markets.

## Harvesting

Mangoes picked for nearby markets may exhibit color changes from dark green to light green or even to light yellow. Fruits are harvested while still dark green if destined for distant markets requiring several days of transit. Harvesting mangoes before external color change makes it difficult to discriminate between fruits that can ripen to an acceptable quality and those that cannot. In general, dark-green mature fruits are harvested when the cheeks have filled out. In some cultivars, maturity is indicated by the degree to which the shoulder extends out at the stem end.

Maturity indices commonly used with other fruits have not proved adaptable to dark-green mangoes. Changes in flesh texture (softening), total soluble solids, and acidity have not proved useful as they occur mostly after the proper harvest time for distant markets. Studies suggest that flesh color, starch content, and specific gravity might be useful indices for some cultivars.

Mangoes are harvested by hand if the pickers can reach them. The fruit is twisted sharply sideways or upward to break the pedicel. To avoid stem punctures, any long pedicels are trimmed flush with the stem end of the fruit. Fruits on high branches are harvested with a picking pole having a cloth bag and cutting knife at the top (fig. 27.16). When the pedicel is severed, the fruit drops into the bag, the picking end of the pole is gently lowered to the ground, and fruits are placed into 18- to 23-kg (40 to 50 lb) field boxes or bins for transport to the packinghouse.

## Packinghouse operations

Mature-green mangoes exude a large amount of latex from the cut stem. This is washed off with water in a tank. The water may contain imazalil or thiabendazole (TBZ) fungicide, mainly to control anthracnose caused by the fungus *Colletotrichum gloeosporioides*. Disease control is more effective if the solution is heated (fig. 27.17). Postharvest fungicidal treatments commonly consist of water at 52°C (125.6°F) containing 0.1 percent TBZ. Mangoes are submersed in the water for 1 to 3 minutes.

Sorters generally examine fruit as it passes on moving belts. Fruits that are judged immature, overmature, or undersized, and those exhibiting limbrub or other defects, are diverted.

Sizing is usually by eye. Individuals, often the packers, select fruits of the desired size from a moving return-flow belt. Use of weight sizers and other mechanical sizers is increasing.

Although various containers are used, mangoes are commonly packed in one or more layers in fiberboard boxes that may have individual compartments for separating fruits.

The use of ethylene dibromide (EDB) is now prohibited on any mango imported into the United States. An alternative hot water dip treatment is now registered for mangos. The duration and temperature of the treatment varies with the country of origin, cultivar and fruit size. Pulp temperature prior to treatment for all mangos must be a minimum of 21.1°C (70°F). Generally the fruit are treated in hot water at 46.4°C (115.5°F) for 65-90 minutes. The fruit pulp temperature must be between 45.6°C (114°F) and 46.1°C (115°F) for 10 minutes during this treatment.

Mature-green mangoes may be subjected to an ethylene treatment—100 ppm for 24 to 48 hours at 20°C (68°F)—to promote faster and more uniform ripening. The treatment may be applied at the shipping point if transit time to market is less than 5 days, or at the destination if transit times are longer.

Fig. 27.16. A picking pole of the type commonly used for mangoes or fruits of other large trees (Colima, Mexico).

Fig. 27.17. Mango packinghouse. Hot-water-thiabendazole bath is in background; sorting table in foreground (Colima, Mexico).

# PINEAPPLES

The pineapple fruit is a multiple structure that evolved, presumably, from a racemose inflorescence. Berry-like fruitlets, generally 100 to 200 in number, and their subtending leafy bracts are together fused to the core, a continuation of the fibrous peduncle. The fruitlets are in a regular spinal pattern on the fruit axis, the pattern usually consisting of two distinct spirals, one turning to the left and the other to the right.

The short, shootlike, leaf-bearing growth at the top of the fruit is called the crown. The crown is a continuation of the original meristem of the plant's main axis extending through the fruit. Crowns are frequently used as planting materials after they are removed from fruits that are to be processed. Slips and suckers developing from axillary buds at the base of the leaves below the fruit are other choices for planting materials.

The most widely grown pineapple cultivar is the Smooth Cayenne, although other cultivars are locally important, especially for the fresh market. These include Red Spanish, Queen, Singapore, Spanish, Selangor Green, Sarawalk, and Maritius. Postharvest handling operations are shown in figure 27.18.

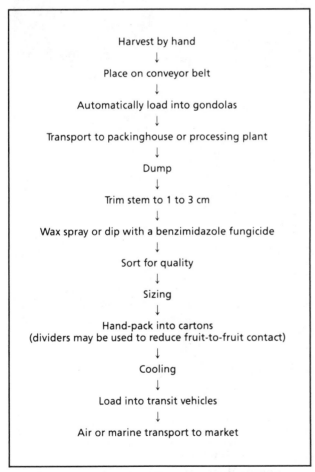

Harvest by hand
↓
Place on conveyor belt
↓
Automatically load into gondolas
↓
Transport to packinghouse or processing plant
↓
Dump
↓
Trim stem to 1 to 3 cm
↓
Wax spray or dip with a benzimidazole fungicide
↓
Sort for quality
↓
Sizing
↓
Hand-pack into cartons
(dividers may be used to reduce fruit-to-fruit contact)
↓
Cooling
↓
Load into transit vehicles
↓
Air or marine transport to market

**Fig. 27.18. Postharvest handling system for pineapples.**

## Maturity

During maturation pineapples increase in weight, flesh soluble solids, and acidity. During ripening, carotenoid pigments and soluble solids of the flesh increase dramatically and the fruit attains its maximum aesthetic and eating quality. During ripening, the shell of the pineapple loses chlorophyll rapidly, starting at the fruit base, in a process similar to that of degreening citrus fruit. Approximately 110 days elapse between the end of flowering and ripeness.

Harvest time is often when the base of the fruit has changed from green to yellow or light brown. Fruits may be harvested for fresh market before striking color changes have occurred. Acceptable quality may develop before color changes occur in the shell. Since the pineapple fruit has no accumulation of starch, there is no reserve for major postharvest quality improvements. As a nonclimacteric fruit, obvious compositional changes after harvest are mostly limited to degreening and a decrease in acidity.

## Harvesting

Pickers select fruit by size, color, or both, and twist it from the stalk. In small operations the harvested fruits may be placed in sacks or baskets that are carried to the end of the row for pickup.

Large-scale production avoids hand carrying by using a harvesting aid consisting of an endless belt extending across a number of rows (fig. 27.19). Fruits picked by hand are placed on the belt and are carried to the machine where they accumulate in a bin or truck gondola. The machine and belts, usually mounted on a truck chassis, move slowly across the field at a speed determined by the pickers.

## Packinghouse operations

Fruits are sorted at the packinghouse to eliminate those that are defective. Sizing may be by eye, but weight sizers are increasingly used (fig. 27.20).

A fungicidal treatment consisting of a TBZ dip or spray is commonly applied before packing to control water blister disease, caused by the fungus *Thiela-*

**Fig. 27.19. A pineapple harvest aid in use (Hawaii).**

Fig. 27.20. Weight-sizing pineapples (Hawaii).

Fig. 27.22. Workers hand pack pineapple fruits.

Fig. 27.21. Wax is applied to pineapple fruits.

Fig. 27.23 Pineapples are packed in shipping cartons.

*viopsis paradoxa*. Uncontrolled, the disease is serious in fruits after harvest. The fungus enters the fruit via wounds or at the stem and grows rapidly throughout the fruit flesh. A very watery soft rot is produced. In the past, the stem surface was smeared with a paste containing a fungicide such as benzoic acid to prevent water blister.

Pineapples are sometimes waxed to reduce physiological disorders (fig. 27.21) and improve appearance. Fruits are packed into full telescoping cartons (figs. 27.22 and 27.23) with inside dimensions of approximately 30.5 cm wide by 45 cm long by 31 cm deep (12 × 17.7 × 12.2 inches). From 8 to 14 or 16 fruits of uniform size are placed in each container.

### Holding and transit

The pineapple is a chilling-sensitive tropical fruit. Chilling injury symptoms include darkening of flesh tissues particularly around the central cylinder. Temperatures below about 6°C (42.8°F) may result in chilling injury. Pineapple fruits are shipped in marine containers (fig. 27.24) or by air.

A physiological condition called endogenous brown spot (EDS) is frequently seen in pineapples (fig. 27.25). This disorder is more common in the winter, and may be associated with chilling in the field.

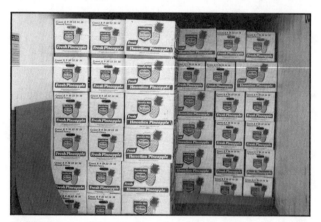

Fig. 27.24. A refrigerated marine container is loaded with pineapples.

### REFERENCES

1. Akamine, E. K. 1967. History of the hot water treatment of papayas. *Hawaii Farm Sci.* 16(3):4.

2. ———. 1975. Hawaii: Papaya and pineapple handling for local and export markets. In *Postharvest physiology, handling and utilization of tropical and subtropical fruits and vegetables*, ed. E. B. Pantastico, 538-41, Westport, CT: AVI Publ. Co.

3. Alvarez, A. M. and W. T. Nishijima. 1987. Postharvest diseases of papaya. *Plant Dis.* 71:681-686.

**Fig. 27.25. Endogenous brown spot disorder of pineapples.**

4. Barmore, C. R. 1974. Ripening mangoes with ethylene and ethephon. *Proc. Fla. State Hortic. Soc.* 87:331-34.

5. Caygill, J. C., R. D. Cooke, D. J. Moore, S. J. Read, and H. C. Passam. 1976. *The mango (Magnifera indica L.). Harvesting and subsequent handling and processing: An annotated bibliography.* London: Trop. Prod. Inst. 124 pp.

6. Cappellini, R. A., M. J. Ceponis, and G. W. Lightner. 1988. Disorders in avocado, mango, and pineapple shipments to the New York market, 1972-1985. *Plant Dis.* 72:270-274.

7. Cappellini, R. A., M. J. Ceponis, and G. W. Lightner. 1988. Disorders in apricot and papaya shipments to the New York market, 1972-1985. *Plant Dis.* 72:366-370.

8. Chen, J. N. and R. E. Paull. 1986. Development and prevention of chilling injury in papaya fruit. *J. Am. Soc. Hortic. Sci.* 111:639-643.

9. Collins, J. L. 1960. *The pineapple: Botany, cultivation and utilization.* New York: Interscience Publications.

10. Couey, H.M. 1989. Heat treatment for control of postharvest diseases and insect pests of fruits. *HortScience* 24:198-202.

11. Eckert, J. W. and J. M. Ogawa. 1985. The chemical control of postharvest diseases: subtropical and tropical fruits. *Annu. Rev. Phytopathol.* 23:421-454.

12. French, C. D. 1972. *Papaya—The melon of health.* New York: Exposition Press. 96 pp.

13. Gortner, W. A., and V. L. Singleton. 1965. Chemical and physical development of the pineapple fruit. III. Nitrogenous and enzyme constituents. *J. Food Sci.* 30:24-29.

14. Krishnamurthy, S., and H. Subramanyam. 1973. Pre- and post-harvest physiology of the mango fruit: A review. *Trop. Sci.* 15:167-93.

15. Marriott, J. 1980. Bananas—Physiology and biochemistry of storage and ripening for optimum quality. *CRC Crit. Rev. Food Sci. Nutr.* 13:41-88.

16. Mendoza, D. B., Jr., and R. B. H. Wills, eds. 1984. *Mango: Fruit development, postharvest physiology, and marketing in ASEAN.* Kuala Lampur, Malaysia: ASEAN Food and Handling Bureau. 111 pp.

17. Mukerjee, P. K. 1972. Harvesting, storage and transport of mango. *Acta Hortic.* 24:251-58.

18. Nagy, S., and P. E. Shaw. 1980. *Tropical and subtropical fruits—Composition, properties and uses.* Westport, CT: AVI Publ. Co. 570 pp.

19. Nagy, S., P.E. Shaw, and W.F. Wardowski, eds. 1990. Fruits of tropical and subtropical origin: composition, properties and uses. Lake Alfred, FL: Florida Science Source, Inc. 391 pp.

20. Pantastico, E. B. 1975. *Postharvest physiology, handling and utilization of tropical and subtropical fruits and vegetables.* Westport, CT: AVI Publ.

21. Paull, R. E. and N. J. Chen. 1989. Waxing and plastic wraps influence water loss from papaya fruit during storage and ripening. *J. Am. Soc. Hortic. Sci.* 114:937-942.

22. Paull, R. E. and K. G. Rohrback. 1985. Symptoms development of chilling injury in pineapple fruit. *J. Am. Soc. Hortic. Sci.* 110:100-105.

23. Rosenbaum, H. 1975. *How to grapple with the pineapple: From planting pineapple tops to baking up-side-down cakes.* New York: Hawthorn Books. 129 pp.

24. Shillingford, C. A. 1978. Postharvest banana fruit rot control with systemic fungicides. *Turrialba* 28:275-78.

25. Simmonds, N. W. 1966. *Bananas.* 2d ed. London: Longmans. 512 pp.

26. Singh, L. B. 1960. *The mango: Botany, cultivation and utilization.* New York: Interscience Publications.

27. Singleton, V. L. 1965. Chemical and physical development of the pineapple fruit. I. Weight per fruitlet and other physical attributes. *J. Food Sci.* 30:98-104.

28. Singleton, V. L., and W. A. Gortner. 1965. Chemical and physical development of the pineapple fruit. II. Carbohydrate and acid constituents. *J. Food Sci.* 30:19-23.

29. Slabaugh, W. R., and M. D. Grove. 1982. Postharvest diseases of bananas and their control. *Plant Dis.* 66:746-50.

30. Spalding, D. H., and W. F. Reeder. 1974. Current state of controlled atmosphere storage of four tropical fruits. *Proc. Fla. State Hortic. Soc.* 87:334-37.

31. Subremanyam, H., N. V. N. Moorthy, S. Lakshminarayana, and S. Krishnamurthy. 1972. Studies on harvesting, transport and storage of mango. *Acta Hortic.* 24:260-64.

32. Thompson, A. K. 1971. Transport of West Indian mango fruits. *Trop. Agric.* (Trinidad) 48:71-77.

# 28

# Postharvest Handling Systems: Tree Nuts

ADEL A. KADER AND JAMES F. THOMPSON

Proper harvesting and postharvest handling are key parts in achieving maximum yield of good quality tree nuts, and that determines marketability and profit. This chapter includes a brief discussion of the steps involved in harvesting and postharvest handling of the three major tree nuts in California—almond, pistachio, and walnut—and their impact on quality and safety attributes.

## Harvesting

### When to harvest

Tree nuts should be harvested as soon as possible after maturation to avoid quality loss and to minimize problems involving fungal attack and infestation with insects, especially the navel orange worm. The following indices determine optimum harvesting dates.

**Almond:** dehiscence (splitting) of the hull; separation of hull from shell; decrease in fruit removal force (development of abscission zone); drying of hull and kernel

**Pistachio:** ease of hull separation from the shell; shell dehiscence (splitting) and color; decrease in fruit removal force; kernel dry weight and crude fat content

**Walnut:** ease of hull removal (hullability); packing tissue browning (when the packing tissue between and around the kernel halves has just turned brown)

Uneven maturation presents a problem in once-over harvesting. Nuts on the tree's periphery usually mature earlier than those at the center. The use of ethephon to overcome this problem or accelerate maturation has produced mixed results that vary with the species. Although ethephon application accelerates almond maturation, it does not improve uniformity of maturity, and it may reduce yield and induce tree gummosis. Consequently, ethephon is not used on almond. Studies with pistachio indicate that ethephon application does not influence nut maturation. Ethephon applied at the time of "packing-tissue browning" allows walnut harvest within 7 to 10 days (instead of 15 to 20 days without ethephon). This treatment is used commercially in walnut harvesting, but it can stress trees.

## Harvesting procedures

Harvest season in California depends on the cultivar, production area, and cultural practices in use. Almonds are harvested between July and October. Pistachios (primarily the Kerman cultivar) are harvested during September. Walnuts are harvested during September and October. Orchard floor management is important in preparing for the harvest of almond and walnut as the nuts are knocked to the ground during harvesting.

Most harvest operations are mechanized. Almonds and walnuts are usually knocked or shaken to the ground by mechanical tree shakers (fig. 28.1), raked into windrows, and then picked up with sweepers (fig. 28.2). Pistachios are harvested with a shake-catch mechanical harvester (fig. 28.3) and the nuts

**Fig. 28.1. Tree shaker used for harvesting almonds and walnuts.**

**Fig. 28.2. Walnut sweeper.**

Fig. 28.3. Shake-catch system used for harvesting pistachios.

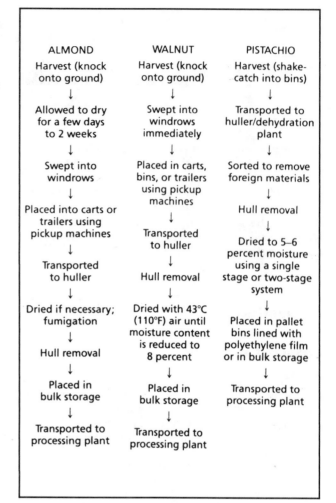

| ALMOND | WALNUT | PISTACHIO |
|---|---|---|
| Harvest (knock onto ground) | Harvest (knock onto ground) | Harvest (shake-catch into bins) |
| ↓ | ↓ | ↓ |
| Allowed to dry for a few days to 2 weeks | Swept into windrows immediately | Transported to huller/dehydration plant |
| ↓ | ↓ | ↓ |
| Swept into windrows | Placed in carts, bins, or trailers using pickup machines | Sorted to remove foreign materials |
| ↓ | ↓ | ↓ |
| Placed into carts or trailers using pickup machines | Transported to huller | Hull removal |
| ↓ | ↓ | ↓ |
| Transported to huller | Hull removal | Dried to 5–6 percent moisture using a single stage or two-stage system |
| ↓ | ↓ | ↓ |
| Dried if necessary; fumigation | Dried with 43°C (110°F) air until moisture content is reduced to 8 percent | Placed in pallet bins lined with polyethylene film or in bulk storage |
| ↓ | ↓ | ↓ |
| Hull removal | Placed in bulk storage | Transported to processing plant |
| ↓ | ↓ | |
| Placed in bulk storage | Transported to processing plant | |
| ↓ | | |
| Transported to processing plant | | |

Fig. 28.4. Handling systems for almonds, walnuts, and pistachio nuts.

are placed in bins (1.2 by 1.2 by 0.6 meter = 4 by 4 by 2 feet). Pistachios should not be shaken to the ground because of their open shells and high moisture content, relative to almonds and walnuts, at harvest.

Hand harvesting (knocking) is used on young trees and where steep terrain makes mechanical harvesting difficult. Tarps are spread under the tree. Pickers then knock the nuts loose by hitting the branches with poles (mauls), and the nuts are collected with the tarps.

## Postharvest Handling

A comparison of the handling systems for almond, walnut, and pistachio is illustrated in figure 28.4. Some important considerations:

**1.** Almonds should be picked up from the orchard floor as soon as they are dry to avoid exposure to adverse weather conditions, especially rain, and to minimize fungal infection and insect infestation.

**2.** Exposure of almonds to wet and hot conditions results in an internal disorder called concealed damage, characterized by a rust-brown to black discoloration of the nut meat and, in extreme cases, an unpalatable off-flavor.

**3.** Walnuts must be picked up, hulled, and dried as soon as possible after harvest. Walnuts left on the ground can deteriorate (the kernels darken) rapidly, especially at high ambient temperatures (e.g., 3 hours at 32°C [90°F] in direct sun).

**4.** Pistachios should be hulled and dried soon after harvest to minimize shell staining and decay. If temporary storage of fresh pistachios at the dehydration plant is necessary, they should be cooled and held before hulling at 0°C (32°F) and a relative humidity lower than 70 percent. Sorting before cold storage to remove defective nuts, leaves, twigs, and other foreign materials (which are usually much more susceptible to decay) minimizes losses during cold storage.

**5.** Fumigation with methyl bromide or phosphine is often used to control insects in stored nuts. It may have to be repeated periodically depending on the conditions and duration of storage.

## Drying

Water is present in plant tissues in three forms: bound water, which is bound with other constituents by strong chemical forces; adsorbed water, which is held by molecular attraction to adsorbing substances; and absorbed water, which is held loosely in the extracellular spaces by the weak forces of capillary action. The absorbed and adsorbed water constitute the "free water," most of which is removed during drying. Bound water is not removed except at very high temperatures that would also decompose some organic matter.

The moisture contents of almonds, pistachios, and walnuts at harvest range between 5 and 15 percent, 40 and 50 percent, and 10 and 30 percent, fresh weight basis, respectively. To improve stability and ensure the safety of the nuts, they should be dried to 5 to 8 percent moisture as soon after harvest as possible.

## Almonds

Most of the final drying of almonds occurs naturally while they are on the orchard floor. However, when rainy conditions prevail during harvesting, heated-air drying may be used to complete their dehydration.

Almonds are usually dried before hulling in a batch dryer with a maximum temperature of 49° to 54°C (120° to 130°F) or in continuous flow dryers with a maximum temperature of 82°C (180°F). Many batch dryers are specially designed, 5-ton capacity wagons with a perforated floor allowing heated air to be distributed underneath the nuts (fig. 28.5). The continuous-flow dryers are either horizontal belt dryers or cross-flow grain dryers. Dried almonds hull more easily than those with wet hulls, allowing hulling equipment to operate at maximum capacity.

## Walnuts

Walnuts are dried to 8 percent final moisture in batch dryers with air temperature not exceeding 43°C (110°F), as higher temperatures induce rancidity. Drying times may be as little as 4 hours for nuts that come from the field nearly dry to 2 days for extremely wet nuts. The wagon dryers used for almonds are also suited to walnut drying, although most walnuts are dried in self-unloading bin dryers. Another system, unique to walnuts, is the pot hole dryer (fig. 28.5). Metal pallet bins with expanded metal floors are filled with nuts and placed on top of an underground air plenum that supplies heated air to the bins.

## Pistachios

Most recently installed pistachio drying facilities use a two-stage process. The hulled nuts are first dried for about 3 hours in a column dryer originally designed for grain (fig. 28.6). Air temperature is kept below about 82°C (180°F) to prevent the shell from opening too far and allowing the kernel to separate from the shell. Nuts are dried to 12 to 13 percent moisture and then transferred to a flat-bottomed grain bin, where they continue to dry to 5 to 6 percent moisture with unheated air or air heated to less than 49°C (120°F). The final drying stage requires 24 to 48 hours. The nuts are left in the bins for storage until they are processed. This system increases the drying capacity of the expensive grain dryers and uses much less energy than complete drying in a high-temperature dryer. Some operations use a rotating drum dryer in place of the grain dryer for the first stage of drying. A few smaller operations use a self-unloading bin dryer for single-stage drying to the final moisture. Air temperatures in these are kept below 60° to 66°C (140° to 150°F) and drying time is about 8 hours.

## Post-drying storage

Dried nuts are usually stored in bins, silos, or other bulk-storage containers for a few weeks or several months before final processing and preparation for market. Optimum storage conditions of 10°C (50°F) or lower and 65 to 70 percent relative humidity must be maintained to minimize deterioration during storage. Protection against insects is also essential during the storage period.

# Preparation for Market

## Quality and safety

**Appearance.** Important factors in a nut's appearance include freedom from defect (shell staining, insect damage, mold, adhering hull, kernel discoloration, shriveling), kernel size relative to nut size (percent edible portion), and cleanliness.

**Texture.** Crispness, chewiness, and other textural quality attributes are influenced by the degree and uniformity of dryness within the nut and among nuts.

**Flavor and nutritive value.** Sweetness, oiliness, and roasted flavor are usually related to good flavor, while rancidity is a major factor in poor quality. Nuts are excellent sources of protein, but they are also very high in fat content.

**Safety.** Aflatoxin contamination must be avoided by protecting the nuts against growth of *Aspergillus flavus* before harvest, and after harvest before drying. Aflatoxin is a highly carcinogenic secondary metabolite produced by the fungi *Aspergillus flavus* and *A. parasiticus*. Often *A. flavus* is used in a collective sense to also include *A. parasiticus*.

Most prone to contamination in the orchard are the nuts poorly protected by hulls. In pistachios, early-split nuts vary between 1 and 5 percent from tree to tree. Early splits are nuts in which the shell splits before the hull has dehisced. Consequently, the entire pericarp splits to expose the kernel to the elements. In one study, about 20 percent of each 50-nut sample of early-split but otherwise good nuts were found to be contaminated with aflatoxin, as opposed to no contamination in regular-split nuts, where the shell had split within the loose, intact hull. Thin-shelled almonds are much more likely to be contaminated than thick-shelled nuts, probably because of poorer sealing of the sutures. Walnuts are less likely to develop aflatoxin than other nuts, but all nut species exposed to moist soil on the orchard floor risk infection.

All nuts infested with navel orange worm or certain other insects risk aflatoxin contamination, but these nuts pose less of a danger because they are

## WAGON DEHYDRATOR

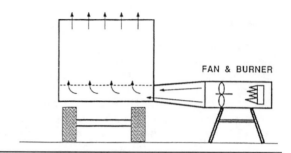

FAN & BURNER

## "POT HOLE" DEHYDRATOR

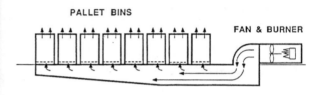

PALLET BINS

FAN & BURNER

## STATIONARY BIN DEHYDRATOR

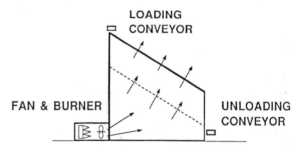

LOADING CONVEYOR

FAN & BURNER

UNLOADING CONVEYOR

**Fig. 28.5. Three types of dehydrators used for drying nut crops.**

easier to eliminate by sorting, and consumers are likely to reject them. In contrast, nuts not infested with insects cannot usually be eliminated by normal sorting methods. Furthermore, consumers will not necessarily reject them.

Early-split nuts not infected in the orchard may become infected during transport and handling. High humidities and temperatures within bulk bins provide ideal conditions for the infection of early-split nuts. The incidence of nuts with aflatoxin increases and aflatoxin levels rise dramatically, until nuts are mycologically stabilized by drying or refrigeration after removal of field heat. Fungicidal sprays in the orchard at the time of early splitting might materially reduce the potential for aflatoxin development, but no tests have demonstrated their effectiveness.

## Processing operations

**1.** Sorting for elimination of defects—nuts with adhering hulls, stained and moldy nuts—using visual or light-reflectance electronic sorting techniques

**2.** Separating empty or partially empty pistachios and walnuts by an air stream, and unsplit pistachios by flotation in water

**Fig. 28.6. Cross-flow dryers used for pistachios.**

3. Cracking shells to extract kernels when desired

4. Sorting to eliminate nuts with possible aflatoxin contamination, a labor-intensive and costly, but necessary, operation (efforts to develop reliable automated sorters continue)

5. Sizing of in-shell or shelled nuts into size categories using mesh screens

6. Sorting by color of shell or kernel; chemical bleaching of in-shell almonds and walnuts to improve shell color

7. Treatment with antioxidants to slow rancidity resulting from oxidation of fatty acids

8. Salting, flavoring, and roasting

9. Packaging of in-shell nuts, shelled nuts, and nutmeats (broken kernels) in various types and sizes of package to protect against insects, provide an effective moisture barrier, exclude oxygen to slow down rancidity, and exclude light to minimize color deterioration of some nuts

10. Nut meats may be pressed for oil extraction and shells are often used for charcoal production or as boiler fuel.

## Marketing and Utilization

Tree nuts are marketed in the shell or as shelled intact kernels or kernel pieces for use as snacks or in confectionary and bakery products. U.S. per capita consumption in 1987 was 0.26 kilogram (0.57 pound) for almonds, 0.08 kilogram (0.18 pound) for pistachios, and 0.23 kilogram (0.51 pound) for walnuts. Export marketing accounts for as much as 50 percent of the U.S. production for some tree nuts.

## Storage of Nuts

Maintenance of quality and storage life of nuts depends on their moisture content, the relative humidity and temperature in storage, the exclusion of oxygen, and the effectiveness of insect control. The role each factor plays in determining the storage potential of nuts is discussed below.

## Moisture content

The relationship between air relative humidity and water activity ($a_w$) of nuts can be expressed as

$$a_w = 0.01 \times RH$$

when RH (relative humidity) is in equilibrium with nut moisture content. Water activity is important in relation to the nuts' susceptibility to fungal attack, including mycotoxin-producing organisms (*Aspergillus flavus* and *A. parasiticus*). FDA regulations for tree nuts define a "safe moisture level" (moisture content that does not support fungal growth) as an $a_w$ that does not exceed 0.70 at 25°C (77°F).

The relationship between $a_w$ and moisture content of selected tree nut meats at 21°C (70°F) is shown below.

| Nut | Moisture content at $a_w$ = 0.2 to 0.8 |
| --- | --- |
| Almond | 3.0–8.7 percent |
| Pistachio | 2.2–8.2 percent |
| Walnut | 2.8–7.0 percent |

## Effect of relative humidity

The equilibrium moisture content of a product as a function of relative humidity (RH) at a given temperature can be illustrated by the sorption curve as shown in fig. 28.7 for walnuts and pistachos. This curve has three sections:

1. The lowest section of the curve, representing bound water, is concave to the RH axis. As the moisture content rises, the water molecules form a monolayer that coincides with about 20 percent RH.

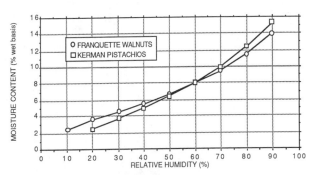

**Fig. 28.7. Relationship between pistachio and walnut moisture content (on a fresh-weight basis) and ambient relative humidity at indicated temperatures.**

**2.** As the RH increases, water molecules form successive layers of diminishing bond to the adsorbing substance of the commodity. This section of the curve (30 to 70 percent RH) is almost linear.

**3.** Above 75 percent RH, the product absorbs water to saturation (large increases in moisture content result from small increases in RH). This encourages deterioration and attack by microorganisms.

A relative humidity of 70 percent at 25°C (77°F) is the lower limit for significant mold growth.

## Effect of temperature

The relationship between moisture content and equilibrium relative humidity (ERH) is temperature-dependent. Between 20 percent and 80 percent ERH for any given moisture content, ERH rises about 3 percent for every 10°C (18°F) rise in temperature. For any given RH, air contains more water vapor at a high temperature than at a low temperature.

Lower temperatures reduce insect activity and retard mold growth and deterioration, including lipid oxidation (which results in rancidity). Temperatures between 0° and 10°C (32° and 50°F) are recommended for tree nuts, depending on expected storage duration. Duration can exceed 1 year; the lower the temperature, the longer the storage life.

Walnuts must be protected from freezing because they have a very short shelf life after thawing.

## Effect of oxygen level

Low oxygen (less than 0.5 percent) is a beneficial supplement to proper temperature management to maintain flavor quality and insect control. Exclusion of oxygen is usually done by vacuum packaging or by packaging in nitrogen.

In-shell almonds and walnuts are much more stable than shelled nuts. The shell is an effective package to protect the kernel from physical damage. The pellicle of a walnut kernel acts as a protective barrier and contains antioxidants.

## Insect control

Fumigation treatments using methyl bromide or phosphine may be used on stored nuts before final processing and packaging. Alternative treatments include controlled atmospheres (especially oxygen levels below 0.5 percent) and irradiation (table 28.1). Temperatures near freezing or between 40°C and 50°C (104°F and 120°F) are also effective for insect control as mentioned above; their use as fumigation substitutes is expected to increase in the future. The use of insect-proof packaging is essential to prevent reinfestation.

## REFERENCES

1. Beuchat, L. R. 1978. Relationship of water activity to moisture content in tree nuts. *J. Food Sci.* 43:754-58.

2. Guadagni, D. G., E. L. Soderstrom, and C. L. Storey. 1978. Effects of controlled atmosphere on flavor stability of almonds. *J. Food Sci.* 43:1077-80.

3. Holmberg, D. 1978. Almond harvest. In *Almond orchard management*, ed. W. Micke, 143-45. Univ. Calif. Div. Agric. Sci. Publ. 4092.

4. Kader, A. A., C. M. Heintz, J. M. Labavitch, and H. L. Rae. 1982. Studies related to the description and evaluation of pistachio nut quality. *J. Am. Soc. Hortic. Sci.* 107:812-16.

5. Labavitch, J. M. 1978. Relationship of almond maturation and quality to manipulations performed during and after harvest. In *Almond orchard management*, ed. W. Micke, 146-50. Univ. Calif. Div. Agric. Sci. Publ. 4092.

6. Labavitch, J. M., C. M. Heintz, H. L. Rae, and A. A. Kader. 1982. Physiological and compositional changes associated with maturation of 'Kerman' pistachio nuts. *J. Am. Soc. Hortic. Sci.* 107:688-92.

7. Lindsey, P. J., S. S. Briggs, A. A. Kader, and K. Moulton. 1989. Methyl bromide on dried fruits and nuts: issues and alternatives. In *Chemical use in food processing and postharvest handling: issues and alternatives*, 41-50. Univ. Calif. Davis. Agric. Issues Center.

8. Martin, G. C., G. S. Sibbett, and D. E. Ramos. 1975. Effect of delays between harvesting and drying on kernel quality of walnuts. *J. Am. Soc. Hortic. Sci.* 100:55-57.

9. Olson, W. H., G. S. Sibbett, and G. C. Martin. 1978. *Walnut harvesting and handling in California*. Univ. Calif. Div. Agric. Sci. Leaf. 21036. 7 pp.

10. Phillips, D. J., S. L. Purcell, and G. J. Stanley. 1980. *Aflatoxins in almonds*. U.S. Dept. Agric. SEA ARM-W-20. 11 pp.

11. Sommer, N. F., J. R. Buchanan, and R. J. Fortlage. 1976. Aflatoxin and sterigmatocystin contamination of pistachio nuts in orchards. *Appl. Environ. Microbiol.* 32:64-67.

12. Thompson, J. F. 1981. *Reducing energy costs of walnut dehydration*. Univ. Calif. Div. Agric. Sci. Leaflet 21257. 11 pp.

13. Wells, A. W., and H. R. Barber. 1959. *Extending the market life of packaged shelled nuts*. U.S. Dept. Agric. Mktg. Res. Rept. 329. 14 pp.

14. Woodroof, J. G. 1979. *Tree nuts: production, processing, products*. Westport, CT: AVI Publ. Co. 712 pp.

Table 28.1. Options and alternatives for insect disinfestation of nuts (Lindsey et al., 1989)

| Characteristic | Chemical fumigation | | Alternatives | |
| | Methyl bromide | Phosphine | Controlled atmospheres (low $O_2$ and/or high $CO_2$) | Irradiation |
| --- | --- | --- | --- | --- |
| Treatment duration | Up to 24 hrs + purging/aeration | 2–3 days + purging/aeration | 3–7 days at 27°C (80.6°F) + purging/aeration | Relatively short exposures |
| Effectiveness | 100% kill rate. | 100% kill rate. | Can achieve 100% kill rate. | Some larvae may survive up to 2 months. |
| Protection against reinfestation | NONE, unless combined with nightly pyrethrin or vapona fogs; requires refumigation. | Same as methyl bromide. | NONE. Requires repeated treatments or continuous maintenance of the CA. | Some limited protection due to possible presence of sterilized males. Reirradiation is not allowed. |
| Quality of product | Good | Good | Good—may be better than with chemical fumigation. | Good at recommended levels. |
| Health effects: | | | | |
| Workers | Highly toxic if directly exposed. | Highly toxic if directly exposed. | Presumed safer than chemical fumigation. | Safe if facility properly shielded and operated. |
| Consumers | Methyl bromide leaves residues that may be carcinogenic. | No known hazards. | No known hazards. | No known hazards. |
| Environmental effect | May involve release of toxic chemicals into environment. | May involve release of toxic chemicals into environment. | Does not pollute environment; somewhat energy-intensive. | Some versions involve transportation and use of radioactive materials. |
| Approximate annual cost of treatment (cents/lb): | | | (low $O_2$ generator) | |
| Almonds | — | <0.1¢ | 0.1–0.25¢ | .36–1.0¢ |
| Walnuts | <0.1¢ | — | 0.25–0.5¢ | .42–3.6¢ |
| Other considerations | Requires little capital investment; same per lb cost for small and large operators; could be withdrawn from market or require Prop. 65 identification in California. | Has some potential insect resistance problems; same per lb cost for any size operator. | Longer kill times could require construction of additional treatment chambers at some locations or alteration of existing storage facilities to make them gas-tight. | Requires large capital investment; more economical for large operators; requires labeling of treated product with potential for consumer resistance. |

# 29

# Postharvest Handling Systems: Fruit Vegetables

Robert F. Kasmire and Marita Cantwell

With the exceptions of peas and broad beans, fruit vegetables are warm-season crops, and with the exception of sweet corn and peas, all are subject to chilling injury. Fruit vegetables are not generally adaptable to long-term storage. Exceptions are the hard-rind (winter) squashes and pumpkin. A useful classification for postharvest discussion of the fruit vegetables is based on the stage of maturity at harvest.

## Immature fruit vegetables

- Legumes: snap, lima, and other beans, snow pea, sugar snap and garden peas
- Cucurbits: cucumber, soft-rind squashes, chayote, bitter melon, luffa
- Solanaceous vegetables: eggplant, peppers, tomatillo
- Others such as okra and sweet corn

## Mature fruit vegetables

- Cucurbits: cantaloupe, honeydew, and other muskmelons; watermelon, pumpkin, hard-rind squashes
- Solanaceous vegetables: mature-green and vine-ripe tomatoes, ripe peppers

This chapter presents the general postharvest requirements and handling systems for this group of commodities. Flow diagrams for tomato and cantaloupe handling are shown in figures 29.1 and 29.2.

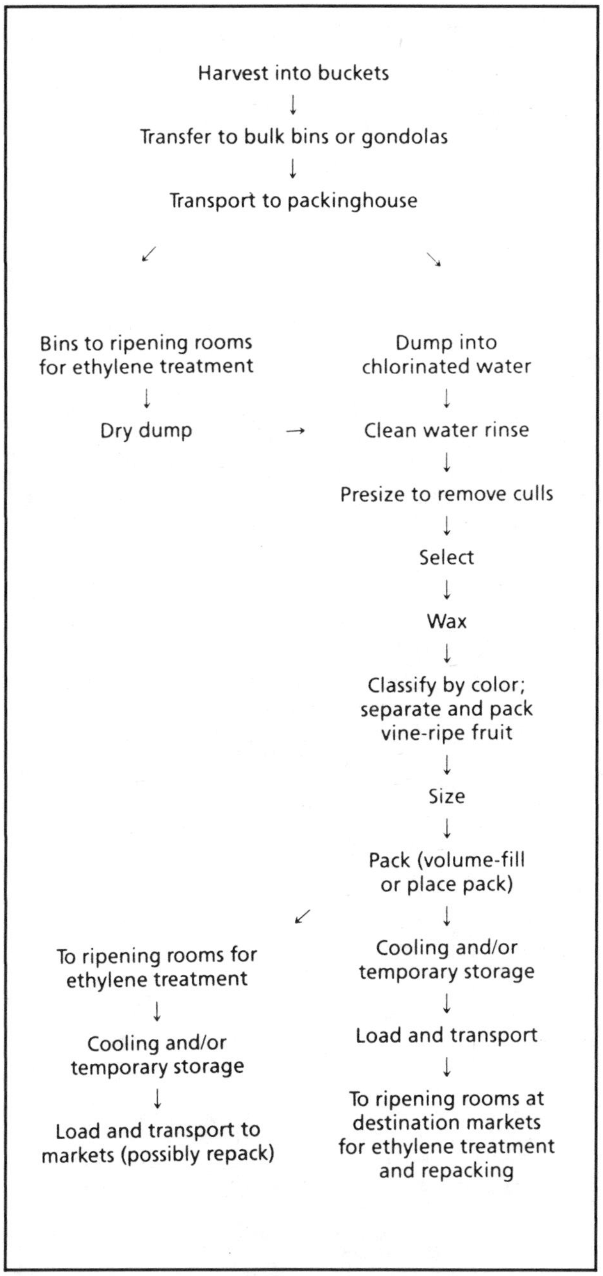

## Field Operations

### Harvesting

The harvest index for most immature fruit vegetables is based principally on size and color. Immature soft-rind squashes, for example, may be harvested at several sizes or stages of development, depending upon market needs. Fruit that are too developed are of inferior internal quality and show undesirable color change after harvest. This also applies to other immature fruit vegetables such as cucumber and bell peppers.

The harvest index for mature fruit vegetables depends on several characteristics, and proper harvest maturity is the key to adequate shelf life and good quality of the ripened fruit. Maturity and ripeness classes for honeydew melon are described in table 29.1 and those for tomato are shown in table 29.2. For cantaloupe, the principal harvest indices are surface color and the development of the abscission zone (fig. 29.3).

Harvest into buckets
↓
Transfer to bulk bins or gondolas
↓
Transport to packinghouse
↙ ↘

Bins to ripening rooms for ethylene treatment | Dump into chlorinated water
↓ | ↓
Dry dump → Clean water rinse
| ↓
| Presize to remove culls
| ↓
| Select
| ↓
| Wax
| ↓
| Classify by color; separate and pack vine-ripe fruit
| ↓
| Size
| ↓
| Pack (volume-fill or place pack)
↙ | ↓
To ripening rooms for ethylene treatment | Cooling and/or temporary storage
↓ | ↓
Cooling and/or temporary storage | Load and transport
↓ | ↓
Load and transport to markets (possibly repack) | To ripening rooms at destination markets for ethylene treatment and repacking

**Fig. 29.1. Postharvest handling systems for fresh market tomatoes.**

Most fruit vegetables are harvested by hand. Some harvest aids may be used, including pickup machines and conveyors for melons. Cantaloupe is also harvested with "sack" crews who empty the melons into

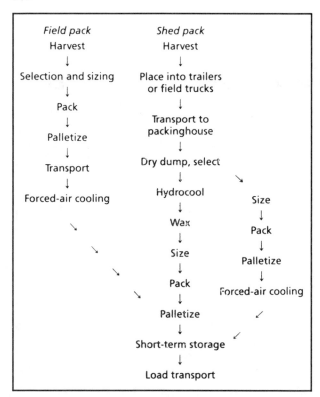

| Field pack | Shed pack |
|---|---|
| Harvest | Harvest |
| ↓ | ↓ |
| Selection and sizing | Place into trailers or field trucks |
| ↓ | ↓ |
| Pack | Transport to packinghouse |
| ↓ | ↓ |
| Palletize | Dry dump, select |
| ↓ | ↓ |
| Transport | Hydrocool |
| ↓ | ↓ |
| Forced-air cooling | Wax |

(Field pack branch converges with shed pack: Size → Pack → Palletize → Forced-air cooling; Wax → Size → Pack → Palletize → Short-term storage → Load transport)

**Fig. 29.2. Postharvest handling systems for cantaloupe.**

**Table 29.1. Maturity and ripeness classes for honeydew melons**

| Class | Characteristics |
|---|---|
| O = Immature | Greenish ground color<br>peel fuzzy/hairy<br>no aroma<br>may be harvested commercially by mistake |
| 1 = Mature, Unripe | Ground color white with greenish aspect<br>peel fuzzy/hairy<br>no aroma<br>melon splits when cut; flesh crisp<br>minimum commercial harvest maturity (10% soluble solids) |
| 2 = Mature, Ripening | Ground color white with trace of green<br>peel less fuzzy; slightly waxy<br>slight to noticeable aroma<br>melon splits when cut; flesh crisp<br>harvested commercially |
| 3 = Ripe | Ground color creamy white to pale yellow<br>peel waxy<br>noticeable aroma<br>abscission zone may begin to form<br>flesh firm when sliced does not split<br>ideal eating, harvested for local market only |
| 4 = Overripe | Ground color yellow<br>soft at blossom end<br>abscission zone fully formed<br>very aromatic<br>flesh soft, somewhat water soaked |

**Table 29.2. Maturity and ripeness classes for fresh market tomatoes**

| Class | Description |
|---|---|
| Mature green 1 | Seeds cut by a sharp knife on slicing the fruit; no jellylike material in any of the locules; fruit is more than 10 days from breaker stage |
| Mature green 2 | Seeds fully developed and not cut on slicing fruit; jellylike material in at least one locule; fruit is 6 to 10 days from breaker stage; minimum harvest maturity |
| Mature green 3 | Jellylike material well developed in locules but fruit still completely green; fruit is 2 to 5 days from breaker stage |
| Mature green 4 | Internal red coloration at the blossom end, but no external color change; fruit is 1 to 2 days from breaker stage |
| Breaker | First external pink or yellow color at the blossom end (USDA Color Stage 2) |
| Turning | More than 10 percent but not more than 30 percent of the surface, in the aggregate, shows a definite change in color from green to tannish-yellow, pink, red, or a combination thereof (USDA Color Stage 3) |
| Pink | More than 30 percent but not more than 60 percent of the surface, in the aggregate, shows pink or red color (USDA Color Stage 4) |
| Light red | More than 60 percent of the surface, in the aggregate, shows pinkish-red or red, but less than 90 percent of the surface shows red color (USDA Color Stage 5) |
| Red | More than 90 percent of the surface, in the aggregate, shows red color (USDA Color Stage 6) |

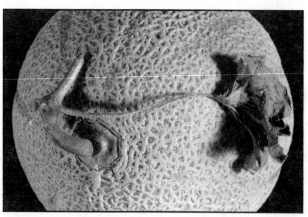

**Fig. 29.3. Cantaloupe with well-developed abscission zone.**

all fresh market tomatoes grown in California are bush-type, and the plants are typically harvested only once or twice. At the time of harvest, 5 to 10 percent of the tomatoes have pink and yellow color and are separated out later on the packing line as vine-ripes.

Immature fruit vegetables generally have very tender skins that are easily damaged in harvest and handling. Special care must be taken in all handling operations to prevent product damage and subsequent decay. Sweet corn, snap beans, and peas may be harvested mechanically or by hand.

bulk-trailers. Crenshaw and other specialty melons are easily damaged and require special care in handling and transport to the packing area. Mature-green tomatoes are usually hand harvested into buckets and emptied into field bins or gondolas. Almost

Many of the mature fruit vegetables are hauled to packinghouses, storage, or loading facilities in bulk bins (hard-rind squashes, peppers, pink tomatoes), gondolas (mature-green tomatoes and peppers), or bulk field trailers or trucks (muskmelons, hard-rind squashes).

Harvesting at night, when products are the coolest, is common for sweet corn and is gaining in use for cantaloupe. Products reach their lowest temperature near daybreak. Night harvest may reduce the time and costs of cooling products, may result in better and more uniform cooling, and helps maintain product quality. Fluorescent lights attached to mobile packing units have permitted successful night harvesting of cantaloupe in California.

### Field packing

The trend is increasing toward field packing of fruit vegetables. Grading, sorting, sizing, packing, and palletizing are carried out in the field. The products are then transported to a central cooling facility. Mobile packing facilities (fig. 29.4) are commonly towed through the fields for cantaloupe, honeydew melon, eggplant, cucumber, summer squashes, and peppers. Field-pack operations entail much less handling of products than in packinghouses. This reduces product damage and, therefore, increases packout yield of products. In melons, for example, field packing means less rolling, dumping, and dropping and thus helps reduce the "shaker" problem, in which the seed cavity loosens from the pericarp wall. It also reduces scuffing of the net which reduces subsequent water loss. Handling costs are also reduced in field pack operations. One difficulty with field packing, however, is the need for increased supervision to maintain consistent quality in the packed product. Field packing is not used for commodities that require classification for both color and size, such as tomato.

**Fig. 29.4. Mobile packing unit for field packing of cantaloupes**

## Packinghouse Operations

### Receiving

Loaded field vehicles should be parked in shade to prevent product warming and sunburning. Products may be unloaded by hand (soft-rind squashes, eggplant, some muskmelons, cucumber, watermelon), dry-dumped onto sloping, padded ramps (cantaloupe, honeydew melon, sweet peppers) or onto moving conveyor belts (tomatoes), or wet-dumped into tanks of moving water to reduce physical injury (honeydew melon, tomatoes, and peppers). Considerable mechanical damage occurs in dry-dumping operations; bruising, scratching, abrading and splitting are common examples. The water temperature in wet-dump tanks for tomatoes should be slightly warmer than the product temperature to prevent uptake of water and decay-causing organisms into the fruits. The dump tank water needs to be chlorinated. An operation may have two tanks separated by a clean water spray to improve overall handling sanitation.

### Preliminary operations

**Presizing.** For many commodities, fruit below a certain size are eliminated manually or mechanically by a presizing belt or chain. Undersize fruit are diverted to a cull conveyor or used for processing.

**Sorting or selection.** The sorting process eliminates cull, overripe, misshapen, and otherwise defective fruit and separates products by color, maturity, and ripeness classes (e.g. tomato and muskmelons). Electronic color sorters are used in some tomato operations.

**Grading.** Fruit are sorted by quality into two or more grades according to U.S. standards, California grade standards, or a shipper's own grade standards.

**Waxing.** Food grade waxes are commonly applied to cucumber, eggplant, sweet peppers, cantaloupe, and tomato and occasionally to some summer squashes. The purpose is to replace some of the natural waxes removed in the washing and cleaning operations, to reduce water loss, and to improve appearance. Waxing may be done before or after sizing, and fungicides may be added to the wax. Application of wax and postharvest fungicides must be indicated on each shipping container. Waxing and fungicides are used only in packinghouse handling of fruit vegetables. European cucumbers are frequently shrink-wrapped rather than waxed.

**Sizing.** After sorting for defects and color differences, the fruit vegetables are segregated into several size categories. Sizing is done manually for many of the fruit vegetables, including the legumes, soft- and hard-rind squashes, cucumber, eggplant, chili peppers, okra, pumpkin, muskmelons, and watermelon. Cantaloupes may be sized by volumetric weights, or diverging roll sizers, sweet peppers are sized commonly by diverging bar sizers, and tomatoes are sized by diameter with belt sizers or by weight.

**Packing.** Mature-green and pink tomatoes, sweet and chili peppers, okra, cucumber, and legumes are commonly weight- or volume-filled into shipping containers. All other fruit-type vegetables and many of the above are place-packed into shipping containers by count, bulk bins (hard-rind squashes, pumpkin, muskmelons, and watermelon) or bulk trucks (watermelon). Fruit-type vegetables that are place-packed are often sized during the same operation.

**Palletizing.** Packed shipping containers of most fruit vegetables in large-volume operations are palletized for shipment. This is a common practice with cantaloupe, muskmelons, sweet peppers, and tomato. Except for sweet corn, the immature fruit vegetables are often handled in low-volume operations, where palletizing is not common because of lack of forklifts. In these cases, the products are palletized at a centralized cooling facility or as they are loaded for transport. Palletizing is usually done after hydrocooling or package-ice cooling, but before forced-air cooling. In field-pack operations, palletizing is generally done in the field.

### Cooling

Various methods are used for cooling fruit vegetables. The most common methods are discussed here.

Forced-air cooling is used for beans, cantaloupe, cucumbers, muskmelons, peas, peppers, soft-rind squashes, and tomato. Forced-air evaporative cooling is used to a limited extent on chilling-sensitive commodities such as squashes, peppers, eggplant, and cherry tomato.

Hydrocooling is used before grading, sizing, and packing of beans, cantaloupe, sweet corn, and okra. Sorting of defective products is done both before and after cooling. Hydrocooling cycles are rarely long enough during hot weather. The need to maintain a continuous, adequate supply of canteloupes to the packers often results in the melons being incompletely cooled. This can be remedied if, after packing and palletizing, enough time is allowed in the cold room to cool the product to recommended temperatures before loading for transport to markets.

Package icing and liquid-icing are used to a limited extent for cooling cantaloupe and routinely as a supplement to hydrocooling for sweet corn.

**Temporary cold storage.** In large-volume operations, most fruit vegetables are placed in cold storage rooms after cooling and before shipment. Cold rooms are less used in small farm operations; the products are often transported to central cooperatively owned or distributor-owned facilities for cooling and short-term storage.

**Loading for transport.** Some tomatoes, cantaloupe, and other muskmelons are shipped in refrigerated railcars, but most fruit vegetables are shipped in refrigerated trucks or container vans. Except for the major volume products such as cantaloupe and tomato, most are shipped in mixed loads, sometimes with ethylene-sensitive commodities. Among the immature fruit-type vegetables, products such as cucumber, legumes, bitter melon, and eggplant are sensitive to ethylene exposure. Among the mature fruit types, watermelon is detrimentally affected by ethylene, resulting in softening of the whole fruit, flesh mealiness, and rind separation.

## Special Treatments

### Ripening

For uniform and controlled ripening, ethylene is often applied to mature-green tomatoes and sometimes to honeydew, casaba, and crenshaw melons. Ethylene treatments may be done at the shipping point or the destination, although final fruit quality is generally considered best if the treatment is applied at the shipping point soon after harvest. Satisfactory ripening occurs at 12.5° to 25°C (55° to 77°F); the higher the temperature, the faster the ripening (table 29.3). Above 30°C (86°F), red color development of tomato is inhibited. An ethylene concentration of about 100 ppm is commonly used. Honeydew melons (usually class 1-2 melons) are sometimes held in ethylene up to 24 hours; tomatoes are usually held at 20°C (68°F) and treated for up to 3 days.

Tomatoes may be ethylene-treated before or after packing, but most are treated after packing. An advantage of treating before packing is that the warmer conditions favor development of any decay-causing pathogens on the fruit, so infected fruit can be eliminated before final packout. Packing after ethylene treatment also permits a more uniform packout. Because most of the mature-green tomatoes produced in California are packed and then treated with ethylene, "checkerboarding" may still occur and make a repack operation necessary.

### Modified atmospheres

Modified atmospheres are seldom used commercially for these commodities, although shipments of melons and tomato under modified atmospheres are being tested for long-distance markets. Consumer

**Table 29.3. Effect of temperature on average ripening rate of mature-green, breaker, turning and pink tomatoes**

| | Days to table ripeness at indicated temperature | | | | | |
|---|---|---|---|---|---|---|
| Ripeness stage | 12.5 (54.5) | 15 (59) | 17.5 (63.5) | 20 (68) | 22.5 (72.5) | 25°C (77)°F |
| Mature-green | 18 | 15 | 12 | 10 | 8 | 7 |
| Breaker | 16 | 13 | 10 | 8 | 6 | 5 |
| Turning | 13 | 10 | 8 | 6 | 4 | 3 |
| Pink | 10 | 8 | 6 | 4 | 3 | 2 |

Source: Kader 1986.

packaging of vine-ripe tomatoes may also involve the use of modified atmospheres. For tomatoes held at recommended temperatures, oxygen levels of 3 to 5 percent slow ripening, with carbon dioxide levels held below 5 percent to avoid injury. Muskmelons have been less studied, but recommended atmospheres under normal storage conditions are 3 to 5 percent oxygen and 10 to 20 percent carbon dioxide.

### Recommended storage/transit conditions

For mature fruit-type vegetables temperature can effectively control the rate of ripening, as illustrated for tomato in table 29.3. Most mature-harvested fruit vegetables are sensitive to chilling injury when held below the recommended storage temperature. Chilling injury is cumulative, and its severity depends on the temperature and the duration of exposure. In the case of tomato, exposure to chilling temperatures below 10°C (50°F) results in lack of color development, decreased flavor, and increased decay (fig. 29.5).

The optimum temperatures for short-term storage and transport are:

- Mature-green tomatoes, pumpkin, and hard-rind squashes: 12.5° to 15°C (55° to 60°F)
- Partially to fully ripe tomatoes, muskmelons (except cantaloupe): 10° to 12.5°C (50° to 55°F).
- Honeydew melons that are ripening naturally or have been induced with ethylene are best held at 5° to 7.5°C (41° to 45°F).
- Watermelon: 7° to 10°C (45° to 50°F)
- Cantaloupe: 2.5° to 5°C (36° to 41°F)

The optimum relative humidity range is 85 to 90 percent for tomato and muskmelons (except cantaloupe), 90 to 95 percent for cantaloupe, and 60 to 70 percent for pumpkin and hard-rind squashes.

### Immature fruit vegetables

- All fruit vegetables harvested immature are sensitive to chilling injury as shown in figure 29.6 for cucumber. Exceptions are the peas and sweet corn, which are stored best at 0°C (32°F) and 95 percent RH.
- The optimum product temperatures with RH at 90 to 95 percent for short-term storage and transport are as follows:

  Eggplant, cucumber, soft-rind squashes, okra: 10° to 12.5°C (50° to 55°F)

  Peppers: 5° to 7°C (41° to 45°F)

  Lima beans, snap beans: 5° to 8°C (41° to 46°F)

### REFERENCES

1. Atherton, J. G., and J. Rudich, eds. 1986. *The tomato crop: a scientific basis for improvement*. New York: Chapman and Hall. 647 pp.

2. Davies, J. N., and G. E. Hobson. 1981. The constituents of tomato fruit—the influence of environment, nutrition and genotype. *CRC Crit. Rev. Food. Sci. Nutr.* 15:205-80.

3. Fahy, J. V. 1976. *How fresh tomatoes are marketed*. U.S. Dept. Agric. Mktg. Bull. 59. 31 pp.

4. Fairbank, W. C., et al. 1987. Night picking. *Calif. Agric.* 41:(1-2)13-16.

5. Gould, W. A. 1974. *Tomato production, processing and quality evaluation*. Westport, CT: AVI Publ. Co. 445 pp.

6. Hawkins, W. 1986. Harvesting and handling of fresh tomatoes. In *US-China Seminar on handling, storage and processing of fruits and vegetables*, ed. M. Bourne, Y. Zonglun, and F.W. Liu, 54-64. Washington, DC: National Academy Press.

7. Isenberg, F. M. R. 1979. Controlled atmosphere storage of vegetables. *Hortic. Rev.* 1:337-94.

8. Kader, A. A. 1986. Effect of postharvest handling procedures on tomato quality. *Acta Hortic.* 190:209-21.

9. Kader, A. A., L. L. Morris, M. A. Stevens, and M. Albright-Holton. 1978. Composition and flavor quality of fresh market tomatoes as influenced by some postharvest handling procedures. *J. Am. Soc. Hortic. Sci.* 103:6-13.

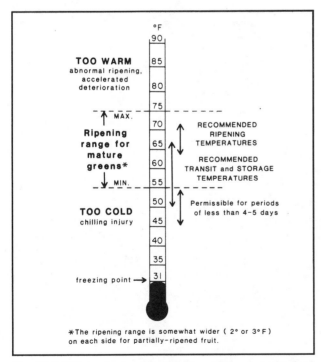

**Fig. 29.5. Effect of temperature on quality and ripening of tomatoes.**

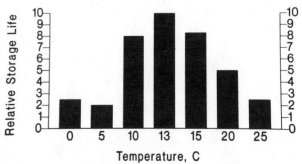

**Fig. 29.6. Storage life of cucumbers held continuously at different temperatures.**

266

10. Kasmire, R. F. 1973. Precooling, refrigeration and post-harvest handling of tomatoes and cantaloupes. *ASHRAE Symp. LO-73-7*, 19-20. Washington, DC: Am. Soc. Heating, Refrig. Air Cond. Eng.

11. Kasmire, R. F., et al. 1981. *Muskmelon production in California*. Univ. Calif. Div. Agric. Sci. Leaflet 2671. 23 pp.

12. Lee, C. Y. 1989. Green peas. In *Quality and preservation of vegetables*, ed. N. A. M. Eskin, 159-83. Boca Raton, FL: CRC Press.

13. McColloch, L. P., H. T. Cook, and W. R. Wright. 1968. *Market diseases of tomatoes, peppers and eggplants*. U.S. Dept. Agric. Handb. 28. 74 pp.

14. Miller, C. H., and T. C. Whener. 1989. Cucumbers. In *Quality and preservation of vegetables*, ed. N. A. M. Eskin, 245-64. Boca Raton, FL: CRC Press.

15. Pratt, H. K. 1971. Melons. In *The biochemistry of fruits and their products*, ed. A. C. Hulme, vol. 2, 207-32. New York: Academic Press.

16. Ryall, A. L., and W. J. Lipton. 1979. *Handling, transportation and storage of fruits and vegetables*. Vol. 1, *Vegetables and melons*, 2nd ed. Westport, CT: AVI Publ. Co. 587 pp.

17. Salveit, M. E., Jr. 1989. A summary of requirements and recommendations for the controlled and modified atmosphere storage of harvested vegetables. In *Proc. 5th Internat. CA Res. Conf.*, ed. J. K. Fellman, 329-54. Moscow: Univ. Idaho.

18. Sherman, M., R. K. Showalter, J. R. Bartz, and G. W. Simone. 1981. *Tomato packinghouse dump tank sanitation*. Univ. Fla. Veg. Crops Fact Sheet VC-31. 4 pp.

19. Sistrunk, W. A., A. R. Gonzales, and K. J. Moore. 1989. Green beans. In *Quality and preservation of vegetables*, ed. N. A. M. Eskin, 185-215. Boca Raton, FL: CRC Press.

20. Stevens, M. A. 1985. Tomato flavor: effects of genotype, cultural practices, and maturity at picking. In *Evaluation of quality of fruits and vegetables*, ed. H. E. Pattee, 367-86. Westport, CT: AVI Publ. Co.

21. Wiley, R. C., F. D. Schales, and K. A. Corey. 1989. Sweet corn. In *Quality and preservation of vegetables*, ed. N. A. M. Eskin, 121-57. Boca Raton, FL: CRC Press.

# 30

# Postharvest Handling Systems: Flower, Leafy, and Stem Vegetables

ROBERT F. KASMIRE AND MARITA CANTWELL

The leafy, stem, and floral vegetables are represented by the following commodities:

- Leafy vegetables: lettuce, cabbage, Chinese cabbage, Brussels sprouts, rhubarb, celery, spinach, chard, kale, endive, escarole, other leafy greens, green onion, Witloof chicory, radicchio, sprouts
- Stem vegetables: asparagus, kohlrabi, fennel
- Floral vegetables: artichoke, broccoli, cauliflower
- Mushrooms

Most of these vegetables are marketed throughout the year since they are harvested from various California production areas. For this reason, no long-term storage is required. In general, these commodities are characterized as very perishable, with high respiration and water loss rates.

## Harvesting

Virtually all leafy vegetables are cut by hand, but harvesting aids may be used with some (Brussels sprouts, celery, and parsley). Mechanical harvesting systems have been developed for crisp-head lettuce, celery, cabbage, Brussels sprouts, and cauliflower, but they are not presently used in California. The determination of horticultural maturity varies with commodity, but in general size is the principal criterion. For others, the solidity of the head determines harvest maturity, as exemplified by the maturity classes used for crisp-head lettuce (table 30.1.).

**Table 30.1. Solidity (maturity) classes of crisp-head lettuce**

| Solidity | Postharvest considerations |
|---|---|
| Soft, no head formation | More susceptible to physical damage; has a higher respiration rate than more mature lettuce; unacceptable for market |
| Fairly firm, slight head formation | Higher respiration rate |
| Firm, good head formation; optimum density | Maximum storage life |
| Hard, maximum density, but no split ribs | More susceptible to russet spotting, pink rib, and other physiological disorders; decreased storage life |
| Extra hard; split mid-ribs common; extreme internal pressure | Has minimum storage and shelf life remaining; most difficult to vacuum cool |

Stem vegetables are also hand harvested. A limited amount of asparagus has been experimentally machine harvested. Asparagus is generally hand cut when spears are at least 23 cm (9 inches) above the soil surface. All floral vegetables are hand harvested, but harvest aids (conveyors) are sometimes used for broccoli. Maturity of floral vegetables is determined by head size and development.

## Field Packing

Field packing is used for all leafy vegetables, except Brussels sprouts (figs. 30.1 to 30.7). The products are selected for maturity and quality, and then cut, trimmed, packed in cartons or crates, transported to cooling facilities, cooled, put into temporary cold storage prior to loading or loaded directly, and transported to market. Field packing generally provides greater marketable yields because of reduced mechanical damage. Wrapped and unwrapped lettuce, celery, cauliflower, broccoli, and spinach are mostly field packed, though the latter three are still packed in packinghouses by a few shippers.

Small celery stalks may be trimmed and packed as hearts in the field or field packed in bulk containers after harvesting and transported to packinghouses for trimming, sorting, prepackaging, and packing as celery hearts. Wrapped lettuce and cauliflower are hand selected, cut, and trimmed, and then placed on mobile field units where they are wrapped and packed into cartons. They are then palletized and transported to the cooling facilities for cooling and subsequent handling. Rough handling in field packing is a major cause of lettuce and cauliflower marketing losses. Keeping the commodity clean is a problem in field packing operations, particularly when fields are muddy.

## Packinghouse Operations

The floral and stem vegetables not packed in the field are selected, cut, placed in bulk containers, then transported to packinghouses for all subsequent handling operations. Compared with field packing, packinghouse handling requires more energy and results in more physical damage to the product, reducing marketable yields.

Packinghouse operations needed to prepare these products for market include:

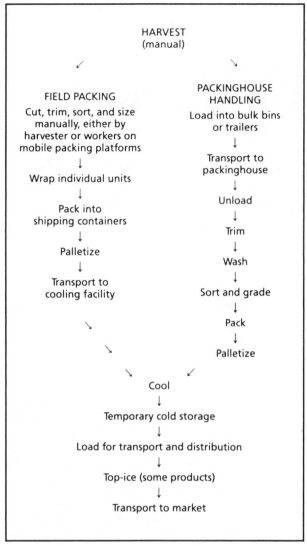

**Fig. 30.1. Postharvest handling of leafy vegetables, such as lettuce, celery, and green onions**

- Trimming and cleaning with chlorinated water (desirable concentration is about 200 ppm chlorine).
- Sorting and grading to eliminate defective products.
- Sizing, in some cases (all sizing is subjective and done by hand).
- Wrapping or tying individual units (cauliflower, broccoli), or in some cases, prepackaging (Brussels sprouts, broccoli, cauliflower florets).
- Packing in shipping containers (often wax-impregnated) or wood crates.

## Cooling

Delays between harvest and cooling should be avoided, especially during warm weather. Different cooling methods may be applied to the same commodity. The most common cooling methods in commercial use are:

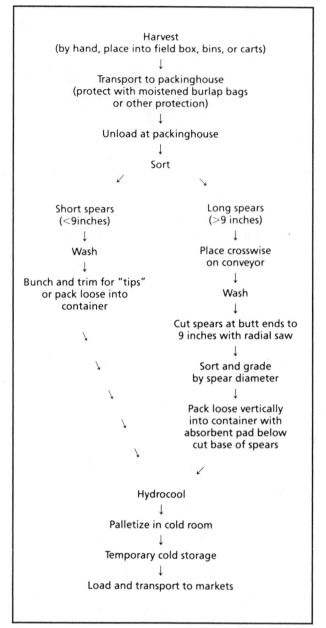

**Figure 30.2. Postharvest handling of asparagus.**

- Vacuum cooling for crisp-head lettuce, leaf lettuce, spinach, cauliflower, Chinese cabbage, bok choy, cabbage and other leafy vegetables, and mushrooms.
- Hydro-Vac cooling (vacuum cooling with injection of water prior to vacuum cycle) for celery and many other leafy vegetables.
- Hydrocooling for artichoke, leaf lettuce, celery, spinach, some green onions, leek, and many other leafy vegetables.
- Package-icing and liquid-icing for broccoli, spinach, parsley, green onions, and Brussels sprouts.
- Room cooling, primarily for artichoke and cabbage, and for the other leafy vegetables in some opera-

Harvest (by hand)

| FIELD PACKING | PACKINGHOUSE HANDLING |

**FIELD PACKING**

Cut, trim and size manually either by harvester or workers on mobile packing unit

↓

Consumer-packaging (wrap heads with film, or band stalks into bunches)

↓

Pack by count into containers (waxed if hydorcooling or in-package icing is used)

↓

Palletize cartons

↓

Transfer to vehicle for transport to cooling facility

↓
↓
↓
↓
↓

**PACKINGHOUSE HANDLING**

Load into bulk bins or trailers

↓

Transport to packinghouse

↓

Unload (mechanically or manually)

↓

(Hydrocool artichoke)

↓

Trim leaves from cauliflower and broccoli, and stems from artichoke

↓

Wash cauliflower with chlorinated water

↓

Tie or band broccoli; wrap cauliflower head

↓

Size (manually for broccoli and cauliflower; mechanically for artichoke)

↓

Pack by count

↓

Palletize

↓

Cool (liquid-ice; forced air)

↓

Temporary cold storage

↓

Load into refrigerated transit vehicles

↓

Top-ice (artichoke and broccoli)

↓

Transport to markets

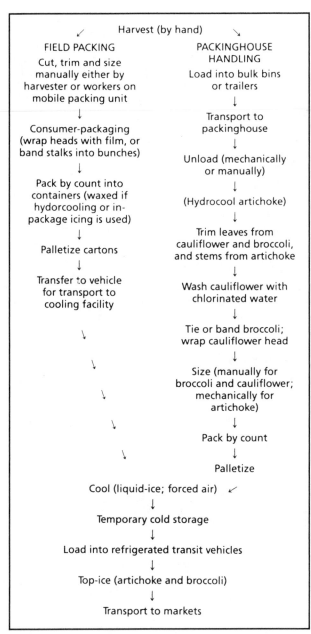

**Fig. 30.3. Postharvest handling of floral vegetables (artichoke, broccoli, and cauliflower).**

**Fig. 30.4. Field packing of crisp-head lettuce.**

**Fig. 30.5. Harvest and field packing of crisp-head lettuce by a mobile packing unit crew.**

tions (not generally recommended for this group of vegetables because it is too slow).

■ Forced-air cooling (sometimes with initial spraying of water), primarily for cauliflower and to a limited extent for other leafy and stem vegetables, including sprouts and mushrooms.

## Recommended Storage Conditions

In general, these products respond best to storage temperatures of 0° to 1°C (32° to 34°F). Freezing must be avoided. These products are frequently loaded into refrigerated trailers and containers immediately after cooling. For temporary storage, a temperature of 0° to 2°C (32° to 36°F) and a relative humidity of 90 to 95 percent is recommended.

Long-term storage is not recommended, except for cabbage, Chinese cabbage, and celery. In storage, air circulation should be minimized to that required for proper temperature control, excess carbon dioxide should be removed, and adequate oxygen levels should be maintained. Exposure to ethylene should be avoided throughout the handling system. Ethylene induces "russet spotting" disorder in lettuce (fig. 30.8) and decreases the shelf life of all green, leafy vegetables. Exposure to light causes undesirable greening in Belgian endive; this can be retarded by maintaining the product at low temperature.

Fig. 30.6. Field packing of crisp-head lettuce into cartons with polyliners.

Fig. 30.7. Field packing of cauliflower on a mobile packing unit.

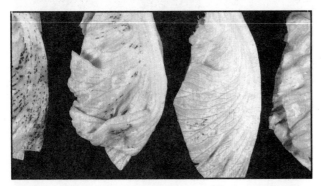

Fig. 30.8. Russet spotting on crisp-head lettuce: no symptoms are apparent on leaf at *left*; severe symptoms appear at *right*.

All the leafy, stem, and floral commodities respond favorably to modified atmospheres, although this technique is used on a limited scale commercially. Atmosphere recommendations for selected commodities are shown in table 30.2. Low oxygen atmospheres (2 to 3 percent $O_2$) favor longer shelf life in all products except asparagus and mushrooms. The recommendations for carbon dioxide modifica-

**Table 30.2. Controlled and modified atmosphere recommendations for leafy, stem, and floral vegetables**

| Product | Temperature (°C) | Temperature (°F) | $O_2$ (%) | $CO_2$ (%) |
|---|---|---|---|---|
| Asparagus | 2 | 36 | 21 | 10–14 |
| Broccoli | 0 | 32 | 2 | 5–10 |
| Cabbage | 0 | 32 | 2–3 | 3–6 |
| Cauliflower | 0 | 32 | 2–3 | 3–4 |
| Celery | 0 | 32 | 1–4 | 3–5 |
| Chinese cabbage | 0 | 32 | 1–2 | 0–5 |
| Green onions | 0 | 32 | 2–3 | 0–5 |
| Lettuce | | | | |
|   Crisp-head | 0 | 32 | 1–3 | 0 |
|   Leafy | 0 | 32 | 1–3 | 0 |
| Mushrooms | 0 | 32 | 21 | 10–15 |
| Spinach | 0 | 32 | 7–10 | 5–10 |
| Witloof chicory | 0 | 32 | 3–4 | 4–5 |

Source: Saltveit 1989.

tion are more variable. Spinach, which is highly perishable, does not tolerate low oxygen atmospheres and is routinely washed and packed in perforated polybags.

## REFERENCES

1. Duvekot, W. S., ed. 1981. Symposium on postharvest handling of vegetables. *Acta Hortic.* 116. 204 pp.

2. Isenberg, F. M. R., ed. 1977. Symposium on vegetable storage. *Acta Hortic.* 62. 361 pp.

3. Kader, A. A., J. M. Lyons, and L. L. Morris. 1974. Quality and postharvest responses of vegetables to preharvest field temperatures. *HortScience* 9:523-27.

4. Lipton, W. J. 1987. Senescence of leafy vegetables. *HortScience* 22:854-59.

5. Lipton, W. J., and E. J. Ryder. 1989. Lettuce. In *Quality and preservation of vegetables*, ed. N. A. M. Eskin, 217-44. Boca Raton, FL: CRC Press.

6. Lipton, W. J., J. K. Stewart, and T. W. Whitaker. 1972. *An illustrated guide to the identification of some market disorders of head lettuce.* U.S. Dept. Agric. Market Res. Rept. 950. 7 pp.

7. Ludford, P. M., and F. M. R. Isenberg. 1987. Brassica crops. In *Postharvest physiology of vegetables*, ed. J. Weichmann, 497-522. New York: Marcel Dekker.

8. Moline, H. E., and W. J. Lipton. 1987. *Market diseases of beets, chicory, endive, escarole, globe artichokes, lettuce, rhubarb, spinach and sweet potatoes.* U.S. Dept. Agric. Handb. No. 155. 86 pp.

9. Morris, L. L., A. A. Kader, and J. A. Klaustermeyer. 1974. Postharvest handling of lettuce. *ASHRAE Trans.* 80:341-49.

10. Phan, C. T., ed. 1985. Postharvest handling of vegetables. *Acta Hortic.* 157. 313 pp.

11. Pritchard, M. K., and R. F. Becker. 1989. Cabbage. In *Quality and preservation of vegetables*, ed. N. A. M. Eskin, 265-84. Boca Raton, FL: CRC Press.

12. Ramsey, G. B., and M. A. Smith. 1961. *Market diseases of cabbage, cauliflower, turnips, cucumbers, melons and related crops.* U.S. Dept. Agric. Handb. 184. 49 pp.

13. Ryall, A. L., and W. J. Lipton. 1979. *Handling, transportation and storage of fruits and vegetables. Vol. 1, Vegetables and Melons.* 2nd ed., chap. 6. Westport, CT: AVI Publ. Co.

14. Saltveit Jr., M. E. 1989. A summary of requirements and recommendations for the controlled and modified atmosphere storage of harvested vegetables. In *Proceedings of the International Controlled Atmosphere Conference*, Wenatchee, WA, ed. J. R. Fellman, vol. 2:329-52.

# 31

# Postharvest Handling Systems: Underground Vegetables (Roots, Tubers, and Bulbs)

ROBERT F. KASMIRE AND MARITA CANTWELL

The edible portions of this group of vegetables develop mostly underground and include several botanical structures.

**Roots:** beet, carrot, celeriac, radish, horseradish, parsnip, turnip, sweet potato, cassava, jicama

**Tubers:** potato, Jerusalem artichoke, yam

**Bulbs:** onion, garlic, shallot

**Others:** ginger rhizomes, taro (dasheen) corms

Vegetables within this group can also be divided into two subgroups based on their postharvest temperature requirements: ·

**Temperate-zone underground vegetables:** beet, carrot, celeriac, radish, horseradish, parsnip, turnip, potato, onion, garlic, shallot, daikon, salsify, water chestnut

**Subtropical and tropical underground vegetables:** sweet potato, yam, cassava, ginger, taro, jicama, malanga

The commodities in this grouping have several common characteristics. They are all storage organs, principally of carbohydrates; they generally have low respiration rates (depends on the stage of development); they are considered relatively nonperishable, especially if the tops are removed; they continue growth after harvest (rooting and sprouting); and they all can be stored for relatively long periods.

## Harvesting

Maturity indices vary with commodity. Many of these products may be harvested and marketed at various stages of development (e.g., "new" or immature potatoes versus mature potatoes, "baby" or immature carrots versus mature carrots). Criteria commonly used to harvest these commodities are as follows:

**Carrot:** size, length of root

**Radish:** days from planting, size

**Potato:** drying of foliage, setting of skins

**Cassava and taro:** drying of foliage begins

**Garlic and onion:** drying and bending over of tops

**Sweet potato:** drying of foliage

Both mechanical and manual harvest are used for this group of vegetables. Most roots and tubers are harvested mechanically and transported in bulk to packinghouses or processing facilities. Garlic and onion for fresh market are mechanically undercut, hand harvested and trimmed, cured in the field, and field-packed or transported to packinghouses. Sweet potato harvesting is still done mostly by hand, although vine cutting and lifting are done mechanically. Vine-killing chemicals may be used on potatoes before mechanical harvesting. Carrots are undercut, lifted by their tops, and detopped during mechanical harvest. Physical damage during the harvesting operations can be extensive and is a major cause of postharvest losses.

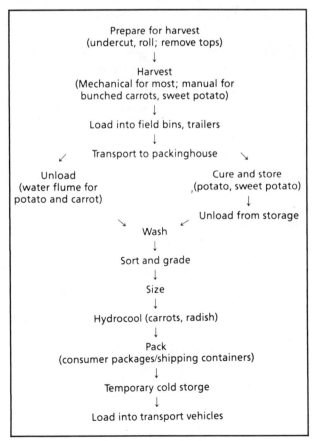

**Fig. 31.1. Harvest and postharvest operations for root and tuber vegetables.**

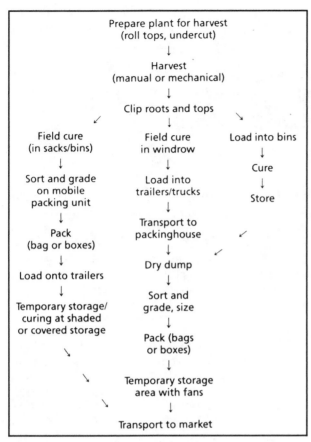

Prepare plant for harvest
(roll tops, undercut)
↓
Harvest
(manual or mechanical)
↓
Clip roots and tops
↓

Field cure          Field cure          Load into bins
(in sacks/bins)     in windrow                ↓
      ↓                  ↓                    Cure
Sort and grade     Load into                  ↓
on mobile          trailers/trucks          Store
packing unit            ↓
      ↓            Transport to
Pack               packinghouse
(bag or boxes)          ↓
      ↓            Dry dump
Load onto trailers      ↓
      ↓            Sort and
Temporary storage/ grade, size
curing at shaded        ↓
or covered storage Pack (bags
                   or boxes)
                        ↓
                   Temporary storage
                   area with fans
                        ↓
                   Transport to market

**Fig. 31.2. Harvest and postharvest operations for bulb vegetables.**

**Table 31.1. Effect of temperature on the wound healing process of potatoes (from Burton 1982)**

| Temperature | | Time (days) necessary to form | |
| --- | --- | --- | --- |
| (°C) | (°F) | Suberin | Periderm |
| 25 | 77 | 1 | 2 |
| 15 | 59 | 2 | 3 |
| 10 | 50 | 3 | 6 |
| 5 | 41 | 5–8 | 10 |
| 2 | 36 | 7–8 | not formed |

## Postharvest Procedures

### Curing

One of the simplest and most effective ways to reduce water loss and decay during postharvest storage of root, tuber, and bulb crops is curing after harvest. In root and tuber crops, curing refers to the process of wound healing with the development and suberization of new epidermal tissue called wound periderm. The effect of temperature on wound healing in the potato is shown in table 31.1. The type of wound also affects periderm formation: abrasions result in the formation of deep, irregular periderm, cuts result in a thin periderm, and compressions and impacts may entirely prevent periderm formation.

In bulb crops, curing refers to the process of drying of the neck tissues and of the outer leaves to form dry scales. Some water loss takes place during curing. Removing decayed bulbs before curing and storage ensures a greater percentage of usable product after storage. Recommended conditions for curing vary according to commodity, as shown in table 31.2.

When onions and garlic are cured in the field, they are undercut, then hand pulled. Sometimes the roots and tops are trimmed and the bulbs then are allowed to dry in field racks or bins from 2 to 7 days or longer (depending on ambient conditions). Sometimes they are pulled and cured before trimming. Curing may be done in windrows with the tops covering the bulbs to prevent sunburn. Where ambient conditions are unfavorable, curing may be done in rooms with warm forced air. Onions develop the best scale color if cured at temperatures of 25° to 32°C (77° to 90°F).

### Storage

Onion, garlic, potato, and sweet potato are often stored after curing and before preparation for market (cleaning, grading, sizing, and packing). These products, and other root crops such as carrot and turnip, may be stored from 3 to 10 months in mechanically refrigerated or ventilated storages.

**Table 31.2. Conditions for curing root, tuber and bulb crops**

| Commodity | Temperature | | Relative humidity (%) | Duration (days) |
| --- | --- | --- | --- | --- |
| | (°C) | (°F) | | |
| Potato | 15–20 | 59–68 | 85–90 | 5–10 |
| Sweet potato | 30–32 | 85–90 | 85–90 | 4–7 |
| Yam | 32–40 | 90–104 | 90–100 | 1–4 |
| Cassava | 30–40 | 86–104 | 90–95 | 2–5 |
| Onion, garlic | 30–45 | 86–113 | 60–75 | 1–4 |

**Fig. 31.3. Mobile packing unit for field packing of onions.**

## Preparation for market

The following operations are commonly used to prepare root, tuber, and bulb crops for market:

**Cleaning.** Dry brush or wash and partially dry, removing excess moisture.

**Sorting.** Eliminate defective products and plant debris.

**Decay control.** Postharvest fungicides are used on some of these commodities such as sweet potato; chlorination of wash and flume waters provides sanitation for carrot and potato.

**Sizing.** Size mechanically or by hand. Mechanical sizers are generally diverging rollers or weight sizers. Modified volumetric sizers are used for potatoes. Carrots present special sizing problems, since they must be sized by diameter (diverging rollers) and by length (manually or with a gravity-length sizer).

**Grading.** Separate into quality grades.

**Packing.** Pack into consumer units (bags, trays) and then pack into master shipping containers; or bulk pack into shipping containers (bags, boxes, and bins).

**Loading into transit vehicles.** Bulk transport to processing plants is sometimes used for onion, potato, and radish. For fresh market, most products are loaded in packed shipping containers (palletized if boxes, manually stacked if bags).

All temperate-zone root crops except potato, onion, and garlic can be hydrocooled. In areas where potatoes are harvested during hot weather, they may also be hydrocooled. Tropical root crops, potato, onion, and garlic are occasionally room-cooled before shipment to market. Potatoes may be cooled to 13° to 16°C (55° to 60°F) before shipment when they

Fig. 31.4. USDA Inspection of field-packed onions.

Fig. 31.5. Curing of bagged onions under a shaded area before shipment.

are harvested, packed, and shipped to market during hot weather.

Potatoes and onions destined for storage are cooled during the early phase of the storage period with cool air forced through storage piles or bins. Cooling may be done with cold ambient air or with air cooled by mechanical refrigeration.

**Special treatments.** For storage, onions and potatoes are generally sprayed with maleic hydrazide (MH) a few weeks before harvest to inhibit sprouting during storage. Aerosol applications of CIPC (3-chloro-isopropyl-N-phenyl carbamate) are often circulated around stored potatoes to further inhibit sprouting. Rodent control is also necessary storages.

**Nonrefrigerated storage methods.** Some growers occasionally store mature potatoes in the ground several weeks before harvest. Ground storage is also used for several of the tropical and subtropical roots, including cassava and jicama. Pits, trenches, and clamps are used for storage of harvested tropical roots and tubers. Pits are occasionally used for short-term, small-scale storage of potatoes in some areas.

Ventilated storage in cellars and warehouses is used for potatoes, sweet potatoes, garlic and onions. Newer facilities with temperature and relative humidity controls provide forced-air circulation through bulk piles of potatoes or onions, or through and around stacks of bulk bins.

## Recommended Storage Conditions

### Temperate-zone root vegetables

In California, temperate-zone root vegetables are not usually stored. When they are stored, the following conditions should be maintained: 0°C (32°F), 95 to 98 percent RH, and adequate air circulation to remove vital heat from the product and prevent $CO_2$ accumulation.

Potatoes can be stored up to 10 months under proper conditions. Most long-term potato storage facilities are in the northern U.S. For fresh market, potatoes should be stored under the following conditions: 4° to 7°C (39° to 45°F), 95 to 98 percent RH, enough air circulation to prevent $O_2$ depletion and $CO_2$ accumulation (about 0.8 cubic feet per minute per 100 pounds of potatoes), and exclusion of light to avoid greening. Greening is due to chlorophyll synthesis and is associated with accumulation of a toxic alkaloid, solanine. For processing (e.g., chipping), the proper conditions are 8° to 12°C (46° to 54°F), 95 to 98 percent RH, adequate ventilation, and exclusion of light. This higher temperature storage retards undesirable sweetening of the potatoes and consequent dark color of the processed products. Seed potatoes are best kept at 0° to 2°C (32° to 36°F), 95 to 98 percent RH, with adequate ventilation.

Garlic should be kept at 0°C (32°F) for long-term storage (6 to 7 months); 28° to 30°C (82° to 86°F) can be used for storage up to 1 month. Ventilation of about 1 cubic meter of air per minute per cubic meter of garlic with 70 percent RH is adequate.

Onions vary in their storage capability. The more pungent types with high soluble solids contents store longer, whereas mild onions with low soluble solids contents are rarely stored for more than 1 month. Storage temperatures should be either 0° to 5°C (32° to 41°F) or 28° to 30°C (82° to 86°F), as intermediate temperatures favor sprouting. Relative humidity should be maintained at 65 to 70 percent, and ventilation rate should be from 0.5 to 1 cubic meter of air per minute for each cubic meter of onions. Avoid light exposure to prevent greening. The storage potential of onions depends on the cultivar.

The use of controlled or modified atmospheres for this group of commodities is negligible. Limited commercial CA storage (3 percent $O_2$, 5 percent $CO_2$) of mild types of onions has been tested recently. Low concentrations of ethylene in the storage environment of carrots and parsnips induce bitterness.

### Tropical-zone root vegetables

The general storage recommendations for tropical-zone root vegetables, summarized in table 31.3, show that most of these products are chilling sensitive. Figure 31.6 illustrates internal discoloration of jicama due to storage at chilling temperatures. Some of these root crops (e.g., cassava) are successfully field-

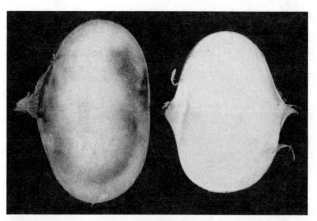

**Fig. 31.6. Internal appearance of jicama root stored 3 weeks at 10°C (50°F) (*left*) and 12.5°C (55°F) (*right*). The internal discoloration is one symptom of chilling injury in jicama. The roots did not differ in external appearance.**

stored but deteriorate rapidly if harvested and held under ambient conditions.

### REFERENCES

1. Booth, R. H. 1974. Postharvest deterioration of tropical root crops: losses and their control. *Trop. Sci.* 16:49-63.

2. Booth, R. H., and R. L. Shaw. 1981. *Principles of potato storage*. Lima: Internat. Potato Center (CIP). 105 pp.

3. Burton, W. G. 1982. *Postharvest physiology of food crops.* New York: Longman. pp. 199-261. 339 pp.

4. Burton, W. G. 1989. *The Potato*. New York: Longman Scientific and Technical co-published with J. Wiley & Sons. 742 pp.

5. Cooke, R. D., J. E. Rickard, and A. K. Thompson. 1988. The storage of tropical root and tuber crops—cassava, yam and edible aroids. *Exper. Agric.* 24:457-70.

6. Dewey, D. D. 1980. *For commercial growers: using temperature and humidity as guides to curing and storing onions.* Mich. State Univ. Ext. Bull. E-1409. File 26.17. 3 pp.

7. Edmond, J. B., and G. R. Ammerman. 1971. *Sweet potatoes: production, processing, marketing.* Westport, CT: AVI Publ. Co. 334 pp.

8. Kay, D. E. 1973. *Root crops* (TPI Crop and Product Digest No. 2). London: Tropical Products Inst. 245 pp.

9. Kushman, L. J., and F. S. Wright. 1969. *Sweet-potato storage.* U.S. Dept. Agric. Handb. 358. 35 pp.

10. Mazza, G. 1989. Carrots. In *Quality and preservation of vegetables*, ed. N. A. M. Eskin, 75-119. Boca Raton, FL: CRC Press.

11. Orr, P. H. 1971. *Handling potatoes from storage to packing line.* U.S. Dept. Agric. Mktg. Res. Rept. 890. 52 pp.

12. Paterson, W. D. 1979. *How onions are marketed.* U.S. Dept. Agric. Mktg. Bull. 65. 22 pp.

13. Purcell, A. E., W. M. Walter, Jr., and L. G. Wilson. 1989. Sweet potatoes. In *Quality and preservation of vegetables*, ed. N. A. M. Eskin, 285-304. Boca Raton, FL: CRC Press, Inc.

14. Rastorski, A., A. van Es et al. 1987. *Storage of potatoes. Postharvest behavior, store design, storage practices, handling.* Wageningen: Pudoc. 453 pp.

15. Ryall, A. L., and W. J. Lipton. 1979. *Handling, transportation, and storage of fruits and vegetables.* Vol. 1, *Vegetables and melons.* 2nd ed. Westport, CT: AVI Publ. Co. 587 pp.

16. Salunkhe, D. K., B. B. Desai, and J. K. Chavan. 1989. Potatoes. In *Quality and preservation of vegetables*, ed. N. A. M. Eskin, 1-52. Boca Raton, FL: CRC Press.

**Table 31.3. Storage conditions recommended for tropical root-type vegetables**

| Commodity | Temperature (°C) | Temperature (°F) | Relative humidity (%) | Storage life |
|---|---|---|---|---|
| Cassava | 5–8 | 41–46 | 80–90 | 2–4 weeks |
| or | 0–5 | 32–41 | 85–95 | ≤6 months |
| Ginger | 12–14 | 54–57 | 65–75 | 6 months |
| Jicama | 12–15 | 54–59 | 65–75 | 3 months |
| Sweet potato | 12–14 | 54–57 | 85–90 | ≤6 months |
| Taro | 13–15 | 55–59 | 85–90 | ≤4 months |
| Yam | 13–15 | 55–59 | near 100 | ≤6 months |
| or | 27–30 | 80–86 | 60–70 | 3–5 weeks |

17. Schouten, S. P. 1987. Bulbs and tubers. In *Postharvest physiology of vegetables*, ed. J. Weichmann, 555-81. New York: Marcel Dekker.

18. Smith, M. A., L. P. McColloch, and B. A. Friedman. 1966. *Market diseases of asparagus, onions, beans, peas, carrots, celery, and related vegetables.* U.S. Dept. Agric. Handb. 303. 65 pp.

19. Smith, O. 1977. *Potatoes: production, storing, processing.* 2nd ed. Westport, CT: AVI Publ. Co. 776 pp.

20. Smith, W. L., Jr., and J. B. Wilson. 1978. *Market diseases of potatoes.* U.S. Dept. Agric. Handb. 479. 99 pp.

21. Smittle, D. A. 1989. Controlled atmosphere storage of Vidalia onions. In J. K. Fellman (ed.), *Proceedings International Controlled Atmosphere Research Conf.*, Vol. 2, 171-177. Moscow: Univ. Idaho.

22. Stoll, K., and J. Weichmann. 1987. Root vegetables. In *Postharvest physiology of vegetables*, ed. J. Weichmann, 541-53. New York: Marcel Dekker.

23. Thompson, A. K. 1982. *The storage and handling of onions.* London: Trop. Products Inst. 14 pp.

24. Uritani, I., and E. D. Reyes, eds. 1984. *Tropical root crops.* Tokyo: Japan Scientific Societies Press. 328 pp.

25. Williams, L. G., and D. L. Franklin. 1971. *Harvesting, handling, and storing yellow sweet Spanish onions.* Univ. Idaho Agric. Exp. Sta. Bull. 526. 31 pp.

# 32

# Postharvest Handling Systems: Minimally Processed Fruits and Vegetables

MARITA CANTWELL

"Minimally processed" horticultural products are prepared and handled to maintain their fresh nature while providing convenience to the user. Producing minimally processed products involves cleaning, washing, trimming, coring, slicing, shredding, and so on. Other terms used to refer to minimally processed products are "lightly processed," "partially processed," "fresh-processed," and "preprepared."

Minimally processed fruits and vegetables include peeled and sliced potatoes; shredded lettuce and cabbage; washed and trimmed spinach; chilled peach, mango, melon, and other fruit slices; vegetable snacks, such as carrot and celery sticks, and cauliflower and broccoli florets (figs. 32.1 and 32.2); packaged mixed salads; cleaned and diced onions (fig. 32.3); peeled

Fig. 32.3. Cleaned and diced onions.

Fig. 32.1. Cauliflower florets.

Fig. 32.2. Broccoli florets.

and cored pineapple; fresh sauces; peeled citrus fruits; and microwaveable fresh vegetable trays.

Whereas most food processing techniques stabilize the products and lengthen their storage and shelf life, light processing of fruits and vegetables increases their perishability. Because of this and the need for increased sanitation, preparation and handling of these products require knowledge of food science and technology and postharvest physiology.

Growth in demand has led to increased marketing of fresh horticultural products in lightly processed form. An industry dedicated to this type of food processing has been established, and the National Association of Fresh Produce Processors was recently formed.

## Physiological Responses

Minimal processing generally increases the rates of metabolic processes that cause deterioration of fresh products. The physical damage or wounding caused by preparation increases respiration and ethylene production within minutes, and associated increases occur in rates of other biochemical reactions responsible for changes in color (including browning), flavor, texture, and nutritional quality (such as vitamin loss). The greater the degree of processing, the greater the wounding response. Control of the wound response is the key to providing a processed product of good quality. The impact of bruising and wounding can be reduced by cooling the product before processing. Strict temperature control after processing is also critical in reducing wound-induced metabolic activity, as shown in the respiration data of intact and shredded cabbage stored at different temperatures (fig. 32.4). Other techniques that substantially reduce damage include use of sharp knives, maintenance of stringent sanitary conditions, and efficient washing and drying (removal of surface moisture) of the cut product.

## Microbiological Concerns

Fruits and vegetables are ecological niches for a diverse and changing microflora, which usually does not include types pathogenic to humans. Intact fruits and vegetables are safe to eat partly because the surface peel is an effective physical and chemical barrier to most microorganisms. In addition, if the peel is damaged, the acidity of the pulp prevents the growth of organisms, other than the acid-tolerant fungi and bacteria that are the spoilage organisms usually associated with decay. On vegetables, the microflora is dominated by soil organisms. The normal spoilage flora, including the bacteria *Erwinia* and *Pseudomonas*, usually have a competitive advantage over other organisms that could potentially be harmful to humans.

Changes in the environmental conditions surrounding a product can result in significant changes in the microflora. The risk of pathogenic bacteria may increase with film packaging (high relative humidity and low oxygen conditions), with packaging of products of low salt content and high cellular pH, and with storage of packaged products at too high temperatures (>5°C or 41°F). Food pathogens such as *Clostridium*, *Yersinia*, and *Listeria* can potentially develop on minimally processed fruits and vegetables under such conditions.

With minimally processed products, the increase in cut-damaged surfaces and availability of cell nutrients provides conditions that increase the numbers and types of microbes that develop. Furthermore, the increased handling of the products provides greater opportunity for contamination by pathogenic organisms.

Microbial growth on minimally processed products is controlled principally by good sanitation and temperature management. Sanitation of all equipment and use of chlorinated water are standard approaches. Low temperature during and after processing generally retards microbial growth but may select for psychrotropic organisms such as Pseudomonads. Moisture increases microbial growth, therefore removal of wash and cleaning water by centrifugation or other methods is critical. Low humidity reduces bacterial growth, although it also leads to drying (wilting and shriveling) of the product. Low oxygen and elevated carbon dioxide levels, often in conjunction with carbon monoxide, retard microbial growth. Plastic film packaging materials modify the humidity and atmosphere composition surrounding processed products and therefore may modify the microbial profile. An example of the effects of different temperatures and packaging conditions is shown in table 32.1 for carrot slices.

## Product Preparation

Minimal processing may occur in a "direct chain" of preparation and handling in which the product is processed, distributed, and then marketed or uti-

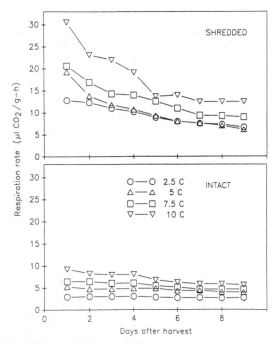

**Fig. 32.4. Respiration rates of intact and shredded cabbage stored at 2.5°C (36.5°F), 5°C (41°F), 7.5°C (45°F), and 10°C (50°F). The intact heads of cabbage had been harvested, cooled, and processed the same day.**

**Table 32.1. Percentage distribution of bacteria on carrot slices packaged in vacuum or in air in heat-sealed packages and stored at three temperatures for 8 days***

| Genus of bacteria | Storage temperature | | | | | |
| | 15°C(59°F) | | 10°C(50°F) | | 5°C(41°F) | |
| | Vac | Air | Vac | Air | Vac | Air |
|---|---|---|---|---|---|---|
| *Erwinia* | 0 | 90 | 20 | 90 | 50 | 80 |
| *Pseudomonas* | 0 | 10 | 0 | 10 | 0 | 10 |
| *Bacillus* | 0 | 0 | 0 | 0 | 10 | 10 |
| *Leuconostoc* | 100 | 0 | 80 | 0 | 40 | 0 |

*From Buick and Damoglou 1987. Carrot slices in the vacuum packages at 15° and 10°C showed anaerobic respiration after 2 and 4 days, respectively. Oxygen levels in the sealed packages were 2, 5, and 18 percent at 15°, 10°, and 5°C, respectively.

lized. Many products are also handled in an "interrupted chain" in which the product may be stored before or after processing or may be processed to different degrees at different locations. Because of this variation in time and point of processing, it would be useful to be able to evaluate the quality of the raw material and predict the shelf life of the processed product.

Minimally processed products may be prepared at the source of production or at regional and local processors; table 32.1 summarizes some of the advantages, disadvantages, and requirements in each case. Whether a product may be processed at source or locally depends on the perishability of the processed form relative to the intact form, and on the quality required for the designated use of the product. Processing has shifted from destination (local) to source processors as improvements in equipment, modified atmosphere packaging, and temperature management have become available.

An example of the operations involved in minimal processing is shown in figure 32.5 for lettuce. In the past, processed lettuce operations often salvaged lettuce remaining in the fields after harvesting for fresh market. It is now recognized that first-cut lettuce should be used for maximum processed product quality. After trimming and coring (fig. 32.6), piece size may be reduced with rotating knives or by tearing

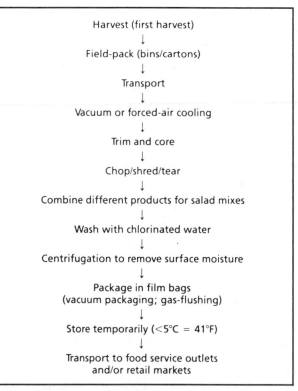

Fig. 32.5. Source preparation of minimally processed lettuce products.

Fig. 32.6. Cleaned and cored lettuce.

**Table 32.2. Advantages, disadvantages, and requirements of minimally processing fresh produce at various locations**

| Location of processing | Advantages, disadvantages, and requirements |
|---|---|
| **Source of production** | Product processed fresh when of the highest quality |
| | Processed product requires a minimum of 14 days postprocessing life (preferably 21 days); good temperature management critical |
| | Economy of scale, avoiding transport of unusable product and portions of product long distances |
| | Vacuum and gas-flushing techniques common in conjunction with differentially permeable films |
| **Regional** | Product processed when of good quality, typically 3 to 7 days after harvest |
| | Reduced need to maximize shelf life; 7 days minimum postprocessing life required; good temperature management vital |
| | Several deliveries weekly to end users; can better respond to short-term demands |
| | Vacuum packaging common; less use of differentially permeable films |
| **Local** | Product of fair to good quality, processed 7 to 14 days after harvest |
| | Relatively short postprocessing life required or expected; good temperature management required, but most often not found |
| | Small quantities processed and delivered; more labor intensive; discard large amounts of unusable product |
| | Simple packaging; little use of vacuum or gas-flushing techniques |

into salad-size pieces. Damage to cells near cut surfaces influences the shelf life and quality of the product. For example, shredded lettuce cut by a sharp knife with a slicing motion has a storage life approximately twice that of lettuce cut with a chopping action. Shelf life of lettuce is less if a dull knife is used rather than a sharp knife.

Washing the cut product removes sugar and other nutrients at the cut surfaces that favor microbial growth and tissue discoloration. Because of differences in composition and release of nutrients with processing, some products such as cabbage are known as "dirty" products. It is desirable to maintain separate processing lines, or thoroughly clean the line before another product follows cabbage. Free moisture must be completely removed after washing. Centrifugation is generally used, although vibration screens and air blasts can also be used. The process should remove at least the same amount of moisture that the product retained during processing. It has been shown that removal of slightly more moisture (i.e., slight desiccation of the product) favors longer postprocessing life.

## Packaging, Modified Atmospheres, and Handling

Polyvinylchloride (PVC), used primarily for overwrapping, and polypropylene (PP) and polyethylene (PE), used for bags, are the films most widely used for packaging minimally processed products. Multilayered films, often with ethylene vinyl acetate (EVA), can be manufactured with differing gas transmission rates. For lettuce processed at source, a 2.5 mil 8 percent EVA co-extruded PE bag has been used. Products are often packaged under partial vacuum (fig. 32.7) or after flushing with different mixtures of gases (oxygen, carbon dioxide, carbon monoxide, and/or nitrogen). Vacuum packaging and gas flushing establish the modified atmosphere quickly and increase the shelf life and quality of processed products. For example, browning of cut lettuce occurs, before a beneficial atmosphere is established by the product's respiration. For other products, such as fast-respiring broccoli florets, impermeable barrier films are used with permeable membrane "patches" to modify the atmosphere through the product's respiration. It is not yet agreed what are the ideal films and atmospheres for minimally processed products. In addition to different atmosphere requirements for different products, the specifics of the handling chains must be taken into account, especially their time delays and temperature fluctuations.

The modified atmospheres that best maintain the quality and storage life of minimally processed products have an oxygen range of 2 to 8 percent and carbon dioxide concentrations of 5 to 15 percent.

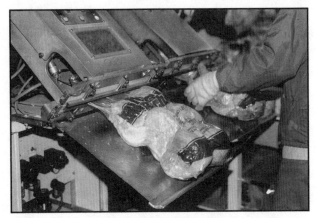

**Fig. 32.7. Vacuum packaging of cleaned and cored lettuce.**

Carbon monoxide concentrations of 5 to 10 percent under low oxygen (<5 percent) conditions retard browning and reduce microbial growth, lengthening shelf life in lettuce and other products. With some nonpermeable barrier-type PE films, an elevated oxygen level (25 to 50 percent) is used with carbon monoxide (3 to 10 percent) to maintain aerobic respiration during the handling period.

The following factors are known to be critical to maintaining quality and shelf life in minimally processed products: using the highest quality raw product, reducing mechanical damage before processing, reducing piece size by tearing or by slicing with sharp knives, rinsing cut surfaces to remove released cellular nutrients and kill microorganisms, centrifugation to the point of complete water removal or even slight desiccation, packaging under a slight vacuum with some addition of CO to retard discoloration, and maintaining product temperature at 1° to 2°C (34° to 36°F) during storage and handling. Temperature maintenance is currently recognized as the most deficient factor.

Other techniques such as irradiation, chemical preservation (dips in ascorbic acid, calcium chloride, and/or citric acid), modification of pH, and reduction of water activity (with sugars/salts) may also control deterioration of processed products, mainly by controlling microbial growth.

## Quality of Minimally Processed Products

The nature of the demand for minimally processed products requires that they be visually acceptable and appealing (fig. 32.8). The products must have a fresh appearance, be of consistent quality throughout the package, and be reasonably free of defects. Field defects such as tipburn on lettuce can reduce the quality of the processed product because the brown tissue is distributed throughout the packaged product.

In mixed salads, the quality of the total product is only as good as that of the most perishable component. This also applies to cleaned and washed spinach and other products where differences in leaf age or physical damage to leaves may yield a product of nonuniform perishability.

Quality assurance programs, long regarded as essential in the processed food industry, are difficult to apply to horticultural crops and the corresponding minimally processed products. Fresh horticultural products have not yet been subjected to the same sanitation, labeling, and shelf-life requirements as other processed foods.

**Fig. 32.8. Shredded lettuce in a bulk 10-pound package.**

## REFERENCES

1. Anonymous. 1990. *Proceedings of Fifth International Conference on Controlled/modified atmosphere/vacuum Packaging.* Princeton, NJ: Schotland Business Research Inc. 260 pp.

2. Ballantyne, A., R. Stark, and J. D. Selman. 1988. Modified atmosphere packaging of shredded lettuce. *Internat. J. Food Sci. Technol.* 23:267-74.

3. ———. 1988. Modified atmosphere packaging of broccoli florets. *Internat. J. Food Sci. Technol.* 23:353-60.

4. Barmore, C. R. 1987. Packing technology for fresh and minimally processed fruits and vegetables. *J. Food Quality* 10:207-17.

5. Berrang, M. E., R. E. Brackett, and L. R. Beuchat. 1989. Growth of *Listeria monocytogenes* on fresh vegetables stored under controlled atmosphere. *J. Food Protection* 52:702-5.

6. Bolin, H. R., A. E. Stafford, A. D. King, Jr., and C. C. Huxsoll. 1977. Factors affecting the storage stability of shredded lettuce. *J. Food Sci.* 42:1319-21.

7. Bolin, H. R., and C. C. Huxsoll. 1989. Storage stability of minimally processed fruits. *J. Food Process. Preserv.* 13:281-92.

8. ———. 1990. Effect of production procedures and storage parameters on the quality retention of salad-cut lettuce. *J. Food Sci.* 56:61-67.

9. Brackett, R. E. 1987. Microbiological consequences of minimally processed fruits and vegetables. *J. Food Quality* 10:195-206.

10. Brocklehurst, T. F., C. M. Zaman-Wong, and B. M. Lund. 1987. A note on the microbiology of retail packs of prepared salad vegetables. *J. Appl. Bacteriol.* 63:409-15.

11. Buick, R. K., and A. P. Damoglou. 1987. The effect of vacuum packaging on the microbial spoilage and shelf-life of "Ready to use" sliced carrots. *J. Sci. Food Agric.* 38:167-75.

12. Couture, R., M. Saltveit, and M. Cantwell. 1990. *Predictors of shelf-life of lettuce processed for salad mixes.* Univ. California, Davis, unpublished report.

13. Giannuzzi, L., N. Rodriguez, and N. E. Zaritzky. 1988. Influence of packaging film permeability and residual sulphur dioxide on the quality of pre-peeled potatoes. *Internat. J. Food Sci. Technol.* 23:147-52.

14. Huxsoll, C. C., and H. R. Bolin. 1989. Processing and distribution alternatives for minimally processed fruits and vegetables. *Food Technol.* 43(2):124-28.

15. Huxsoll, C. C., H. R. Bolin, and A. D. King, Jr. 1989. Physiochemical changes and treatments for lightly processed fruits and vegetables. In *Quality factors of fruits and vegetables— Chemistry and technology*, ed. J. J. Jen, 203-15. Washington, DC: American Chemical Society.

16. Kader, A. A. 1987. Effects of adding CO to controlled atmospheres. In *Postharvest physiology of vegetables*, ed. J. Weichman, 277-84. New York: Marcel Dekker.

17. King, Jr., A. D., and H. R. Bolin. 1989. Physiological and microbiological storage stability of minimally processed fruits and vegetables. *Food Technol.* 43(2):132-35, 139.

18. Klein, B. P. 1987. Nutritional consequences of minimal processing of fruits and vegetables. *J. Food Quality* 10:179-93.

19. Manvell, P. M., and M. R. Ackland. 1986. Rapid detection of microbial growth in vegetable salads at chill and abuse temperatures. *Food Microbiology* 3:59-65.

20. McDonald, R. E., L. A. Risse, and C. R. Barmore. 1990. Bagging chopped lettuce in selected permeability films. *HortScience* 25:671-73.

21. Myers, R. A. 1989. Packaging considerations for minimally processed fruits and vegetables. *Food Technol.* 43(2):129-31.

22. O'Beirne, D., and A. Ballantyne. 1987. Some effects of modified- atmosphere packaging and vacuum packaging in combination with antioxidants on quality and storage life of chilled potato strips. *Internat. J. Food Sci. Technol.* 22:515-23.

23. Ponting, J. D., R. Jackson, and G. Watters. 1971. Refrigerated apple slices: Effects of pH, sulfites and calcium on texture. *J. Food Sci.* 36:349-50.

24. Priepke, R. E., L. S. Wei, and A. I. Nelson. 1976. Refrigerated storage of prepackaged salad vegetables. *J. Food Sci.* 41:379-82.

25. Rolle, R. S., and G. W. Chism III. 1987. Physiological consequences of minimally processed fruits and vegetables. *J. Food Quality* 10:157-77.

26. Ronk, R. J., K. L. Carson, and P. Thompson. 1989. Processing, packaging and regulation of minimally processed fruits and vegetables. *Food Technol.* 43(2):136-39.

27. Rosen, J. C., and A. A. Kader. 1989. Postharvest physiology and quality maintenance of sliced pear and strawberry fruits. *J. Food Sci.* 54:656-59.

28. Sapers, G. M., and F. W. Douglas, Jr. 1987. Measurement of enzymatic browning at cut surfaces and in juice of raw apple and pear fruits. *J. Food Sci.* 52:1258-62, 1285.

29. Senter, S. D., J. S. Bailey, and N. A. Cox. 1987. Aerobic microflora of commercially harvested, transported and cryogenically processed collards (*Brassica oleracea*). *J. Food Sci.* 52:1020-21.

30. Shewfelt, R. L. 1987. Quality of minimally processed fruits and vegetables. *J. Food Quality* 10:143-56.

31. Watada, A. E., K. Abe, and N. Yamuchi. 1990. Physiological activities of partially processed fruits and vegetables. *Food Technol.* 44(5):116, 118, 120-22.

# 33

# The Extension Link: Getting the Message Across

## I. Extension Methods

ROBERT F. KASMIRE

There is a great need for effective extension programming in postharvest technology to help solve the industry's problems and to disseminate research results and other information useful to the fresh market horticultural-crops shipping, marketing, transportation, and distribution. To fill this need, extension personnel must be highly motivated professionals, well trained and equipped. They must be located where they can readily and effectively interact with both researchers and the shipping, distribution, and marketing industry. In this sense all postharvest workers are extension workers, whether they work for universities, government agencies, or private industry.

## The Role of Extension

The overall objectives of an extension postharvest program are (1) to improve the quality and value of horticultural crops available to consumers, (2) to reduce marketing losses of horticultural crops, and (3) to improve marketing efficiency. Specific objectives focus on solving particular problems affecting one commodity or a group of commodities.

To plan a postharvest extension program, one must first identify the problem, then identify and understand the clientele being addressed. Sometimes the audience's composition, background, and distribution are major constraints. The fresh produce shipping, marketing, and distribution industry is very heterogeneous and widespread, often including handlers at distant points in the same country or in more than one country. Initially this clientele may know little, if anything, about extension's role and objectives and may be suspicious of any attempts to change their handling practices, techniques, or facilities. To gain their attention and confidence, a program must first demonstrate that it can benefit them, most often economically. Once this is done, they can be as eager and cooperative as any other extension audience.

The clientele of the postharvest extension worker consists of local (county) extension workers (farm advisors or county agents) and industry personnel involved in preparation, shipping, marketing, and distribution of fresh-market horticultural products. It also includes students, via teaching, guidance, and counseling. The local extension worker's clientele consists primarily of industry personnel within a specific county, district, or region.

After learning about the audience, the following steps can be helpful:

**1.** Identify the problems to be resolved and determine their priorities. It is not possible to work on all problems at one time. Determine which are the most important and which can be realistically solved, and then assign priorities in both long- and short-range plans.

**2.** Develop the long- and short-range (1-year) objectives of the program. A long-range objective might be to improve the nutritional quality of fresh fruits and vegetables sold to consumers; a short-range objective might be to improve cooling practices and facilities used before loading.

**3.** Determine the elements needed to conduct the long- and short-range programs. Compare the manpower, equipment, and facilities that will be needed with those that are currently available. It is best to involve cooperators in programs whenever possible.

## Extension Methods

Many extension methods are available for conducting useful programs. The relative effectiveness of these methods depends upon the involvement of research workers, extension workers, and industry clients. Below are the methods most commonly used in postharvest extension programs.

**Applied research.** Many handling problems and practices in the industry cause increased product deterioration and marketing losses. Some problems are relatively simple and can be studied easily. Most applied research studies the causes or magnitude of deterioration or losses and develops and evaluates possible corrective measures. While some research must be conducted in laboratories and other research can be conducted in industry facilities, all must use scientifically sound methods and procedures. It is often necessary, even desirable, to cooperate with other researchers from universities, government agencies, and industry. This makes researchers aware of industry problems and adds their scientific input, and it makes industry aware of the problem-solving assistance available from scientists.

**Consultations.** Requests for consultations most often originate with industry leaders who want to improve their operations. Consultations include individuals, companies, or other groups. Consultations generally deal with specific subjects (e.g., cooling methods). Repeat requests for consultations are sure indications of the client's satisfaction with the technique.

**Group meetings.** Meetings with groups allow the extension worker to present information to larger audiences than through consultations. Group meetings also encourage audience participation and discussion.

**Demonstrations.** Extension workers can use demonstrations to show how to use a new practice, procedure, or facility or to illustrate its results. Demonstrations are often used to extend the results of applied research. Careful attention to the equipment, facilities, and visual aids used in demonstrations can increase their effectiveness. Tutorial audiovisual programs such as slide sets, videotapes, or movies can be valuable.

**Short courses.** Audiences can benefit from intensive, broad coverage of specific subjects in short courses through classroom lectures, laboratory demonstrations, or tours. The Postharvest Technology Short Course offered annually by the University of California at Davis is an example. Others are conducted by individual marketing firms, for their own personnel, or by trade association specialists. Cooperative Extension postharvest specialists can make their own programs more effective by participating in other industry short courses. Short courses may be from 1 day to about 2 weeks long. They may be used to refresh the audience with previously learned information and to provide updated or new information on a subject. Effective short courses require much planning, considerable professional involvement and input, proper facilities, and follow-up to evaluate their effectiveness. Inadequate preparation reduces a course's effectiveness and can result in complete failure. Printed material, including a syllabus, is generally provided to each participant in a short course.

**Workshops.** Workshops can improve the skills of individuals or groups. For example, in postharvest technology, one might conduct workshops on grading, sorting, packing, cooling, or careful handling. Workshops can last for one or more hours, and can meet only once or, as a series, over a period of time.

**Tours.** Facilities and operations are often easier to understand after a well-run tour. A tour can be an effective way to introduce a new subject to an audience. Tours can be a part of a demonstration, short course, or workshop. Industry cooperation is essential for a successful tour.

**Publications.** Publishing in professional journals, trade publications, newsletters, and other media outlets helps the extension worker to extend information to a greater, more distant audience than can be reached through personal contacts. Effective articles present information clearly, succinctly, and in an appealing manner; they are addressed to specific audiences and for a specific purpose, and they do not impose the writer's personal bias upon the information presented. A publication can be written in a variety of formats—technical, semitechnical, or popular. The choice depends on the intended audience. Following are examples of types of publications:

**Technical and semitechnical.** These include articles on applied research published in professional societies' publications and in university and government technical reports. They are written primarily for the benefit of professional workers; few industry representatives read these publications.

**Progress reports.** These publications extend current information to cooperators, research sponsors, industry personnel, and other interested persons—an example might be results of a just-completed preliminary study on the effects of a questionable, presently used industry practice. They are usually brief reports, no more than a few pages long. Their main advantages are timeliness, brevity, and directness, and they keep cooperators informed about the results.

Brief, single-subject guides. This type of publication addresses a single specific subject or development. The University of Florida Fact Sheets are examples.

**Newsletters.** These periodical publications extend information to broad audiences on a regular basis, typically four to six times a year. A good newsletter is an effective route for extension of brief, pertinent reports and articles to the postharvest industry and to fellow extension and research workers. The *University of California Perishables Handling—Horticultural Crops Newsletter* has been an effective informational tool since its founding in 1962. Issues usually contain a review article on a specific subject related to postharvest handling, brief articles on recent research results, and a list of recent postharvest publications and reports. The readership for this newsletter includes about 50 percent professionals (extension, research, libraries, and government agencies) and about 50 percent industry recipients. The University of Florida postharvest newsletters for industry, *Packinghouse Newsletter* and *Handling Florida Vegetables*, and Washington State University's *Postharvest Pomology Newsletter* are other effective newsletters. The industry also has its own postharvest newsletters.

**Trade publications.** Trade articles are most effective for extending information to the industry handlers, who regularly read them. Magazines and bulletins produced by grower-shipper associations—*The Western Grower and Shipper*, published by the Western

Growers Association; the monthly *PMA Bulletin*, published by the Produce Marketing Association; and *Fresh Outlook*, published quarterly by the United Fresh Fruit and Vegetable Association—and weekly newspapers of the fresh produce industry, *The Packer* and *Produce News*, are included in this group. Articles by extension workers in trade publications extend information to a broad, otherwise hard-to-reach audience. These trade associations and their publications can be very effective media for extending postharvest information.

Results from specific studies or other relevant information are often published in two or more types of publications to reach a broader audience and achieve the maximum effect. For example, the scientific details of a study might be reported in a professional society journal, the semitechnical aspects published in a university or government report, and a popular report of the study, showing its relevance to the postharvest industry, might be published in an extension newsletter and one or more trade publications.

**Committees and programs.** Extension postharvest technology programs can be effectively enhanced through participation in professional society and trade association programs and committees. This means more work for extension personnel, but it can also mean greater program effectiveness. People in the University of California postharvest programs are active in the American Society for Horticultural Science and work and participate actively with several industry trade associations, both as program participants and as working committee members (e.g., United Fresh Fruit and Vegetable Association, Produce Marketing Association, California Grape and Tree Fruit League, Western Growers' Association, the American Society of Heating, Refrigeration and Air Conditioning Engineers, and the Refrigerated Transpor-

tation Foundation). Extension personnel also work closely with committees and advisory boards for fresh-market horticultural crops established by state marketing order programs. These groups can help extend information more effectively and widely than is possible by working alone.

## Personal Program Development

Postharvest extension workers must maintain a high degree of professionalism if their programs are to be effective. They should continue to take courses and short courses when possible, and take sabbatical leaves to advance their capabilities. They must be familiar with modern techniques and instrumentation. Extension workers need to use creativity and initiative in program approaches, techniques, and skills. They also must be highly motivated. They must seek out postharvest handling problems and the groups that face them, because industry handlers generally will not reveal their problems until extension gains their confidence. Finally, extension workers must communicate effectively in speech and in writing, and they must be good listeners, able to take advantage of any constructive criticism made about their program or skill. Criticism can simply mean that people are listening to what the speaker is saying and are observing extension programs. By listening to criticism, extension workers involve more people in their programs, and that helps program effectiveness.

## Program Coordination

The key to success in extension programs is to maintain effective coordination of research and industry personnel. There are several ways to accomplish this coordination, and one is discussed in the following section of this chapter.

# II. Extension and the California Fruit Industry

F. GORDON MITCHELL

In the United States, extension and local research programs are conducted under the auspices of each state's land grant university. The U.S. Department of Agriculture also operates a research program designed to study problems of national or regional scope, and it maintains offices to coordinate research and extension activities among the states.

In California, agricultural research and extension responsibilities rest with the University of California. Agricultural faculties on the various campuses and in field facilities have responsibilities for research and teaching, and personnel with Cooperative Extension appointments are responsible for all extension activities.

## California's Deciduous Fruit Industry

Total fruit and nut production in California covers about 2.2 million acres and has an annual product value of about $3.1 billion. The program described here involves the pomology portion and excludes grapes (over 740,000 acres) and subtropicals (over 340,000 acres), which have their own programs. Pomological crops involve over 1.1 million acres in California and have an annual product value exceeding $1.5 billion, including 20 important fruit and nut crops. California produces 90 percent to 100 percent of the U.S. total for 11 of these crops.

Some deciduous fruit production occurs in most of California's 58 counties, although the biggest part of production is centered in the large, fertile Central Valley and in several smaller coastal valleys. Although typically of a Mediterranean climate, considerable climatic differences exist among fruit-growing areas within the state.

## University Research

The pomological research responsibility rests with the Pomology Department of the University of California, Davis. Considerable research is also conducted by other University departments in such areas as pest management, engineering, food science, soils, and water. The academic research staff of the Pomology Department is now at about 20, and research involves various areas of plant improvement, production physiology, and postharvest handling. Additional research responsibilities rest with extension personnel.

## Extension

The agricultural portion of the University of California Cooperative Extension is composed of farm advisors and statewide extension specialists. The farm advisor offices are located in most California counties, and the specialists are members of a research department on one of the University campuses. They are housed either with that department or on a field station. About 30 farm advisors have part- or full-time responsibility for pomological crops, and 9 specialists have full or partial appointments in the Pomology Department.

As academic staff members of the University of California, all farm advisors are expected to develop their own programs, using those techniques that are best suited to the needs of the fruit industry within their area. Thus, program emphasis varies considerably up and down the state. Most local programs involve some combination of newsletters, local publications, radio broadcasts, short courses, field meetings, field demonstrations, demonstration plots, problem-solving research, and individual problem diagnosis.

There are some "area" or "cross-county" assignments for farm advisors. These sometimes allow the farm advisors to specialize in narrower commodity or subject orientations and have eliminated duplication of effort within an area. Usually, cooperative agreements between counties are established so that each county program receives its full allotment of extension resources.

Research activities by farm advisors also receive increasing emphasis. Awareness of the value of extension involvement in research has grown over the years, and now most farm advisors conduct some local research. Depending on the advisor and the needs of the county or area, this can involve a substantial dedication of time. Research can be conducted jointly with departmental research personnel, other advisors, and specialists. Such activities provide important local solutions to pressing industry problems and help the advisors to become recognized authorities in their subject matter specialty.

**Extension specialist programs.** Over the years, the specialist program has changed dramatically in California, and now all agricultural specialists are members of and housed with the relevant subject matter department. Many also have researcher and lecturer titles and are involved in graduate student training, and some have joint Cooperative Extension and Experiment Station appointments.

Extension specialists typically maintain active programs in cooperation with farm advisors throughout the state. This involves meetings, conferences, training sessions, publications, and individual consultation as needed. The specialist coordinates the field research being conducted by a number of farm advisors, maintains contact with statewide organizations within the fruit industry, and makes the departmental researchers aware of industry research needs. The pomology production specialists divide their responsibilities largely along commodity lines. Some specialists work in less traditional areas that are of limited concern to production farm advisors, and some do much work with segments of the agricultural industry that spread across county lines.

Each extension specialist is expected to maintain a personal research program, with the research emphasis dependent upon the program needs. Research may be conducted jointly with departmental research personnel and farm advisors. Projects are expected to involve problem solving or adaptive studies, the results of which are directly applicable to industry needs.

**The Extension professional.** The strength of the extension program is in large measure determined by the quality of its personnel. In California, extension personnel are full academic staff members of the University of California. While salaries may not match some in the private sector, they are high enough to keep employee turnover to a minimum. Considering how many years it takes for an extension worker to reach peak effectiveness, this is important. Extension personnel are entitled to sabbatical leave privileges and are expected to develop sabbatical study programs designed to improve their effectiveness. They are also expected to participate in professional societies and other programs to maintain their professional competence. The research requirement and the close contact with research personnel helps to maintain that professionalism.

**Pomology-Extension continuing conference.** Pomology Department research and extension personnel and farm advisors with pomology responsibility coordinate their efforts through an ongoing program called the Pomology-Extension Continuing Conference to improve their overall response to the California fruit industry's needs. As the name implies, this is a continuing program, and it has several parts and several purposes. About once each year all personnel meet for 2 or 3 days to report, review, and plan. Smaller commodity working groups coordinate statewide research and extension activities within individual commodity areas. These groups often include people from other departments or agencies and may invite fruit industry participation. The groups assign research priorities, discuss plans, and project the needs for publications or other activities. New cooperative projects have emerged from some of the discussions. While working groups select their own chairs, the commodity specialist retains a coordination function.

This program is voluntary, and participation has been good. Many participants have responded to group decisions by starting, stopping, or modifying research projects. The varied elements of the pomology program have been better coordinated, and the conferences' deliberations have helped project future extension and research staffing needs.

**Pomology short courses.** Though certainly not a new concept, the short course has emerged as an significant activity closely linking extension, research, and the fruit industries. Pomology short courses are scheduled through the Pomology-Extension Continuing Conference, generally no more than two each year. To date, pomology short courses have been held on walnut, almond, prune, and stone fruit production; kiwifruit production and handling; and stone-fruit handling.

These short courses combine classroom lectures and field or laboratory demonstrations in a 3- to 5-day program. This format allows more in-depth coverage of the subject matter than is possible in shorter meetings. The courses are intended to train and refresh industry personnel who have active responsibilities in the subjects covered.

Short courses provide a special opportunity for research personnel and extension personnel (both farm advisors and specialists) to work together in planning, preparation, and presentation. Faculty selection for a short course is based entirely on subject-matter competence and teaching ability. Faculty are asked to prepare written material for a course syllabus, and whenever possible, syllabi are reworked into permanent form as comprehensive published manuals on the subject covered. To date, these short courses have been well attended and well received by fruit industry personnel.

**One program.** From the point of view of the California fruit and nut industries, the University's activities in research and extension are a single program. The agricultural audience does not distinguish between research and extension personnel; it is simply concerned with obtaining the help and information it needs from an appropriate authority. The University of California's Pomology-Extension Continuing Conference is an attempt to effect a single, coordinated program that serves these needs.

# Index